AF440685

LOGIC, ARTIFICIAL INTELLIGENCE AND ROBOTICS

Frontiers in Artificial Intelligence and Applications

Series Editors: J. Breuker, R. López de Mántaras, M. Mohammadian, S. Ohsuga and W. Swartout

Volume 71

Previously published in this series:

Vol. 70, In production

Vol. 69, N. Baba et al. (Eds.), Knowledge-Based Intelligent Information Engineering Systems & Allied Technologies

Vol. 68, J.D. Moore et al. (Eds.), Artificial Intelligence in Education

Vol. 67, H. Jaakkola et al. (Eds.), Information Modelling and Knowledge Bases XII

Vol. 66, H.H. Lund et al. (Eds.), Seventh Scandinavian Conference on Artificial Intelligence

Vol. 65, In production

Vol. 64, J. Breuker et al. (Eds.), Legal Knowledge and Information Systems

Vol. 63, I. Gent et al. (Eds.), SAT2000

Vol. 62, T. Hruška and M. Hashimoto (Eds.), Knowledge-Based Software Engineering

Vol. 61, E. Kawaguchi et al. (Eds.), Information Modelling and Knowledge Bases XI

Vol. 60, P. Hoffman and D. Lemke (Eds.), Teaching and Learning in a Network World

Vol. 59, M. Mohammadian (Ed.), Advances in Intelligent Systems: Theory and Applications

Vol. 58, R. Dieng et al. (Eds.), Designing Cooperative Systems

Vol. 57, M. Mohammadian (Ed.), New Frontiers in Computational Intelligence and its Applications

Vol. 56, M.I. Torres and A. Sanfeliu (Eds.), Pattern Recognition and Applications

Vol. 55, G. Cumming et al. (Eds.), Advanced Research in Computers and Communications in Education

Vol. 54, W. Horn (Ed.), ECAI 2000

Vol. 53, E. Motta, Reusable Components for Knowledge Modelling

Vol. 52, In production

Vol. 51, H. Jaakkola et al. (Eds.), Information Modelling and Knowledge Bases X

Vol. 50, S.P. Lajoie and M. Vivet (Eds.), Artificial Intelligence in Education

Vol. 49, P. McNamara and H. Prakken (Eds.), Norms, Logics and Information Systems

Vol. 48, P. Návrat and H. Ueno (Eds.), Knowledge-Based Software Engineering

Vol. 47, M.T. Escrig and F. Toledo, Qualitative Spatial Reasoning: Theory and Practice

Vol. 46, N. Guarino (Ed.), Formal Ontology in Information Systems

Vol. 45, P.-J. Charrel et al. (Eds.), Information Modelling and Knowledge Bases IX

Vol. 44, K. de Koning, Model-Based Reasoning about Learner Behaviour

Vol. 43, M. Gams et al. (Eds.), Mind Versus Computer

Vol. 41, F.C. Morabito (Ed.), Advances in Intelligent Systems

Vol. 40, G. Grahne (Ed.), Sixth Scandinavian Conference on Artificial Intelligence

Vol. 39, B. du Boulay and R. Mizoguchi (Eds.), Artificial Intelligence in Education

Vol. 38, H. Kangassalo et al. (Eds.), Information Modelling and Knowledge Bases VIII

Vol. 37, F.L. Silva et al. (Eds.), Spatiotemporal Models in Biological and Artificial Systems

Vol. 36, S. Albayrak (Ed.), Intelligent Agents for Telecommunications Applications

ISSN: 0922-6389

Logic, Artificial Intelligence and Robotics

LAPTEC 2001

Edited by

Jair Minoro Abe

SENAC – College of Computer Science and Technology, São Paulo, Brazil

and

João Inácio da Silva Filho

SENAC – College of Computer Science and Technology, São Paulo, Brazil

IOS
Press

Ohmsha

Amsterdam • Berlin • Oxford • Tokyo • Washington, DC

ISBN 1 58603 206 2 (IOS Press)
ISBN 4 274 90476 8 C3000 (Ohmsha)
Library of Congress Control Number: 2001095319

Publisher
IOS Press
Nieuwe Hemweg 6B
1013 BG Amsterdam
The Netherlands
fax: +31 20 620 3419
e-mail: order@iospress.nl

Distributor in the UK and Ireland
IOS Press/Lavis Marketing
73 Lime Walk
Headington
Oxford OX3 7AD
England
fax: +44 1865 75 0079

Distributor in the USA and Canada
IOS Press, Inc.
5795-G Burke Centre Parkway
Burke, VA 22015
USA
fax: +1 703 323 3668
e-mail: iosbooks@iospress.com

Distributor in Germany, Austria and Switzerland
IOS Press/LSL.de
Gerichtsweg 28
D-04103 Leipzig
Germany
fax: +49 341 995 4255

Distributor in Japan
Ohmsha, Ltd.
3-1 Kanda Nishiki-cho
Chiyoda-ku, Tokyo 101
Japan
fax: +81 3 3233 2426

Preface

It constitutes a great honor for us to present the approved and invited papers of the 2nd Congress of Logic Applied to Technology – LAPTEC'2001 held in São Paulo, Brazil, from Nov. 12th to 14th, 2001. LAPTEC began its life in 2000, with diverse backgrounds which included Artificial Intelligence, Robotics, Informatics, Technology, and correlated themes, having Logic (classical and non-classical) as underlying common matter.

Although LAPTEC was formerly thought as a national congress, LAPTEC'2000 had significant foreign contributions. So, the present version became officially an international congress, offering to participants opportunity to share a wide variety of domains in the fields above mentioned.

Over the past decade, Informatics in general has experienced amazing and unexpected directions. A significant number of young researchers has joined the academical community contributing with new knowledge, living knowledge, knowledge that reaches the frontiers of science.

LAPTEC'2001 had as Chairs:

General Chairs:

Jair Minoro Abe (Brazil) (Chair)
João Inácio da Silva Filho (Brazil) (Vice-Chair)

We would like to express our gratitude to the members of the following committees:

Honorary Chairs:

Edgar G. K. Lopez-Escobar - (USA), Heinz-Dieter Ebbinghaus - (Germany), Kiyoshi Iséki - (Japan), Lotfi A. Zadeh - (USA), Newton C. A. da Costa - (Brazil), and Patrick Suppes - (USA)

Scientific Committee:

Artibano Micali - (Brazil), Atsuyuki Suzuki - (Japan), Bráulio Coelho Ávila - (Brazil), Daniel Dubois - (Belgium), David Poole - (Canada), Don Pigozzi - (USA), Edgar G. K. Lopez-Escobar - (USA), Eduardo Massad - (Brazil), Germano Lambert Torres - (Brazil), Germano Resconi - (Italy), Heinz-Dieter Ebbinghaus - (Germany), Helmut Thiele - (Germany), Hiroakira Ono - (Japan), João Inácio da Silva Filho (Brazil), Jochen Pfalzgraf - (Austria), Kazumi Nakamatsu - (Japan), Kiyoshi Iséki - (Japan), Lotfi A. Zadeh - (USA), Luis Fernandez Lopez - (Brazil), Mara Marly Gomes Barreto - (Brazil), Marcel Guillaume - (France), Marcello M. Veiga - (Canada), Marcelo Finger - (Brazil), Maria Carolina Monardi - (Brazil), Masafumi Yamashita - (Japan), Masahiro Inuiguchi - (Japan), Michiro

Kondo - (Japan), Nelson Fávilla Ebecken - (Brazil), Newton C. A. da Costa - (Brazil), Norman Foo - (Australia), Paulo Veloso - (Brazil), Patrick Suppes - (USA), Piotr Koszmider - (Brazil), Ricardo Bianconi - (Brazil), Sandra Sandri - (Brazil), Seiki Akama - (Japan), Setsuo Arikawa - (Japan), Setsuo Ohsuga - (Japan), Sheila Veloso - (Brazil), Shotaro Tanaka - (Japan), Tadashi Shibata - (Japan), Tetsuya Murai - (Japan), Tru H. Cao - (United Kingdom), and Tsutomu Date - (Japan)

Organizing Committee:

Alexandre Scalzitti - (Brazil) (Vice-Chair), Bráulio Coelho Ávila - (Brazil), Cláudio Rodrigo Torres - (Brazil) (Chair), Eduardo Massad - (Brazil), Eliane D'Ippolito - (Brazil), João Eduardo Kögler Jr. - (Brazil), Kazumi Nakamatsu (Japan), Marcos Roberto Bombacini (Brazil), Neli Ortega (Brazil), Piotr Koszmider - (Brazil), Ricardo Bianconi - (Brazil), Ricardo Rodrigues - (Brazil), Rita Maria Lino Tarcia - (Brazil), and Toshi-ichi Tachibana - (Brazil)

Local Organization Committee:

Luiz Gonzaga Xavier de Barros (Brazil) (Chair) and Rosana Martins das Neves (Brazil) (Vice-Chair)

Secretariat:

Ligia Pappone (Brazil) and Rosimara Freitas de Souza (Brazil)

Technical Support:

Elton de France Amorim (Brazil) and Renan Rocha de Moraes (Brazil)

Also our special gratitude to the following additional scholars who helped us in refereeing papers: Adriano D. Pila (Brazil), Alexandre Rasi Aoki (Brazil), Andre Carvalho (Brazil), Caetano Traina Junior (Brazil), Claudia A. Martins (Brazil), Claudia Milare (Brazil), Flávio Soares Corrêa da Silva (Brazil), Gustavo Batista (Brazil), Ines de Castro Dutra (Brazil), Jonathan Lawry (United Kingdom), Jonathan Rossiter (United Kingdom), Luis Alfredo de Carvalho (Brazil), Luiz Eduardo Borges da Silva (Brazil), João C. P. da Silva (Brazil), Mario R. F. Benvides (Brazil), Solange O. Rezende (Brazil), Steve McCoy (United Kingdom), and Valmir Barbosa (Brazil).

Many people were involved in the practical organization of this congress and gave a very effective support: Alessandra Aparecida de Souza, Ana Lúzia Magalhães Carneiro, Beatriz Varella, Carlos Glaujor, Carlos Henrique Leibholz, Elaine de Souza Silva, Eliane Pereira, Elton de França Amorim, Ernane Melo Jr., Fabiana Domanoski Gurniak, Flávia Gabanela da Costa, Gizela Carvalho da Fonseca, Luciana Perederko, Marlon Disney R. Pontes, Neiva Augusta da Silva, Renan Rocha de Moraes, Renata Mieko Matsumura, Renata Boni Ruschel, Rita Maria Lino Tarcia, Vanessa Maria Gomes, and Vera Lúcia F. Zeitune Santos. We greatly appreciate their efforts.

We would like to thank the numerous sponsors, particularly the SENAC-College of Computer Science and Technology, São Paulo, Brazil, especially its Dean, Prof. Ricardo Castillo Molina, who has given conditions for that LAPTEC'2001 could be hosted at

SENAC-SP. Also we would like to acknowledge the following entities: FAPESP, IEEE, CNPq, Institute For Advanced Studies – University of São Paulo, Universtity of São Paulo – Campus São Carlos, Himeji Institute of Technology – Japan, Shizuoka University – Japan, Teikyo Heisei University – Japan, Escola de Engenharia de Itajubá – Brazil, Universidade do Brasil (Federal University of Brazil), Sociedade Brasileira de Computação, Sociedade Brasileira para o Progresso da Ciência, ABJICA – Brazil, and IOS Press, the publisher of this volume.

Last but not least, we would like wish to express our appreciation for the work of Cláudio Rodrigo Torres and Alexandre Scalzitti, Chair and Vice-Chair of the Organizing Commitee, respectively. Their efforts were the main organizing force behind the LAPTEC'2001.

Jair Minoro Abe
João Inácio da Silva Filho
Chair and Vice-Chair – LAPTEC'2001

Contents

Preface   v

Annotated Logics $Q\tau$ and Model Theory, *J.M. Abe*   1

The Degree of Inconsistency in Paraconsistent Logics, *S. Akama and J.M. Abe*   13

ParaFrame: A Paraconsistent Frame System, *B.C. Ávila and F. Hembecker*   23

Some Model Theory of Ordered Structures and Undefinability Results,
     R. Bianconi   31

Evolutionary Approach to Design of Artificial Neural Networks,
     L.M. Lima de Campos, M. Roisenberg and G.A. Lima de Campos   35

Fuzzy Conceptual Graphs for the Semantic Web, *T.H. Cao*   43

Logic and Valuations, *N.C.A. da Costa*   52

Emmy: A Paraconsistent Autonomous Mobile Robot, *J.I. da Silva Filho*
     and J.M. Abe   53

A Treatment of Limit Figures of Quadratic Transformations as One-way Functions
     in Authentication for Pictures or Documents of Digital Contents, *T. Da-te,*
     T. Yoshikawa, H. Shioya and H. Noanaka   62

Toward a Domain-Theoretic Modelling of Measuring Processes, *G.P. Dimuro*
     and A.C. da Rocha Costa   68

Logic of Incursive Synchronization Applied to the Anticipation of a Chaotic
     Epidemic, *D.M. Dubois*   76

Rules from Supervised Neural Networks in Data Mining Tasks, *N.F.F. Ebecken*
     and E.R. Hruschka   84

Applications of Product Boolean Algebras in Cluster Analysis, *C.G. González*
     and E.C. Maia   101

On Finite BCK with Condition (S), *K. Iséki*   109

Some Fundamental Theorems on BCK, *K. Iséki*   123

Note on the Structure of Weak Interlaced Bilattice $\kappa(L)$, *M. Kondo*   129

Decision-Making System based on Fuzzy and Paraconsistent Logics,
     G. Lambert-Torres, C.I. de Almeida Costa and H. Gonzaga   135

A Polysynthetic Theory of Sets, *E.G.K. López-Escobar*   147

Reasons and Ways to Cope with a Spectrum of Logics, *A. Martini, U. Wolter*
     and E.H. Haeusler   148

The Logic of Medical Diagnosis, *E. Massad and N.R.S. Ortega*   156

Representation of Incompleteness and Inconsistency in Possible-Worlds Semantics,
     T. Murai, G. Resconi and Y. Sato   166

Applications of EVALP Based Reasoning, *K. Nakamatsu, J.M. Abe and A. Suzuki*   174

Logic – A Key of New Information Technology, *S. Ohsuga*   186

Non Local Computation by Semantic Field (Semantic Web), *G. Resconi*
     and T. Murai   200

Ehrenfeucht Theorem for the Annotated Logics $Q\tau$, *A. Scalzitti and J.M. Abe*   208

A Formalization for Signal Analysis of Information in Annotated Paraconsistent
     Logics, *A. Scalzitti, J.I. da Silva Filho and J.M. Abe*   215

Rule and Agent Oriented Software Architecture for Controlling Automated
Manufacturing Systems, *J.M. Simão, P.R.O. da Silva, P.C. Stadzisz
and L.A. Künzle* 224
What Can Be Inferred Without Embracive World-Knowledge, *P. Smrž* 232
Semantic Computation by Humans, Computers and Robots, *P. Suppes* 238
A New Approach to Type-2 Fuzzy Sets, *H. Thiele* 255
A Concurrent Algorithm for Logical Subsumption, *I. Tonin and G. Bittencourt* 263
On Interpolation and Modularity for Ultrafilter Logic, *P.A.S. Veloso* 270
On a Logical Framework for 'Generally', *S.R.M. Veloso and P.A.S. Veloso* 279

Author Index 287

Logic, Artificial Intelligence and Robotics
J.M. Abe & J.I. da Silva Filho (Eds.)
IOS Press, 2001

Annotated Logics $Q\tau$ and Model Theory

Jair Minoro ABE
Institute for Advanced Studies, University of São Paulo, São Paulo, Brazil
SENAC-College of Computer Science and Technology, São Paulo, Brazil

Abstract. Annotated logics are a category of non-classical logics (paraconsistent, paracomplete, and non-alethic logics). In this work we study the annotated first-order logics $Q\tau$ and their model theory, showing that practically all the main basic results in classical model theory can be reproduced in these systems.

1. Introduction

Annotated logics are a family of non-classical logics initially proposed by SUBRAHMANIAN (1987). Subsequently they were studied from the point of view of logic programming by BLAIR & SUBRAHMANIAN (1987), and from a foundational viewpoint by N.C.A. da Costa, V.S. Subrahmanian, J.M. Abe, S. Akama, and others (see DA COSTA, ABE & SUBRAHMANIAN (1991), and ABE (1992, 1997)). Simultaneously, KIFER & LOZINSKII (1989) extended the annotated logic framework by the addition of new logical symbols. This extended framework has also been applied to the development of a declarative semantics for inheritance networks (cf. KIFER & KRISHNAPRASAD (1989)) and object-oriented databases (cf. KIFER & WU (1989)) Other applications are summaried in ABE(1997). In view of the applicability of annotated logics to these differing formalisms in computer science, it has become essential to study these logics more carefully, mainly from the foundational point of view.
In general, annotated logics are a kind of paraconsistent, paracomplete, and non-alethic logic. The latter systems are among the most original and imaginative systems of non-classical logic developed in our century. Nowadays, paraconsistent logic has established a distinctive position in a variety of fields of knowledge.
We study the completeness of the logics $Q\tau$ and their model theory, showing that practically all classical basic results can be adapted to these systems.

2. Paraconsistent, paracomplete, and non-alethic logics

In what follows, we sketch the non-classical logics discussed in the paper, establishing some conventions and definitions.
Let T be a theory whose underlying logic is L. T is called inconsistent when it contains theorems of the form A and $\neg A$ (the negation of A). If T is not inconsistent, it is called *consistent*. T is said to be *trivial* if all formulas of the language of T are also theorems of T. Otherwise, T is called *non-trivial*. When L is classical logic (or one of several others, such as intuitionistic logic), T is inconsistent iff T is trivial. So, in trivial theories the extensions of the concepts of formula and theorem coincide. A *paraconsistent logic* is a logic that can be used as the basis for inconsistent but non-trivial theories. A *theory* is called *paraconsistent* if its

underlying logic is a paraconsistent logic. Issues such as those described above have been appreciated by many logicians. In 1910, the Russian logician Nikolaj A. Vasil'év (1880-1940) and the Polish logician Jan Lukasiewicz (1878-1956) independently glimpsed the possibility of developing such logics. Nevertheless, Stanislaw Jaskowski (1996-1965) was in 1948 effectively the first logician to develop a paraconsistent system, at the propositional level. His system is known as 'discussive propositional calculus'. Independently, some years later, the Brazilian logician Newton C.A. da Costa (1929-) constructed for the first time hierarchies of paraconsistent propositional calculi C_i, $1 \leq i \leq \omega$ of paraconsistent first-order predicate calculi (with and without equality), of paraconsistent description calculi, and paraconsistent higher-order logics (systems NF_i, $1 \leq i \leq \omega$). Another important class of non-classical logics are the paracomplete logics. A logical system is called *paracomplete* if it can function as the underlying logic of theories in which there are formulas such that these formulas and their negations are simultaneously false. Intuitionistic logic and several systems of many-valued logics are paracomplete in this sense (and the dual of intuitionistic logic, Brouwerian logic, is therefore paraconsistent).

As a consequence, paraconsistent theories do not satisfy the principle of non-contradiction, which can be stated as follows: of two contradictory propositions, i.e., one of which is the negation of the other, one must be false. And, paracomplete theories do not satisfy the principle of the excluded middle, formulated in the following form: of two contradictory propositions, one must be true.

Finally, logics which are simultaneously paraconsistent and paracomplete are called *non-alethic logics.*

Problems of various kinds give rise to these non-classical logics: for instance, the paradoxes of set theory, the semantic antinomies, some issues originating in dialectics, in Meinong's theory of objects, in the theory of fuzziness, and in the theory of constructivity. However, one of the most amazing applications was obtained in recent years in Artificial Intelligence (as noted in §1), in particular with expert systems, belief, knowledge, truth, among others.

Throughout this paper, $A \cap B$ and $A \cup B$ indicate the set-theoretical intersection and union, respectively; and if $(A_i)_{i \in I}$ is a family of sets, its union is indicated by $\bigcup_{i \in I} A_i$.

$A \subseteq B$ means that A is a subset of B. $\#A$ indicates the cardinal number of A. $\mathbb{N}$ indicates the set of natural numbers and $\mathbb{N}^* = \mathbb{N} - \{0\}$. Some other usual conventions and notions of set theory are assumed without extensive comments.

3. The logics $Q\tau$

$Q\tau$ is a family of first-order logics, called annotated first-order predicate calculi. They are defined as follows: throughout this work, $\tau = \langle |\tau|, \leq, \sim \rangle$ will be some arbitrary, but fixed, finite lattice of truth values. The least element of τ is denoted by $\bot$, while its greatest element by $\top$. We also assume that there is a fixed unary operator $\sim: |\tau| \to |\tau|$ which constitutes the "meaning" of our negation. $\vee$ and $\wedge$ denote, respectively, the least upper bound and the greatest lower bound operators (of τ) .

The language L of $Q\tau$ is a first-order language (with equality) whose primitive symbols are the following:

1. Individual variables: a denumerably infinite set of variable symbols.

2. Logical connectives: $\neg$, (negation), $\wedge$ (conjunction), $\vee$ (disjunction), and $\rightarrow$ (conditional).

3. For each n, zero or more n-ary function symbols (n is a natural number).

4. For each $n \neq 0$, zero or more n-ary predicate symbols.

5. Quantifiers: $\forall$ (for all) and $\exists$ (there exists).

6. The equality symbol: $=$.

7. Annotated constants: each member of τ is called an annotational constant.

8. Auxiliary symbols: parentheses and commas.

For each n, the number of n-ary function symbols may be zero or non-zero, finite or infinite. A 0-ary function symbol is called a *constant*. Also, for each n-1, the number of n-ary predicate symbol may be finite or infinite.

In the sequel, we suppose that $Q\tau$ possesses at least one predicate symbol.

We define the notion of *term* as usual. Given a predicate symbol p of arity n, an annotational constant λ and n terms $t_1, \dots, t_n$, an *annotated atom* is an expression of the form $p_\lambda t_1 \dots t_n$. In addition, if t_1 and t_2 are terms whatsoever, $t_1 = t_2$ is an *atomic formula*. We introduce the general concept of *formula* in the standard way. Among several intuitive readings, an annotated atom $p_\lambda t_1 \dots t_n$ can be read is *it is believed that*

$p_\lambda t_1 \dots t_n$'*s truth value is at least* λ.

In general, the syntactical notions, as well as the terminology, the notations, etc. are those of SHOENFIELD(1967) with obvious adaptations. We will employ them without extensive comments.

Definition 1. Let A and B formulas of L. We put

$$(A \leftrightarrow B) =_{\text{Def.}} ((A \rightarrow B) \wedge (B \rightarrow A)) \text{ and } (\neg\!\!\!\top A) =_{\text{Def.}} (A \rightarrow ((A \rightarrow A) \wedge (A \rightarrow A))).$$

The symbol '$\leftrightarrow$' is called the *biconditional* and '$\neg\!\!\!\top$ ' is called *strong negation*.

Let A be a formula. $\neg^0 A$ indicates A, $\neg^1 A$ indicates $\neg A$, and $\neg^n A$ indicates $(\neg(\neg^{n-1} A))$, $(n \geq 1)$. Also, if $\mu \in \tau$, $\sim^0 \mu$ indicates μ, $\sim^1 \mu$ indicates $\sim\mu$, and $\sim^n \mu$ indicates $(\sim(\sim^{n-1}\mu))$, $(n \geq 1)$.

Definition 2. Let $p_\lambda t_1 \dots t_n$ be an annotated atom. A formula of the form $\neg^k p_\lambda t_1 \dots t_n$ ($k \geq 0$) is called a *hyper-literal*. A formula other than hyper-literal is called a *complex formula*.

We now introduce the concept of structures for L.

Definition 3. A *structure* S for L consists of the following objects:

1. A non-empty set $|S|$, called the *universe* of S. The elements of $|S|$ are called *individuals* of S.

2. For each n-ary function symbol f of L an n-ary function f_S: $|S|^n \rightarrow |S|$. (In particular, for each constant e of L, e_S is an individual of A.)

3. For each n-ary predicate symbol p of L an n-ary function p_S: $|S|^n \rightarrow |\tau|$.

Let A be a structure for L. As in SHOENFIELD(1967), the *diagram language* L_S is obtained as usual.

If a is a free-variable term, we define the individual $S(a)$ of S. We use i and j as syntactical variables which vary over names.

We define a truth value $S(A)$ for each closed formula A in L_S.

1. If A is $a = b$

$S(A) = 1$ iff $S(a) = S(b)$; otherwise $S(A) = 0$.

2. If A is $p_\lambda t_1 \dots t_n$

$S(A) = 1$ iff $p_S(S(t_1) \dots S(t_n)) \geq \lambda$

$S(A) = 0$ iff it is not the case that $p_S(S(t_1) \dots S(t_n)) \geq \lambda$

3. *If A is $B \wedge C$, or $B \vee C$, or $B \rightarrow C$, we let*

$S(B \wedge C) = 1$ iff $S(B) = S(C) = 1$.

$S(B \vee C) = 1$ iff $S(B) = 1$ or $S(C) = 1$.

$S(B \rightarrow C) = 0$ iff $S(B) = 1$ and $S(C) = 0$.

If A is $\neg^k p_\lambda t_1 \dots t_n$ $(k \geq 1)$, then $S(A) = S(\neg^{k-1} p_{\neg\lambda} t_1 \dots t_n)$.

4. If A is a complex formula, then,

$S(\neg A) = 1 - S(A)$.

5. If A is $\exists x B$, then

$S(A) = 1$ iff $S(B_x[i]) = 1$ for some i in L_S.

6. If A is $\forall x B$, then

$S(A) = 1$ iff $S(B_x[i]) = 1$ for all i in L_S.

A formula A of L is said to be *valid in S* if $S(A') = 1$ for every S-instance A' of A. A formula A is called *logically valid* if it is valid in every structure for L. In this case, we symbolize it by $\vDash A$. If Γ is a set of formulas of L we say that A is a *semantic consequence* of Γ if for any structure S in what $S(B) = 1$ for all $B \in \Gamma$, it is the case that $S(A) = 1$. We symbolize this fact by $\Gamma \vDash A$. Note that when $\Gamma = \varnothing$, $\Gamma \vDash A$ iff $\vDash A$.

Lemma 4. We have:

1. $\vDash p_\perp t_1 \dots t_n$

2. $\vDash \neg^k p_\lambda t_1 \dots t_n \leftrightarrow \neg^{k-1} p_{\neg\lambda} t_1 \dots t_n)$ $(k \geq 1)$

3. $\vDash p_\lambda t_1 \dots t_n \rightarrow p_\mu t_1 \dots t_n, \lambda \geq \mu$

4. $\vDash p_{\mu 1} t_1 \dots t_n \wedge p_{\mu 2} t_1 \dots t_n \wedge \dots \wedge p_{\mu m} t_1 \dots t_n \rightarrow p_{\mu \underset{i=1}{\overset{m}{\vee}}} t_1 \dots t_n$

Now, we shall describe an axiomatic system which we call $A\tau$ whose underlying language is L: A, B, C are any formulas whatsoever, F, G are complex formulas, and $p_\lambda t_1 \dots t_n$ an annotated atom. $A\tau$ consists of the following postulates (axiom schemes and primitive rules of inference), with the usual restrictions:

$(\rightarrow_1)$ $A \rightarrow (B \rightarrow A)$

$(\rightarrow_2)$ $(A \rightarrow (B \rightarrow C) \rightarrow ((A \rightarrow B) \rightarrow (A \rightarrow C))$

$(\rightarrow_3)$ $((A \rightarrow B) \rightarrow A \rightarrow A)$

$(\rightarrow_4)$ $\dfrac{A,\ A \rightarrow B}{B}$ (Modus Ponens)

$(\wedge_1)$ $A \wedge B \rightarrow A$

$(\wedge_2)$ $A \wedge B \rightarrow B$

$(\wedge_3)$ $A \rightarrow (B \rightarrow (A \wedge B))$

$(\vee_1)$ $A \rightarrow A \vee B$

$(\vee_2)$ $B \rightarrow A \vee B$

$(\vee_3)$ $(A \rightarrow C) \rightarrow ((B \rightarrow C) \rightarrow ((A \vee B) \rightarrow C))$

$(\neg_1)$ $(F \rightarrow G) \rightarrow ((F \rightarrow \neg G) \rightarrow \neg F)$

$(\neg_2)$ $F \rightarrow (\neg F \rightarrow A)$

$(\neg_3)$ $F \vee \neg F$

$(\exists_1)$ $A(t) \to \exists x A(x)$

$(\exists_2)$ $\dfrac{A(x) \to B}{\exists x A(x) \to B}$

$(\forall_1)$ $\forall x A(x) \to A(t)$

$(\forall_2)$ $\dfrac{B \to A(x)}{B \to \forall x A(x)}$

(τ_1) $p_\perp t_1 \ldots t_n$

(τ_2) $\neg^k p_\lambda t_1 \ldots t_n \to \neg^{k-1} p_{\sim\lambda} t_1 \ldots t_n, \, k \geq 1$

(τ_3) $p_\lambda t_1 \ldots t_n \to p_\mu t_1 \ldots t_n \, , \, \lambda \geq \mu$

(τ_4) $p_{\lambda 1} t_1 \ldots t_n \wedge p_{\lambda 2} t_1 \ldots t_n \wedge \ldots \wedge p_{\lambda m} t_1 \ldots t_n \to p_\lambda t_1 \ldots t_n, \text{ where } \lambda = \overset{m}{\underset{i=1}{\vee}} \lambda_i$

$(=_1)$ $x = x$

$(=_2)$ $x = y \to (A[x] \leftrightarrow A[y])$

Theorem 5 - 3.5. In $Q\tau$, the operator $\sim$ has all properties of the classical negation. For instance, we have:

1. $\vdash A \vee \neg A$
2. $\vdash \neg(A \wedge \neg A)$
3. $\vdash (A \to B) \to ((A \to \neg B) \to \neg A)$
4. $\vdash A \to \neg\neg A$
5. $\vdash \neg A \to (A \to B)$
6. $\vdash (A \to \neg A) \to B$

among others, where A, B are any formulas whatsoever .

Corollary 6. In $Q\tau$ the connectives $\neg$, $\wedge$, $\vee$, and $\to$ together with the quantifiers $\forall$ and $\exists$ have all properties of the classical negation, conjunction, disjunction, conditional and the universal and existential quantifiers, respectively. If A, B, C are formulas whatsocver, we have, for instance:

1. $(A \wedge B) \leftrightarrow \neg(\neg A \vee \neg B)$
2. $\neg \forall A \leftrightarrow \exists x \neg A$
3. $\exists x B \vee C \leftrightarrow \exists x(B \vee C)$
4. $B \vee \exists x C \leftrightarrow \exists x(B \vee C)$

Theorem 7. If A is a complex formula, then $\vdash \neg A \leftrightarrow \neg A$

Definition 8. We say that a *structure S* is *non-trivial* if there is a closed annotated atom $p_\lambda t_1 \ldots t_n$ such that $S(p_\lambda t_1 \ldots t_n) = 0$.

Hence a structure S is non-trivial iff there is some closed annotated atom that is not valid in S.

Definition 9. We say that a *structure A* is *inconsistent* if there is a closed annotated atom $p_\lambda t_1 \ldots t_n$ such that

$S(p_\lambda t_1 \ldots t_n) = 1 = S(p_\lambda t_1 \ldots t_n)$.

So, a structure S is inconsistent iff there is some closed annotated atom such that it and its negation are both valid in S.

Definition 10. A *structure S* is called *paraconsistent* if S is both inconsistent and non-trivial. The *system $Q\tau$* is said to be *paraconsistent* if there is a structure S for $Q\tau$ such that S is paraconsistent.

Definition 11. A *structure S* is called *paracomplete* if there is a closed annotated atom $p_\lambda t_1 \ldots t_n$ such that $S(p_\lambda t_1 \ldots t_n) = 0 = S(p_\lambda t_1 \ldots t_n)$.

The *system* $Q\tau$ is said to be *paracomplete* if there is a structure S such that S is paracomplete.

Theorem 12. $Q\tau$ is paraconsistent iff $\#|\tau| \geq 2$.

Proof. Suppose that $\#|\tau| \geq 2$. There is at least one predicate symbol p. Let IAI be a non-empty set which $\#|\tau| \geq 2$. Let us define $p_S:|S|^n \to |\tau|$ setting $p_S(a_1, \ldots , a_n) = \perp$ and $p_S(b_1, \ldots , b_n)$ $= \top$ where $(a_1, \ldots , a_n) \neq (b_1, \ldots , b_n)$.

Then, $S(p_\perp i_1 \ldots i_n) = 1$, where i_j the name of b_j, $j = 1, \ldots , n$, and $S(\neg p_\perp i_1 \ldots i_n) = 1$. Likewise, $S(p_\top j_1 \ldots j_n) = 0$, where j_i is the name of a_i, $i = 1, \ldots , n$. So, $Q\tau$ is paraconsistent. The converse is immediate.

Theorem 13. For all τ there are systems $Q\tau$ that are paracomplete; and also systems that are not paracomplete. If $Q\tau$ is paracomplete, then $\#|\tau| \geq 2$.

Proof. Similar to the proof of the preceding theorem.

Definition 14. A *structure* S is called *non-alethic* if S is both paraconsistent and paracomplete. The *system* $Q\tau$ is said to be *non-alethic* if there is a structure S for $Q\tau$ such that S is non-alethic.

Theorem 15. For all $\#|\tau| \geq 2$ there are systems $Q\tau$ that are non-alethic; and also systems that are not non-alethic. If $Q\tau$ is non-alethic, then $\#|\tau| \geq 2$. Given a structure S, we can define the theory $\text{Th}(S)$ associated with S to be the set $\text{Th}(S) = C_n(\Gamma)$, where Γ is the set of all annotated atoms which are valid in S $C_n(\Gamma)$ indicates the set of all semantic consequences of elements of Γ.

Theorem 16. Given a structure S for $Q\tau$, we have:
1. $\text{Th}(S)$ is a paraconsistent theory iff S is a paraconsistent structure. 2. $\text{Th}(A)$ is a paracomplete theory iff S is a paracomplete structure. 3. $\text{Th}(S)$ is a non-alethic theory iff S is a non-alethic structure.

In view of the preceding theorem, $Q\tau$ is, in general, a paraconsistent, paracomplete and non-alethic logic.

4. Completeness

We give a Henkin-type proof of the completeness theorem for the logics $Q\tau$.

Definition 4.1. A theory T based on $Q\tau$ is said to be *complete* if for each closed formula A we have $\vdash_T A$ or $\vdash_{T\neg} A$.

Lemma 17. Let $\lambda_0 = \vee\{\lambda \in |\tau|: \vdash_T p_\lambda t_1 \ldots t_n \}$. Then $\vdash_T p_{\lambda_0} t_1 \ldots t_n$. Now let T be a non-trivial theory containing at least one constant. We shall define a structure S that we call the *canonical structure* for T. If a and b are variable-free terms of T, then we define aRb to mean $\vdash_T a = b$. It is easy to check that R is an equivalence relation. We let $|S|$ be the quotient set F/R, where F indicates the set of all formulas of L. The equivalence class determined by a is designed by a°. We complete the definition of S by setting $f_S(a_1^{\,0}, \ldots , a_m^{\,0}) = (f_S(a_1, \ldots , a_m))^0$ and

$$p_S(a_1^{\,0}, \ldots , a_n^{\,0}) = \vee\{\lambda \in |\tau|: \vdash_T p_\lambda a_1 \ldots a_n\}$$

It is straightforward to check the formal correctness of the above definitions.

Theorem 18. If $p_\lambda a_1 \ldots a_n$ is a variable-free annotated atom, then

$$S(p_\lambda a_1 \ldots a_n) = 1 \text{ iff } \vdash_T p_\lambda a_1 \ldots a_n.$$

Proof. Let us suppose that $S(p_\lambda a_1 \ldots a_n) = 1$. Then $p_S(a_1^{\,0}, \ldots , a_n^{\,0}) \geq \lambda$.

But $p_S(a_1^0, \ldots, a_n^0) = \vee\{\lambda \in |\tau|: \vdash_T p_\lambda a_1 \ldots a_n\}$; so, $\vdash_T p_{\lambda_0} a_1 \ldots a_n$ by the preceding lemma. As $\lambda_0 \geq \lambda$, it follows that $\vdash_T p_\lambda a_1 \ldots a_n$ by axiom (τ_3).

Conversely, let us suppose that $\vdash_T p_\lambda a_1 \ldots a_n$. Then $\vee\{\mu \in |\tau|: \vdash_T p_\mu a_1 \ldots a_n\}$. Let $\lambda_0 = \vee\{\mu \in |\tau|: \vdash_T p_\mu a_1 \ldots a_n\};)$. Then it follows that $\lambda_0 \geq \lambda$. But $\lambda_0 = p_S(a_1^0, \ldots, a_n^0)$, and so $p_S(a_1^0, \ldots, a_n^0) \geq \lambda$; hence $S(p_\lambda a_1 \ldots a_n) = 1$.

A formula A is called *variable-free* if A does not contain free variables.

Theorem 19. Let $a = b$ be a variable-free formula. Then $S(a = b) = 1$ iff $\vdash_T a = b$.

We define the Henkin theory as in the classical case. Now, suppose that T is a Henkin theory and S the canonical structure for T.

Theorem 20. Let $\neg^k p_\lambda a_1 \ldots a_n$ be a variable-free hyper-literal. Then

$$S(\neg^k p_\lambda a_1 \ldots a_n) = 1 \text{ iff } \vdash_T \neg^k p_\lambda a_1 \ldots a_n$$

Proof. By induction on k taking into account the axiom (τ_2) and theorem 18.

Theorem 21. Let T be a complete Henkin theory, S the canonical structure for T, and A a closed formula. Then, $S(A) = 1$ iff A.

Proof. By induction on the length of A.

Corollary 22. Under the conditions of the theorem the canonical structure for T is a model of T.

We construct Henkin theories as in the classical case.

Theorem 23. (*Lindenbaum's theorem*). If T is a non-trivial theory, then T has a complete simple extension.

Theorem 24. (*Completeness theorem*). A theory (consistent or not) is non-trivial iff it has a model.

Theorem 25. Let Γ be a set of formulas. Then $\Gamma \vdash A$ iff $\Gamma \vDash A$.

Theorem 26. A formula A of a theory T is a theorem of T iff it is valid in T.

5. Some metatheorems

In this section we establish some necessary syntactical results for the development of the model theory of $Q\tau$.

Theorem 27.(*Equivalence theorem*). Let $B_1, \ldots, B_n$ and $B'_1, \ldots, B'_n$ be complex formulas. Let A' be obtained from A by replacing some occurrences of $B_1, \ldots, B_n$ by $B'_1, \ldots, B'_n$ respectively. If $\vdash B_i \leftrightarrow B'_i$, $i = 1, \ldots, n$, then $\vdash A \leftrightarrow A'$.

Proof. By induction on the length of A, as in the classical corresponding theorem. Nonetheless, let us consider the case when A is $\neg C$. There are two possibilities:
1. an occurrence of a formula in A is the whole of A or
2. it is entirely contained in C.

The first case is immediate. In the second one, A' is $\neg C'$, where C' results from C by replacements of the type described in the statement of the theorem. By the induction hypothesis, $\vdash C \leftrightarrow C'$, as C and C' are complex formulas, it follows that $\vdash \neg C \leftrightarrow \neg C'$, and so $\vdash A \leftrightarrow A'$, by theorem 7.

Let us recall some standard syntactical notions. A formula A is called *open* if A does not contain quantifiers. A formula A is said to be in *prenex-form* if it has the form $Qx_1 \ldots$

Qx_nB, where each Qx_i is either $\forall x_i$ or $\exists x_i$; $x_1, \ldots, x_n$ are distinct, and B is open. We call $Qx_1 \ldots Qx_nB$ the *prefix* and B the *matrix* of A. We allow the prefix to be empty.

Using the equivalence theorem we can deduce that for some formulas (but not for all) there are equivalent formulas in prenex form.

We say that a formula A is *logically equivalent* to a formula B iff $\vdash A \leftrightarrow B$. A formula A is said to be *existential* (*universal*) iff it is logically equivalent to a formula in prenex form, where all the quantifiers in its prefix are existential (universal). A formula is called *universal-existential* iff it is equivalent to a formula in prenex form, such that all the universal quantifiers precede all the existential quantifiers in the prefix. We abbreviate this saying simply a $\forall\exists$-formula.

Each existential formula and each universal formula is degenerately universal-existential.

Theorem 28. (*Reduction theorem*). Let Γ be a set of formulas in the theory T, and let A be a formula of T. Then A is a theorem of $T[\Gamma]$ iff there is a theorem of T of the form $B_1 \to B_2 \to \ldots \to B_n \to A$, where each B_i is the closure of a formula in Γ.

Theorem 29. (*Reduction theorem for trivialization I*). Let Γ be a non-empty set of formulas in the theory T. Then $T[\Gamma]$ is trivial iff there is a theorem of T that is a disjunction of strong negations of closures of distinct formulas in Γ.

Proof. Similar to the corresponding classical theorem, using strong negation instead of the weak negation, and 'trivialization' instead of 'inconsistency'.

Theorem 30. (*Reduction theorem for trivialization II*). Let Γ be a non-empty set of complex formulas in the theory T. Then $T[\Gamma]$ is trivial iff there is a theorem of T that is a disjunction of negations of closures of distinct formulas in Γ.

Proof. Consequence of the theorem 7 and of the preceding theorem.

Corollary 31. Let A' be the closure of A. Then A is a theorem of T *iff* $T[\neg A']$ is trivial.

Corollary 32. Let A' be the closure of a complex formula A. Then A is a theorem of T iff $T[\neg A']$ is trivial.

6. The theory of models

As an immediate consequence of the completeness theorem, we have:

Theorem 33. (*Compactness theorem*). A formula A in a theory T is valid in T iff it is valid in some finitely axiomatized part of T.

Corollary 34. A theory T has a model iff every finitely axiomatized part of T has a model.

Definition 35. Let A and B be structures for L. An *isomorphism between* A and B is a bijective function $\varphi : |A| \to |B|$ such that

$(I_1)\, f_B(\varphi(a_1), \ldots, \varphi(a_n)) = \varphi(f_A(a_1, \ldots, a_n))$ and

$(I_1)\, p_B(\varphi(a_1), \ldots, \varphi(a_n)) = \lambda \Leftrightarrow p_A(a_1, \ldots, a_n), \forall\lambda\in|\tau|$ and

hold for all function symbols f, for all predicate symbols p of L, and all $a_1, \ldots, a_n \in |A|$.

Let A and B be structures for L and let $\varphi : |A| \to |B|$ be a function. If i is the name of an individual $a \in |A|$, we use i^φ to designate the name of the individual $\varphi(a) \in |B|$. If u is an expression of $L(A)$, u^φ is the expression obtained from u by replacing each name i by i^φ.

Theorem 36. Let φ be an isomorphism between A and B. Then $\varphi(A(a)) = B(a^\varphi)$ for every variable-free term a of $L(A)$ and $A(\neg^k p_\lambda a_1 \ldots a_n) = B((\neg^k p_\lambda a_1 \ldots a_n)^\varphi)$

for every closed hyper-literal $\neg^k p_\lambda a_1 \ldots a_n$ of $L(A)$.

Proof. The first part is proved by induction on the length of a. For the second part we proceed by induction on k. If $k = 0$, we have

$$A(p_\lambda a_1 \ldots a_n) = 1 \Leftrightarrow p_A(A(a_1, \ldots , a_n)) \geq \lambda \Leftrightarrow p_B(\varphi(A(a_1), \ldots , \varphi(a_n))) \geq \lambda \Leftrightarrow$$
$$p_B(B(a_1{}^\varphi), \ldots , B(a_n{}^\varphi)) \geq \lambda \Leftrightarrow B(p_\lambda a_1 \ldots a_n) = 1$$

Now, let us suppose the theorem valid for $k > 0$. Then

$$A(\neg^{k+1} p_\lambda a_1 \ldots a_n) = 1 \Leftrightarrow A(\neg^k p_{\sim\lambda} a_1 \ldots a_n) = 1 \text{ (by axiom } (\tau_2))$$
$$\Leftrightarrow B(\neg^k p_{\sim\lambda} a_1 \ldots a_n) = 1 \text{ (by induction hypothesis)}$$
$$B(\neg^{k+1} p_\lambda a_1 \ldots a_n) = 1 \text{ (by axiom } (\tau_2))$$

Corollary 37. Under the conditions of the theorem $A(A) = B(A^\varphi)$ for every closed formula A of $L(A)$.

Proof. By induction on the length of A. If A is a hyper-literal, then the result follows from the theorem. If A is $a = b$, then it is easy to see that $A(A) = 1 = B(a^\varphi = b^\varphi)$. If A is a conjunction, disjunction, or conditional or is of the form $\exists xB$ or $\forall xB$, the result follows as in the classical case.

Two structures A and B are called *elementarily equivalent* if the same formulas of L are valid in A and B. This implies that A and B are models of the same theories. An *embedding* of A in B is an injective mapping $\varphi : |A| \to |B|$ such that

$(E_1) f_B(\varphi(a_1), \ldots , \varphi(a_n)) = \varphi(f_A(a_1, \ldots , a_n))$ and

$(E_1) p_B(\varphi(a_1), \ldots , \varphi(a_n)) = \lambda \Leftrightarrow p_A(a_1, \ldots , a_n), \forall \lambda \in |\tau|$ and

hold for all function symbols f, for all predicate symbols p of L, and all $a_1, \ldots , a_n \in |A|$.

If $|A| \subseteq |B|$ and the identity mapping from $|A|$ to $|B|$ is an embedding of A in B, then A is called a *substructure* of B and B is an *extension* of A. If A and B are models of some theory, we sometimes say *submodel* for substructure.

Theorem 38. Let A and B be structures for L and $\varphi : |A| \to |B|$ be a mapping. Then φ is an embedding of A in B iff $A(A) = B(A^\varphi)$ for every variable-free formula A of L.

Proof. The proof that the conditions holds if φ is an embedding is just like the proof of the preceding theorem. Now suppose that the condition holds. Let $a_1, \ldots , a_n$ be individuals of A, and let $i_1, \ldots , i_n$ be the names of $a_1, \ldots , a_n$ respectively. Then

$A(p_\lambda i_1 \ldots i_n) = B(p_\lambda i_1{}^\varphi \ldots i_n{}^\varphi)$, for each $\lambda \in |\tau|$. Now let $p_A(A(i_1), \ldots , A(i_n)) = \mu$ and

$p_B(B(i_1{}^\varphi), \ldots , B(i_n{}^\varphi)) = \theta$

It is clear that $A(p_\lambda i_1 \ldots i_n) = 1$; so $B(p_\lambda i_1{}^\varphi \ldots i_n{}^\varphi) = 1$, that is, $p_B(B(i_1{}^\varphi), \ldots , B(i_n{}^\varphi)) \geq \mu$ or, $\theta \geq \mu$. In a similar way, $\mu \geq \theta$. Therefore, $\theta = \mu$. This implies (E_2). The proof of (E_1) is just like the classical one.

Corollary 39. Let A and B be structures for L such that $|A| \subseteq |B|$. Then A is a substructure of B iff $A(A) = B(A)$ for every variable-free formula A of $L(A)$.

Let Γ be a set of formulas of L and A a structure for L. $\Gamma(A)$ designates the set of A-instances of formulas in Γ. Thus if Γ is the set of all open formulas in L then $\Gamma(A)$ is the set of all variable-free formulas in $L(A)$; if Γ is the set of all formulas in L then $\Gamma(A)$ is the set of all closed formulas of $L(A)$. Let A and B be structures for L such that $|A| \subseteq |B|$, and let Γ be a set of formulas in L. We say that A is a Γ-*substructure* of B and that B is a Γ-*extension* of A if for every formula $A \in \Gamma(A)$, $A(A) = 1$ implies $B(A) = 1$.

Theorem 40. Let Γ be the set of open formulas of L. Then the strong negation of every formula in Γ is in Γ. Moreover, the same is true for $\Gamma(A)$, where A is a structure for L.

Proof. It is enough to consult the definition 1 and the definition of $\Gamma(A)$.

Theorem 41. Let Γ be the set of open formulas of L. Then a Γ-substructure is simply a substructure and a Γ-extension is simply an extension.

Proof. In effect, let A be a Γ-substructure of B, and A a formula in $\Gamma(A)$. If $A(A) = 0$, then $A(\neg A) = 1$; so $B(\neg A) = 1$ and therefore $B(A) = 0$.

If F is the set of all formulas in L, we say *elementary structure* for F-substructure and *elementary extension* for F-extension.

Theorem 42. If $x = e$ is in Γ and A is a Γ-substructure of B, then $A(e) = B(e)$.

Let A and B be. structures for L such that $|A| \subseteq |B|$. We expand B to a structure B_A for $L(A)$ as follows: if i is the name of an individual a of A, then B_A assigns a to i.

The notion of Γ-diagram of A is the same as in the classical case, and denoted by $D_\Gamma(A)$.

If Γ if the set of open formulas, we write $D(A)$ for $D_\Gamma(A)$; if Γ is the set of all formulas, we write $D_F(A)$, for $D_\Gamma(A)$.

Theorem 43. (*Diagram lemma*). Let Γ be a set of formulas in L, and let A and B be structures for L such that $|A| \subseteq |B|$. Then A is a Γ-substructure of B iff B_A is a model of $D_\Gamma(A)$.

Let Γ be a set of formulas in L. We say that Γ is *regular* if every formula of the form $x = y$ or $\neg(x = y)$ is in Γ, and if for every formula $A \in \Gamma$, every formula of the form $A[x_1, \ldots, x_n] \in \Gamma$.

Theorem 44. (*Model extension theorem I*). Let A be a structure for L, T a theory in the language L, and Γ a regular set of formulas in L. Then A has a Γ-extension that is a model of T iff every theorem of T that is a disjunction of strong negations of formulas in Γ is valid in A.

Proof. Similar to the classical case, using the reduction theorem for trivialization I.

Theorem 45. (*Model extension theorem II*). Let A, L, T, and Γ be as in the preceding theorem. Then A has a Γ-extension that is a model of T iff every theorem of T that is a disjunction of negations of complex formulas in r is valid in A.

Proof. Similar to the classical case, using the reduction theorem for trivialization II.

Corollary 46. Let Γ a regular set of formulas in L and let Δ be a set of formulas containing every formula $\forall x_1 \ldots \forall x_n A$, where A is a disjunction of strong negations of formulas in Γ. If A is a structure for L and B is a Δ-extension of A, then there is a Γ-extension C of B that is an elementary extension of A.

Corollary 47. Let Γ and L be as in the preceding corollary. Let Δ be a set of formulas containing every formula $\forall x_1 \ldots \forall x_n A$, where A is a disjunction of negations of complex formulas in Γ. If A is a structure for L and B is a Δ-extension of A, then there is a Γ-extension C of B that is an elementary extension of A.

When Γ is either the set of open formulas or the set of existential formulas, the preceding results provide some conditions on the models of T that determine T is equivalent to a theory whose non-logical axioms are in Γ.

Theorem 48. Let Γ be a set of formulas in L and let Γ' be the set of formulas in Γ that are theorems of T. If every structure for $L(T)$ in which all the formulas of Γ' are valid is a model of T, then T is equivalent to a theory whose non-logical axioms are in Γ.

Theorem 49. (*Los-Tarski theorem*). A theory T is equivalent to an open theory iff every substructure of a model of T is a model of T.

Proof. As in the corresponding classical case, employing our version of the model extension theorem (I or II).

A sequence $(A_i)_{i \in \mathbb{N}^*}$ of structures for L is called a *chain* if for each n, A_{n+1} is an extension of A_n. Given a chain, we define a structure A called the *union* of the chain.

The universe of A is $|A| = \bigcup_{i=1}^{\infty} |A_i|$. If $a_1, \ldots, a_k \in |A|$, then there is an n such that all $a_1, \ldots, a_k \in |A|$. We then set $f_A(a_1, \ldots, a_k) = f_{An}(a_1, \ldots, a_k)$ and

$p_A(a_1, \ldots, a_k) = \lambda \Leftrightarrow p_{An}(a_1, \ldots, a_k) = \lambda, \forall \lambda \in |\tau|$

It is immediate that this definition is well defined. An *elementary chain* is a chain $(A_i)_{i \in \mathbb{N}*}$ such that for each n, A_{n+1} is an elementary extension of A_n.

Theorem 50. (*Tarski*). If $(A_i)_{i \in \mathbb{N}*}$ is an elementary chain, then the union

$A = \bigcup_{i=1}^{\infty} A_i$ is an elementary extension of each A_i.

Proof. By induction on the length of A, where A is a closed formula in $.L(A_i)$ in order to proof that $A_i(A) = A(A)$, using the corollary 39.

Theorem 50. (*Chang-Los-Suszko theorem*). A theory is equivalent to a theory whose non-logical axioms are existential iff every union of a chain of models of T is a model of T.

Proof. The proof that the condition holds if T is equivalent to a theory T', whose non-logical axioms are existential, is formally the same as in the classical case. So, suppose that every union of a chain of models of T is a model of T. We shall construct a chain $(A_i)_{i \in \mathbb{N}*}$ such that $A_1 = A$, A_{2i} is a model of T, and A_{2i+3} is an elementary extension of A_{2i+1}. To construct the chain, we use the model extension theorem (I or II). We need to establish that, if A is a disjunction of strong negations of universal formulas, then A is equivalent to a formula B in prenex form, that is existential. But this is true in virtue of corollary 6.

Given such a chain, we let B be $\bigcup_{i=1}^{\infty} A_i$. Then B is the union of the chain $(A_{2i})_{i \in \mathbb{N}*}$ of models of T, and hence, is a model of T. But B is also the union of the elementary chain $(A_{2i+1})_{i \in \mathbb{N}*}$. Hence by Tarski's theorem, B is an elementary extension of $A_1 = A$, and therefore is elementarily equivalent to A. It follows that A is a model of T.

Theorem 51. If a closed formula A is $\forall\exists$-formula valid in each structure of a chain $(A_i)_{i \in \mathbb{N}*}$, then A is valid in each union of the chain.

The classical results on cardinality of models like the cardinality theorem of Tarski and the Löwenheim-Skolem theorem, can be immediately extended, since they are theorems of a structural character.

Now let T and T' be theories. The *union* of T and T', designated by $T \cup T'$, is the theory whose non-logical symbols are the non-logical symbols of T', and whose non-logical axioms are all the non-logical axioms of both T and of T'.

Theorem 52. (*Craig-Robinson theorem I*). Let T and T' be theories. Then $T \cup T'$ is trivial iff there is a closed formula A such that $\vdash_T A$ and $\vdash_{T'} \rceil A$.

Proof. Similar to the classical case, using strong negation and the notion of trivial theory.

Theorem 53. (*Craig-Robinson theorem II*). Let T and T' be theories. Then $T \cup T'$ is trivial iff there is a closed complex formula A such that $\vdash_T A$ and $\vdash_{T'} \neg A$.

Corollary 54. (*Craig interpolation theorem*). Let T and T' be theories, and let A, B formulas such that $\vdash_{T \cup T'} A \to B$ where A is a formula of T and B is a formula of T'. Then there is a formula C such that $\vdash_T A \to C$ and $\vdash_{T'} C \to B$.

We can define also the definability of a symbol in terms of a set of symbols. By using the concept of u-isomorphism, we have the following application of the interpolation theorem:

Theorem 55. (*Beth's definability theorem*). Let Q be a set of non-logical symbols in T, and u be a non-logical symbol of T that is not in Q . Then u is definable in terms of Q in T iff for every pair of models A and B of T and every bijective mapping $\varphi: |A| \to |B|$ that is a v-isomorphism for every $v \in Q$, φ is a u-isomorphism.

7. Concluding remarks

1. The results presented in this work show us a little of the reach of the model theory of annotated systems. Practically all standard results in classical model theory have a version for these systems. Naturally, other important basic topics, such as applications of complete theories (e.g. some version of the quantifier elimination theorem) and saturated model theory remain to be studied.

2. Formulas generated by the equality symbol can be also be 'annotated'. We can also develop versions of first-order languages of $Q\tau$ without equality, and study their model theory, as well as develop some version of annotated higher-order logic (v.g. in the form of type theory or set theory). In DA COSTA, ABE & SUBRAHMANIAN (1991) and ABE (1992) we showed that a version of annotated set theory encompasses fuzzy set theory, and some of its generalizations.

3. The systems $Q\tau$ can be envisaged as extensions of classical first-order logic (see DA COSTA, ABE & SUBRAHMANIAN (1991) and ABE (1992)).

We intend to develop all those questions in forthcoming papers.

References

1. ABE, J. M. (1992), *Fundamentos da Lógica Anotada* (Foundations of Annotated Logics), Ph.D. Thesis, University of São Paulo, São Paulo.
2. ABE, J.M. (1997), Some Aspects of Paraconsistent Systems and Applications, *Logique et Analyse*, 157, 83-96.
3. ABE, J.M. & S. AKAMA (1997), Annotated logics $Q\tau$ and ultraproducts, *Logique et Analyse* 160, 335-343.
4. BLAIR, H. A. and V. S. SUBRAHMANIAN (1987), *Paraconsistent Logic Programming*, Proc. *7th* Conference on Foundations of Software Tech- nology and Theoretical Computer Science, Lecture Notes in Computer Science, vol. 287, 340-360, Springer- Verlag.
5. DA COSTA, N.C.A., V.S. SUBRAHMANIAN & C. VAGO (1991), 'The paraconsistent logics $P\tau$, *Zeitschrift fur Math. Logik und Grund. der Math.* 37,137-148.
6. DA COSTA, N.C.A., J.M. ABE & V.S. SUBRAHMANIAN (1991), 'Remarks on annotated logic', *Zeitschr. f. Math. Logik und Grundlagen d. Math.*, 37: 561-570.
7. KIFER, M. and E. LOZINSKII (1989), *RI: A Logic for Reasoning with Inconsistency*, LICS-89.
8. KIFER, M. and T. KRISHNAPRASAD (1989), *An Evidence Based for a Theory of Inheritance*, Proc. 11 *th* International Joint Conf. on Arti- ficial Intelligence, 1093-1098, Morgan-Kaufmann.
9. KIFER, M. and J. WU (1989), *A logic for Object Oriented Logic Programming*, Proc. *8th* ACM Symp. on Principles of Database Systems, 379-393.
10. SHOENFIELD, J.R. (1967), *Mathematical Logic*, Addison Wesley, Reading.
11. SUBRAHMANIAN, V.S. (1987), *On the Semantics of Quantitative Logic Programs*, Proc. *4th* IEEE Symposium on Logic Programming, Com- puter Society Press, Washington D.C., 173-182.
12. SYLVAN, R. & J.M. ABE (1996), On general annotated logics, with an introduction to full accounting logics, *Bulletin of Symbolic Logic*, 2, 118-119.

Logic, Artificial Intelligence and Robotics
J.M. Abe & J.I. da Silva Filho (Eds.)
IOS Press, 2001

The Degree of Inconsistency in Paraconsistent Logics

Seiki Akama

Computational Logic Laboratory, Department of Information Systems,
Teikyo Heisei University, 2289 Uruido, Ichihara-shi,
Chiba, 290-0193, Japan
e-mail akama@thu.ac.jp

Jair Minoro Abe

Department of Informatics, ICET, Paulista University,
R. Dr. Bacelar 1212, 04026-002, Sao Paulo, SP, Brazil
e-mail jairabe@uol.com.br

Abstract
One of the starting points in paraconsistent logics such as da Costa's (1974) C_1 is to localize inconsistency. However, we may be faced with global inconsistency, which cannot be localized. Existing paraconsistent logics are aimed at dealing only with local inconsistency. We propose to extend paraconsistent logics with the mechanism of representing the (necessity) degree of global inconsistency by introducing the idea of possibilistic logic of Dubois, Lang and Prade (1992). The resulting logic called possibilistic annotated logics $PP\tau$ can be formulated as fusing Subrahmanian's (1987) annotated logics and possibilistic logics. The new logic can be seen as an ideal system capable of treating inconsistency in a uniform manner.

Keywords
local inconsistency, global inconsistency, paraconsistent logics, annotated logics, possibilistic logics, possibilistic annotated logics.

1   Introduction

It was often claimed that the treatment of inconsistency in AI systems is crucial from a practical point of view. *Paraconsistent logics* are capable of constructing inconsitent but non-trivial theories. This means that we cannot derive all formulas from an inconsistent theory in paraconsistent logics. One of the starting points in paraconsistent logics such as da Costa's (1974) C_1 is to localize inconsistency. However, we may be faced with global inconsistency, which cannot be localized. In other words, there are two kinds of inconsistencies, i.e. *local inconsistency* and *global inconsistency*. Existing paraconsistent logics are aimed at dealing only with local inconsistency. We should be able to handle both local and global inconsistency in a single framework.

We propose to extend paraconsistent logics with the mechanism of representing the (necessity) degree of global inconsistency by introducing the idea of *possibilistic logic* of Dubois, Lang and Prade (1994). The resulting logic called possibilistic annotated logics $PP\tau$ can be formulated as fusing Subrahmanian's (1987) annotated logics and possibilistic logics. The new logic can be seen as an ideal system capable of treating inconsistency in a uniform manner. Furthermore, we relate possibilistic annotated logics $PP\tau$ to fuzzy annotated logics $FP\tau$ of Akama and Abe (1990).

The structure of this paper is as follows. In section 2, we give the outline of annotated logics $P\tau$. Section 3 introduces possibilistic logics. In section 4, we develop possibilistic annotated logics $PP\tau$ with the axiomatization and the semantics. We related $PP\tau$ to fuzzy annotated logics $FP\tau$ in section 5. We conclude this paper with some remarks on future work in section 6.

2 Annotated Logics $P\tau$

In this section, we review the paraconsistent annotated propositional logics $P\tau$. A detailed account is to be found in da Costa, Subrahmanian and Vago (1991), da Costa, Abe and Subrahmanian (1991) and Abe (1992). Here, we assume that if S is a set then $\sharp S$ indicates the cardinality of S. Other usual terminology of naive set theory will be employed without major comments.

The symbol $\tau = \langle |\tau|, \leq, \sim \rangle$ denotes some fixed finite lattice called *lattice of truth-values*. We use the symbol $\leq$ to denote the ordering in which τ is a complete lattice, $\bot$ and $\top$ to denote, respectively, the bottom element and top element of τ. Also, $\wedge$ and $\vee$ indicate, respectively, the greatest lower bound and the least upper bound operators with respect to subsets of $|\tau|$. We also fix an operator $\sim : |\tau| \to |\tau|$ which will work as the "meaning" of the negation of the system $P\tau$. The language of $P\tau$ has the following (denumerable) primitive symbols.

1. Propositional symbols: $p, q, r, \dots$.
2. Connectives, i.e. $\neg$ (negation), $\wedge$ (conjunction), $\vee$ (disjunction) and $\to$ (implication).
3. Each member of τ is called an *annotated constant*: $\lambda, \mu, \theta, \dots$
4. Auxiliary symbols: (,).

Formulas are now defined as follows:

1. If p is a propositional symbol and $\lambda \in \tau$ is an annotated constant, then p_λ is an atomic formula.
2. A and B are formulas, then $\neg A, A \wedge B, A \vee B, A \to B$ are formulas.
3. Only those expressions are formulas that are determined to be so by means of conditions 1 and 2.

The atomic formula p_λ can be read "it is believed that p's truth-value is at least λ". Let A be a formula. Then, $\neg^0 A$ is A, $\neg^1 A$ is $\neg A$, and $\neg^k A$ is $\neg(\neg^{k-1}A)$, where $k \geq 0$ is the natural number. The convention is also used for $\sim$. If p is a propositional symbol and λ is an annotated constant, then the formula of the form $\neg^k p_\lambda$ is called a *hyper-literal*. A formula which is not a hyper-literal is called a *complex formula*.

Definition 2.1

Let A and B be formulas. Then we put
$$A \leftrightarrow B =_{def} (A \to B) \wedge (B \to A)$$
and $\neg_* A =_{def} A \to (A \to A) \wedge \neg(A \to A)$.

The formula $A \leftrightarrow B$ is read, as usual, the *equivalence* of A and B. The operator $\neg_*$ is called *strong negation*, so $\neg_* A$ must be read the *strong negation* of A.

The postulates (axiom schemata and primitive rules of inference) of P_τ are the following: A, B, and C are formulas whatsoever, F and G are complex formulas, p is a propositional symbol, and λ, μ, λ_j are annotated constants.

(A1) $A \to (B \to A)$
(A2) $(A \to (B \to C)) \to ((A \to B) \to (A \to C))$
(A3) $((A \to B) \to A) \to A$
(A4) $A, A \to B / B$
(A5) $A \wedge B \to A$
(A6) $A \wedge B \to B$
(A7) $A \to (B \to (A \wedge B))$
(A8) $A \to A \vee B$
(A9) $B \to A \vee B$
(A10) $(A \to C) \to ((B \to C) \to ((A \vee B) \to C))$
(A11) $(F \to G) \to ((F \to \neg G) \to \neg F)$
(A12) $F \to (\neg F \to A)$
(A13) $F \vee \neg F$
(A14) $p_\perp$
(A15) $\neg^k p_\lambda \to \neg^{k-1} p_{\sim\lambda}, k \geq 1$
(A16) $p_\lambda \to p_\mu, \lambda \geq \mu$
(A17) $p_{\lambda_1} \wedge p_{\lambda_2} \wedge ... \wedge p_{\lambda_m} \to p_\lambda$, where $\lambda = \bigvee_{i=1}^{m} \lambda_i$

Theorem 2.2

In P_τ all valid formulas of the classical positive propositional calculus are valid.

Theorem 2.3

In P_τ, $\neg_*$ has all properties of the classical negation. For instrance, we have:

1. $\vdash A \vee \neg_* A$
2. $\vdash \neg_*(A \wedge \neg_* A)$
3. $\vdash (A \to B) \to ((A \to \neg_* B) \to \neg_* A)$
4. $\vdash A \leftrightarrow \neg_* \neg_* A$
5. $\vdash \neg_* A \to (A \to B)$
6. $\vdash (A \to \neg_* A) \to B$

Corollary 2.4

In P_τ, the connectives $\neg_*, \wedge, \vee$, and $\to$ have all properties of the classical negation, conjunction, disjunction and implication, respectively.

Theorem 2.5

$P\tau$ is non-trivial.

Definition 2.6

Let Γ be a set of formulas The *syntactical consequence* of Γ, symbolized by $\bar{\Gamma}$, is the set $\bar{\Gamma} = \{A \in \mathbf{F} \mid \Gamma \vdash A\}$. If $\Gamma = \bar{\Gamma}$, then Γ is called a *theory*. A set of formulas Γ is called *trivial* if $\bar{\Gamma} = \mathbf{F}$. Otherwise Γ is *non-trivial*.

Theorem 2.7

Let Γ be a non-trivial set of formulas. Then Γ can be extended to a maximal (with respect to inclusion of sets) non-trivial set with respect to $\mathbf{F}$.

Theorem 2.8

Let Γ be a maximal non-trivial set of formulas. Then

1. If A is an axiom of $P\tau$, then $A \in \Gamma$.
2. $A, B \in \Gamma$ iff $A \wedge B \in \Gamma$.
3. $A \vee B \in \Gamma$ iff $A \in \Gamma$ or $B \in \Gamma$.
4. If $p_\lambda, p_\mu \in \Gamma$, then $p_\theta \in \Gamma$. where $\theta = max(\lambda, \mu)$.
5. $\neg^k p_\mu \in \Gamma$ iff $\neg^{k-1} p_{\sim\mu}$.
6. If $A, A \to B \in \Gamma$ then $B \in \Gamma$.
7. $A \to B \in \Gamma$ iff $A \notin \Gamma$ or $B \in \Gamma$.

Next, we describe a semantics for $P\tau$; see Abe (1992) for details. We denote by $\mathbf{P}$ the set of propositional symbols and by $\mathbf{A}$ the set of atomic formulas, and by $\mathbf{2}$ the set $\{0, 1\}$.

Definition 2.9

An *interpretation* (or $P\tau$-*interpretation*) is a function $I : \mathbf{P} \to \mid \tau \mid$. For an interpretation I we can associate a valuation $V_I : \mathbf{F} \to \mathbf{2}$, inductively defined by:

1. If $p \in \mathbf{P}$ and $\mu \in |\tau|$, then
1.1 $V_I(p_\mu) = 1$ iff $I(p) \geq \mu$.
1.2 $V_I(p_\mu) = 0$ iff it is not the case that $I(p) \geq \mu$.
1.3 $V_I(\neg^k p_\mu) = V_I(\neg^{k-1} p_{\sim\mu}), k \geq 1$.
2. If A and B are formulas, then
2.1 $V_I(A \wedge B) = 1$ iff $V_I(A) = V_I(B) = 1$.
2.2 $V_I(A \vee B) = 1$ iff $V_I(A) = 1$ or $V_I(B) = 1$.
2.3 $V_I(A \to B) = 1$ iff $V_I(A) = 0$ or $V_I(B) = 1$.
3. If F is a complex formula, then
3.1 $V_I(F) = 1 - V_I(\neg_* F)$.

If $V_I(A) = 1$ for a formula A, then we say that V_I *satisfies* A; similarly if $V_I(A) = 0$, then we say that V_I *does not satisfy* A. If $V_I(A) = 1$ for any I, then we say that A is *valid*, written $\models A$.

Theorem 2.10

Let p be a propositional symbol and $\lambda, \mu, \rho \in |\tau|$. We have

1. $\models p_\perp$.
2. $\models p_\lambda \to p_\mu$, if $\lambda \geq \mu$.
3. $\models p_\lambda \wedge p_\mu \to p_\rho$, where $\rho = \lambda \vee \mu$.

Lemma 2.11

Let $\lambda_0 = \bigvee\{\mu \in |\tau| \,\| \vdash p_\mu\}$. Then, $\vdash p_{\lambda_0}$.

Abe (1992) proved the completeness theorem for $P\tau$.

Theorem 2.12 (Abe (1992))
$P\tau$ is complete.

3　Possibilistic Logics

Possibilistic logic was developed by Dubois and Prade (1987) as a logic of uncertainty inspired by *possibility theory* of Zadeh (1976). It can deal with possibility and necessity measure of a formula in some logical system. Here, necessity measure can express certainty degree in the sense of measureing the certainty of a formula to be true. In addition, possibilistic logic can describe *partial* inconsistencies. And it seems to be closely related to the ideas in paraconsistent logics.

Now, we provide a formal sketch of possibilistic logic PL. Let L be the langauge of classical propositional logic. Let Ω be the set of interpretations of L. A *possibility distrubution* on Ω is a function π from Ω to $[0,1]$. π is called *normalized* iff $\exists \omega \in \Omega$ such that $\pi(\omega) = 1$. The quantity $SN(\pi) = 1 - sup\{\pi(\omega) \mid \omega \in \Omega\}$ is called *subnormalized degree* of π, and is equal to 0 iff π is normalized.

A possibility distribution π on Ω provides two types of functions on the set L'　　　　　　L to $[0,1]$, called *possibility measure* Π and *necessity measure* N, which is defined as follows:

$$\Pi : L' \to [0, 1],$$
$$\forall p \in L'(\Pi(p) = sup\{\pi(\omega) \mid \omega \models p\}),$$
$$N : L' \to [0, 1],$$
$$\forall p \in L'(N(p) = inf\{1 - \pi(\omega) \mid \omega \models \neg p\} = 1 - \Pi(\neg p)\})$$

Here, we note that possibilisitic logic should be distinguished by *fuzzy logic* in that fuzzy logic is a logic of vagueness.

In PL, we use a *necessity-valued formula* of the form $(p\ \alpha)$, where p is a propositional formula and α is a valuation of $[0, 1]$. The necessity-valued formula $(p\ \alpha)$ states that $N(p) \geq \alpha$, namely the satisfaction of p is at least α-necessity. Thus, $(p\ 1)$ expresses that $N(p) \geq \alpha$, namely p must be absolutely satisfied. And, $(p\ 0)$ expresses that p is not useful, namely we know nothing about p.

We turn to the semantics of PL. The standard formula can be interpreted in the usual way. For necessity-valued formulas, the satisfaction relaiton $\models$ is defined as follows:

$$\pi \models (p\ \alpha) \iff N(p) \geq \alpha$$

where π is a possibility distribution on Ω and N is the necessity masure induced by π. Let F be the set of necessity-valued formulas $\{(p_1,\ \alpha_1), ..., (p_n\ \alpha_n)\}$. Then, we can define the following:

$$\pi \models F \iff \forall i \in \{1, ..., n\}, \pi \models (p_i\ \alpha_i)$$

The notion of logical consequence is defined as:

$$F \models (p\ \alpha) \iff \forall \pi (\pi \models F \text{ imply } \pi \models (p\ \alpha)$$

The Hilbert-style axiomatization of PL is presented by the axiom schemata and the rules of inference:

Axiom Schemata
(PL1) $(A \to (B \to A)\ 1))$
(PL2) $((A \to (B \to C) \to ((A \to B) \to (A \to C))\ 1)$
(PL3) $((\neg A \to \neg B) \to ((\neg A \to B) \to A)\ 1)$

Rules of Inference
(GMP) $(A\ \alpha), (A \to B\ \beta) \vdash (B\ min(\alpha, \beta))$
(S) $(p\ \alpha) \vdash (p\ \beta)$, where $\beta \leq \alpha$

Here, $\vdash$ is the provability relation in PL. For the axiom schema, we can use aribitrary axiomatics of the classical propositional logic PC. (GMP) is the *graded modus ponens*, which is a possibilistic generalization of *modus ponens* in PC. (S) is a rule for the necessity measure, which is similar to the necessitation in modal logic.

We can prove the completeness theorem of PL; see Dubois, Lang and Prade (1994). One of the distinguished features of PL is to measure the consistency of formulas. The *consistency degree* of F, denoted $Cons(F)$ is defined as:

$$Cons(F) = Sup_{\pi \models F} Sup_{\omega \models \Omega} \pi(\omega)$$

The dual of $Cons(F)$ is the *inconsistency degree*, denoted $Incons(F)$, defined as follows:

$$Incons(F) = 1 - Sup_{\pi \models F} Sup_{\omega \models \Omega} \pi(\omega)$$

We here observe that PL has the paraconsistent feature by extending the classical logic in the followsing sense. For necessity-valued formula F, if F is consistent then $Incons(F) = 0$ and F is inconsistent then $Incons(F) = 1$. Moreover, we can grade the inconsistency of formulas. F is *completely consistent* if $Incons(F) = 0$. F is *completely inconsistent* if $Incons(F) = 1$. F is *partially inconsistent* if $0 < Incons(F) < 1$. Thus, we can measure the global inconsistency of the formula F using $Incons(F)$.

4 Possibilistic Annotated Logics

It is reasonable to amalgamate paraconsistent and possibilistic logics to formalize both local and global inconsistency. The idea was first explored by Besnard and Lang (2000), who claimed that non-classical logics can be strengthened by possibilistic logic. As a case study, they treated a possibilistic extension of da Costa's C_1 (1974) and developed a notion of *graded paraconsistency*. The choice of C_1 seemed to be made by the historical reason that C_1 is one of the well known paraconsistent logics. However, they discussed their approach could be applied to any paraconsistent logics.

In this regard, to develop a possibilistic annotated logic $PP\tau$ is very interesting. This is because annotated logics are computational logic. We also note that annotated logics employ the device of annotation and share similar ideas with possibilistic logics. Below, we work out the new system and discuss the advantages over other approaches.

The language of PP_τ is that of P_τ with the necessity-valued formulas of the form: $(\alpha\ p)$, where $\alpha \in N$ is a valuation of $[0, 1]$ and p is a formulas in P_τ. It is therefore possible to give a possibilistic interpretation of the formulas of P_τ. Here, N is a P_τ-necessity function from L of the language of P_τ to $[0, 1]$ satisfying the following three conditions:

$(Taut)$ if $\vdash_{P_\tau} A$ then $N(A) = 1$
(Eq_N) if $\vdash_{P_\tau} A \leftrightarrow B$ then $N(A) = N(B)$
$(Conj)$ $N(A \wedge B) = min(N(A), N(B))$

As a consequence, we can also provde a P_τ-possibility function Π satisfying the following:

$(Contr)$ if $\vdash_{P_\tau} \neg_* A$ then $\Pi(A) = 0$
(Eq_Π) if $\vdash_{P_\tau} A \leftrightarrow B$ then $\Pi(A) = \Pi(B)$
$(Disj)$ $\Pi(A \vee B) = max(\Pi(A), \Pi(B))$

Observe here that $(Contr)$ holds only for strong negation $\neg_*$. It is easily verified that P_τ-necessity function satisfies the following theorem.

Theorem 3.1
Let N be a P_τ-necessity function. Then, N satisfies:

$(N1)$ if $\vdash_{P_\tau} A \rightarrow B$ then $N(B) \geq N(A)$
$(N2)$ $N(B) \geq min(N(A), N(A \rightarrow B))$
$(N3)$ if $A_1, ..., A_n \vdash_{P_\tau} B$ then $N(B) \geq min(N(A_1), ..., N(A_n))$

(Proof): For (N1), $\vdash_{P_\tau} A \leftrightarrow (A \wedge B)$ if $\vdash_{P_\tau} A \rightarrow B$. From (Eq_N), $N(A) = N(A \wedge B)$ holds. Next, by $(Conj)$, we have $N(A \wedge B) = min(N(A), N(B))$. Thus, $N(B) \geq N(A)$.

For (N2), the following two formulas are provable from (A5) and (A6).

(1) $\vdash_{P_\tau} A \wedge (A \rightarrow B) \rightarrow A$
(2) $\vdash_{P_\tau} A \wedge (A \rightarrow B) \rightarrow (A \rightarrow B)$

By (A2), we have:

(3) $\vdash_{P_\tau} (A \wedge (A \rightarrow B) \rightarrow (A \rightarrow B)) \rightarrow ((A \wedge (A \rightarrow B) \rightarrow A) \rightarrow (A \wedge (A \rightarrow B) \rightarrow B))$

From (2), (3), by using (A4)

(4) $\vdash_{P_\tau} (A \wedge (A \rightarrow B) \rightarrow A) \rightarrow (A \wedge (A \rightarrow B) \rightarrow B)$

is obtained. From (1) and (4), we get:

(5) $\vdash_{P_\tau} A \wedge (A \rightarrow B) \rightarrow B$

By (A4). From (5), $N(B) \geq N(A \wedge (A \rightarrow B))$ holds by (N1). Next, by applying
$(Conj)$ to it, we can derive $N(B) \geq min(N(A), N(A \rightarrow B))$.

For (N3), from $A_1, ..., A_n \vdash_{P_\tau} B$, we have $\vdash_{P_\tau} A_1 \wedge ... \wedge A_n \rightarrow B$ by deduction
theorem. By $(N1)$, $N(B) \geq N(A_1 \wedge ... \wedge A_n)$ holds. Next, $(Conj)$ enables us to
deduce $N(A_1 \wedge ... \wedge A_n) = min(N(A_1), ..., N(A_n))$. Finally, by $(N1)$, we obtain
$N(B) \geq min(N(A_1), ..., N(A_n))$.

Next, we give a semantics for PP_τ. The interpretation $I : \mathbf{P} \rightarrow |\tau|$ and the
associated valuation $V_I : \mathbf{F} \rightarrow \mathbf{2}$ are defined as in P_τ. In addition, we must
specify the valuation $V_N : N(\mathbf{F}) \rightarrow \mathbf{2}$ for necessity-valued formulas as follows:

$$V_N((A \ \alpha)) = 1 \text{ if } N(A) \geq \alpha$$
$$V_N((A \ \alpha)) = 0 \text{ otherwise}$$

The axiomatization of PP_τ is obtainable from the axiomatization of P_τ by
replacing each axiom ϕ of P_τ by $(\phi \ 1)$ except (A4). These are designated as
(PA1)-(PA3), (PA5)-(PA17). Additionally, we need the following two rules:

(PA4) $(A \ \alpha), (A \rightarrow B \ \beta)/(B \ min(\alpha, \beta))$
(PA5) $(A \ \alpha)/(A \ \beta)$, where $\beta \leq \alpha$

Here, (PA4) is (GMP) and (PA5) is (S) in the sense of possibilistic logic.

When τ is a finite lattice, we can easily prove the completeness of PP_τ from
the results of Abe (1992) and Dubois, Lang and Prade (1994).

PP_τ extends P_τ in several respects. Since P_τ is a paraconsistent logic, it
can torelate local inconsistency at the level of annotation, namely hyper-literals.
But, PP_τ can also express global inconsistency at all levels of formulas by means
of the necessity measure. Consequently, we can grade the level of inconsistency
$(A \wedge \neg_* A)$. In the next section, we compare PP_τ with fuzzy annotated logics
FP_τ, proposed by Akama and Abe (2000), which represents both local and
global inconsistency in a *single* annotation.

5   Relation to Fuzzy Annotated Logics

The use of annotation plays an important role in fuzzy reasoning. Akama and
Abe (2000) developed the idea by presenting *fuzzy annotated logcs* FP_τ. FP_τ
extends P_τ with several aspects. One is that FP_τ allows an annotation in
all levels of formulas. Another is to use the notion of *threshold of certainty*
for interpreting the truth of a formula. Because possibilistic annotated logics
PP_τ are based on annotated and possibilistic logics, it is of special interest to
compare them here.

Following Akama and Abe (2000), we outline FP_τ. The logical connectives
of FP_τ are conjucntion, disjunction and implication, but negation is not included due to the use of annotation. Any formula can be given an annotation.
If p is an atomic formula and $\lambda \in [0, 1]$, then λp is a *fuzzy formula*. λp reads
"the uncertainty of p is at least λ". The interpretation is similar to that of
necessity-valued formulas in possibilistic logic. If A and B are formulas and
$\lambda \in [0, 1]$, then $\lambda : A \wedge B, \lambda : A \vee B, \lambda : A \rightarrow B$ are formulas.

To interpret fuzzy formulas, we need the threshold of certainty α. It gives rised to the semantics for fuzzy formulas as follows:

$$v(\lambda p) = 1 \text{ iff } I(p) \geq \lambda > \alpha$$
$$v(\lambda p) = 0 \text{ iff } I(p) < \lambda < \alpha$$

Here, $I : \mathbf{P} \to \tau$ and $V : \mathbf{F} \to [0,1]$. In the valuation of fuzzy formulas, the threshold provides two kinds of interpretations for each logical connectives, which are based on de Morgan like laws.

$$\lambda > \alpha :$$
$$\lambda : A \wedge B = \lambda : A \wedge \lambda : B$$
$$\lambda : A \vee B = \lambda : A \vee \lambda : B$$
$$\lambda : A \to B = \lambda : A \to \lambda : B$$
$$\lambda < \alpha :$$
$$\lambda : A \wedge B = \lambda : A \vee \lambda : B$$
$$\lambda : A \vee B = \lambda : A \wedge \lambda : B$$
$$\lambda : A \to B = (1 - \lambda) : A \wedge \lambda : B$$

Seen from the above, the annotation on fuzzy formulas behave like *generalized negation*. In addition, in $FP\tau$, both the law of non-contradiction and excluded middle do not hold. This means that $FP\tau$ is both paraconsistent and paracomplete.

The axiomatization and the completeness proof of $FP\tau$ may be found in Akama and Abe (2000). The essential difference of $PP\tau$ and $FP\tau$ is that the former can associate the degree of inconsistency and incompleteness by means of necessity-valued formulas, and the latter can torelate both inconsistency and incompleteness. In this sense, the latter is stronger than the former in that inconsistency (incompletenss) can be ranked by the necessity measure. To relate both systems, the base system of $FP\tau$ must have the negation connective, and the annotation should be equated to the necessity measure. In addition, the necessity function N, which corresponds to the annotation λ, should be read $N : \mathbf{F} \to \{[0, n] \subseteq \mathcal{N}\}$, where $\mathcal{N}$ is a distributed lattice whose least element and greatest element are $\mathbf{0}$ and $\mathbf{1}$, respectively. In this way, fuzzy annotated logics can be viewed as a version of possibilistic annotated logics.

6　Conclusions

We have sketched possibilistic annotated logics $PP\tau$ as a fusion of possibilistic and annotated logics. The system was inspired by the work in Besnard and Lang (2000). The proposed logics are more powerful and computational than the corresponding system based on da Costa's C_1 in the sense that we dispense with the sophisticated treatment of paraconsistent negation and that they have a computational adequacy.

The future work includes the following. First, we need a modal extension of possibilitic annotated logics along the line of Akama and Abe (1998). Such an extension may be fruitful to knoweldge representation. The extension of annotated logics with other negation connectives may be interesting in the context of the present work (cf. Kifer and Subrahmanian (1992)). The second theme addresses the way the proposed systems can be expanded for other types of reasoning, e.g. probabilistic reasoning. Thirdly, we should make a detailed comparison of the proposed system and existing approaches worked on approximated reasoning.

References

Abe, J. M. (1992): *On the Foundations of Annotated Logics* (Portuguese), Ph.D. Thesis, University of São Paulo, Brazil.

Akama, S. and Abe, J. M. (1998): Many-valued and annotated modal logics, *Proc. of ISMVL'98*, 114-119, Fukuoka, Japan.

Akama, S. and Abe, J. M. (2000): Fuzzy annotated logics, *Proc. of IPMU'2000*, 504-508, Madrid, Spain.

Besnard, P. and Lang, J. (2000): Graded paraconsistency - reasoning with inconsistent and uncertain knowledge, D. Batens, C. Moetensen, G. Priest and J.-P. Van Bendegem (eds.), *Frontiers of Paraconsistent Logic*, 75-94, Research Studies Press, Baldock.

da Costa, N. (1974): On the theory of inconsistent formal systems, *Notre Dame Journal of Formal Logic 15*, 497-510.

da Costa, N., Subrahmanian, V. and Vago, C. (1991): The paraconsistent logic $P\tau$, *Zeitschrift für mathematische Logik und Grundlagen der Mathematik 37*, 139-148.

da Costa, N., Abe, J. and Subrahmanian, V. (1991): Remarks on annotated logic, *Zeitschrift für mathematische Logik und Grundlagen der Mathematik 37*, 561-570.

Dubois, D., Lang, L. and Prade, H. (1994): Possibilistic logic, D. Gabbay, C. Hogger, and J.A. Robinson (eds.), *Handbook of Logic in Artificial Intelligence and Logic Programming*, 439-513, Oxford University Press, Oxford.

Kifer, M. and Subrahmanian, V. (1992): Theory of generalized annotated logic programming and its applications, *Journal of Logic Programming 12*, 335-367.

Subrahmanian, V. (1987): On the semantics of quantitative logic programs, *Proc. of the 4th IEEE Symposium on Logic Programming*, 173-182.

Zadeh, L. (1976): Fuzzy sets as a basis for a theory of possibility, *Fuzzy Sets and Systems*, 1, 3-28.

Logic, Artificial Intelligence and Robotics
J.M. Abe & J.I. da Silva Filho (Eds.)
IOS Press, 2001

ParaFrame: A Paraconsistent Frame System

Bráulio Coelho Ávila

Fernanda Hembecker

Pontifical Catholic University of Paraná, Curitiba - PR, Brazil

Abstract. Frames are a scheme for the representation of knowledge which allows the description of complex objects. However, there is a gap between the knowledge represented by Frame Systems and the knowledge in the real world. Some Frame Systems have been designed in order to reduce this gap. Nevertheless, they do not deal with tasks such as exceptions and the inconsistency adequately. Inconsistency is a natural phenomenon arising from the description of the real world. This phenomenon may be encountered in several situations. Nevertheless, human beings are capable of reasoning adequately. The automation of such reasoning requires the development of formal theories. Paraconsistent Logic was proposed by N.C.A. da Costa to provide tools to reason about inconsistencies. In this work an inheritance reasoner has been implemented. It represents knowledge through paraconsistent frames and performs inferences over the tangled hierarchies based upon the degree of inconsistency/underdeterminedness. Moreover, its main characteristic is that it does not eliminate, *ab initio*, contradictions. The inheritance reasoner has provided a more adequate treatment to exceptions and inconsistent information of Frame Systems with multiple inheritance.

1 Introduction

Although no general consensus exists as to what is knowledge representation, many schemes have been proposed to represent and store knowledge. This work concentrates on the Frames approach [12]. A frame is a representation of a complex object. It is identified by a *name* and consist in a set of slots. Each frame possesses at least one hierarchically superior frame, thus providing the basis for the inheritance mechanism. A special frame is the root of this inheritance hierarchy.

The inheritance hierarchy is a consequence of the classical taxonomic hierarchy notion as a way of organizing knowledge. The two main types of existing inheritance systems are: those not admitting exceptions to inherited properties and those admitting exceptions to inherited properties [6].

It is easy to describe the semantics of the first inheritance system in first order Classical Logic, in which frames may be interpreted as unitary predicates and slots may be interpreted as binary predicates. The description of the semantics of the second inheritance system in first order Classical Logic is much more difficult, because exceptions introduce nonmonotonicity.

Several nonmonotonic formalisms have been proposed since the late 70's. Among the most widespread are: Clark's *predicate completion*, Reiter's *default logic*, McDermott and Doyle's *nonmonotonic logic I*, McCarthy's *circumscription technique*, McDermott's *nonmonotonic logic II*, and Moore's *autoepistemic logic*. None of these formalisms, however, adequately handles issues such as the inconsistency phenomenon.

The first attempts to treat exceptions systematically later proved to be incorrect in the presence of redundant links and inconsistencies. The first comprehensive definition to solve such problems was Touretzky's [16]. Since then, many other equally correct systems have been proposed. Nevertheless, despite more than one decade of studies, with the increasing subtle examples and counter-examples being considered, a consensus concerning multiple inheritance treatment with exceptions in such hierarchies is still to emerge [14][1]. It must also be stressed that none of these schemes handle adequately issues such as the inconsistency phenomenon, despite such phenomena being increasingly common in computing environments.

2   Paraconsistent Multiple Inheritance Reasoner

The work by Ávila *et al.* [3] proposes an extension of the ParaLog [9] logic programming language, called ParaLog_e, allowing inconsistency to be handled directly. The development of the ParaLog_e language is based on an *annotated paraconsistent logic* [1], infinitely valued, where the truth-values are members of $\{x \in \Re \mid 0 \le x \le 1\} \times \{x \in \Re \mid 0 \le x \le 1\}$. Therefore, an infinite $T = \langle |T|, \le \rangle$, where

$$|T| = \{x \in \Re \mid 0 \le x \le 1\} \times \{x \in \Re \mid 0 \le x \le 1\}.$$

This lattice possesses a minimum element — $[0,0]$ — and a maximum element — $[1,1]$. The minimum element corresponds to indefinite — underdetermined — and the maximum element corresponds to inconsistent — overdetermined. Intuitively, $[1,0]$ and $[0,1]$ correspond, respectively, to *true* and *false* in bi-valued logic. The underlying $\le$ order is represented by Hasse's diagram in Figure 1.

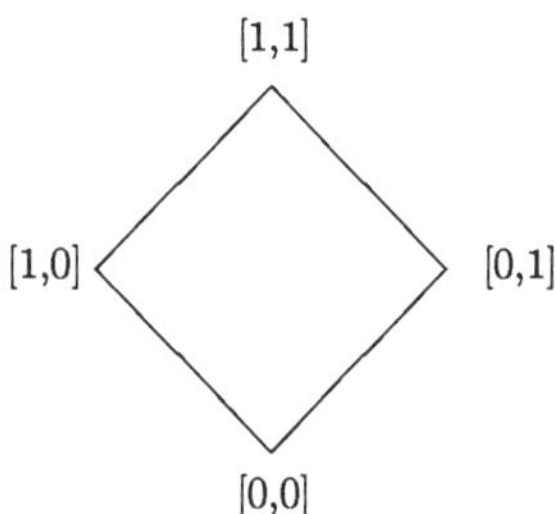

Figure 1: An infinitely valued evidential lattice.

The annotations of this logic system may be considered as *points* in a unitary square on the Cartesian plane, as shown in Figure 2. Thus, a p atom is considered *perfectly* defined when positioned over line $x + y - 1 = 0$; a p atom is considered *overdefined*

---

[1]As there is no formal semantics for inheritance reasoning with exceptions so far, the integrity of a scheme is currently determined by checking its behavior by means of a set of inheritance network examples.

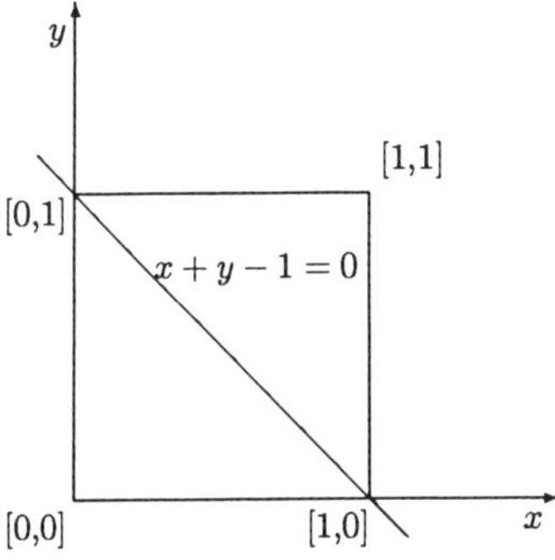

Figure 2: Unitary square on the Cartesian plane.

when positioned above line $x + y - 1 = 0$; and, a p atom is considered *underdefined* when positioned down line $x + y - 1 = 0$. The degree of inconsistency of atom p is $(\alpha + \beta - 1) \times 100$, where $\alpha + \beta \geq 1$. The degree of under-determinedness of atom p is $(\alpha + \beta - 1) \times 100$, where $\alpha + \beta \leq 1$.

According to Subrahmanian [15], a two-evidence annotation may be thought of as: a *favorable evidence* to p and a *contrary evidence* to p. No restriction is placed on these evidences, except that they be within the $\{x \in \Re \mid 0 \leq x \leq 1\}$ interval. The use of both these evidences increases the expression power of the information represented.

Thus it may be seen that by using the Paralog_e language, the representation of information in frame systems may be performed more naturally. Languages such as ParaLog_e, capable of merging Classical Logic Programming concepts with those of inconsistency, widen the scope of Logic Programming applications in environments presenting conflicting beliefs and contradictory information.

The implementation of a paraconsistent inheritance reasoner in ParaLog_e is described below, allowing handling exceptions and inconsistencies in multiple inheritance frame systems.

2.1 Description of the Paraconsistent Inheritance Reasoner

The development of an inheritance reasoner for frames requires three major decisions to be drawn, concerning the way of representing: the inheritance network, the frame knowledge, and the reasoner inference-making.

Inheritance Network Representation

In the first decision, one can observe that almost all the inheritance systems found in literature represent knowledge through semantic nets [13] [16] [11]. This paper represents knowledge by means of paraconsistent frames. Frames are a more complex knowledge representation scheme than semantic nets [5].

The paraconsistent frame inheritance network — frame network — of the paraconsistent reasoner is a *tangled hierarchy*. That is, it allows frame data to possess multiple ascending frames, and thereby multiple inheritance. In this paper, ascending links between two frames may be:

- *strict* — represented by relationships ako^2 with perfectly defined evidences $[1.0, 0.0]$ and $[0.0, 1.0]$. That is, evidence $[1.0, 0.0]$ represents that a frame A *ako* frame B. On the other hand, evidence $[0.0, 1.0]$ represents that a frame A *not-ako* frame B. Strict links ako and not-ako are also known as *positive links* and *negative links*, respectively, in *bipolar inheritance systems*.

- *defeasible* — represented by ako relationships with evidences belonging to the $T = \{x \in \Re \mid 0 \leq x \leq 1\} \times \{x \in \Re \mid 0 \leq x \leq 1\}$ lattice, excepting $[1.0, 0.0]$ and $[0.0, 1.0]$. That is, for instance, a $[0.9, 0.3]$ evidence represents that a frame A, generally, ako frame B, but with a 20% inconsistency/under-determinedness degree[3].

This frame network also admits the existence of redundant ascending links between two frames, because the network and the reasoner are based on ParaLog_e, and after the closure transformation[4], the resulting link is the *supreme* of redundant link.

Knowledge Representation in Frames

In the second decision, knowledge, as said before, is represented by Paraconsistent Frames. The way of representing such frames is similar to the standard frame representation. However, the information represented in paraconsistent frames has greater expression power, since it is associated with evidences. In ParaLog_e a frame may be represented as a fact, that is, a fact for each facet of one slot, as shown next.

$$< frame > (< slot >, < facet >, < value >) : [< e_f >, < e_c >].$$

For instance, *c(color,value,blue)* : $[0.9, 0.3]$. specifies that a frame *c* possesses a *color* slot with *blue* value — 90% *favorable evidence* and 30% *contrary evidence* — with an inconsistency/under-determinedness degree of 20%.

This paper admits two types of facets: value and exception. The first facet specifies the *value* assumed by the slot. The value facets may be:

- *strict* — have perfectly defined evidences $[1.0, 0.0]$ and $[0.0, 1.0]$. For instance,

$$a(color, value, blue) : [0.0, 1.0].$$

specifies that a frame *a* has a *color* slot of *non-blue* value.

- *defeasible* — possess evidences pertaining to lattice $T = \{x \in \Re \mid 0 \leq x \leq 1\} \times \{x \in \Re \mid 0 \leq x \leq 1\}$, excepting $[1.0, 0.0]$ and $[0.0, 1.0]$. For instance,

$$a(color, value, blue) : [0.3, 0.6].$$

specifies that a frame *a* has a *color* slot of *blue* value, but with a 10% degree of inconsistency/under-determinedness.

---

[2] a kind of

[3] The inconsistency/under-determinedness degree is obtained according to [15]

[4] The closure transformation is described in [4]

Value facets also admit the existence of redundant value facets. After the closure transformation, the resulting value facet is the supreme of redundant value facet.

The last facet specifies the *exceptions* assumed by the slot. There might be one or more exception facets for each slot. Exception facets may be:

- *strict* — have perfectly defined evidences $[1.0, 0.0]$ and $[0.0, 1.0]$. For instance,

$$c(color, exception, red) : [1.0, 0.0].$$
$$c(color, exception, black) : [1.0, 0.0].$$

specify that a frame c possesses a *color* slot with *red* and *black* exceptions.

- *defeasible* — possess evidences pertaining to lattice $\mathcal{T} = \{x \in \Re \mid 0 \leq x \leq 1\} \times \{x \in \Re \mid 0 \leq x \leq 1\}$, excepting $[1.0, 0.0]$ and $[0.0, 1.0]$. For instance,

$$c(color, exception, red) : [0.2, 0.5].$$
$$c(color, exception, black) : [0.1, 0.3].$$

specify that a frame c possesses a *color* slot with *red* and *black* exceptions, but with a 30% and 60% degree of inconsistency/under-determinedness, respectively.

Exception facets also admit the existence of redundant exception facets. After the closure transformation, the resulting exception facet is the supreme of redundant exception facet.

Reasoner Inference-Making

Finally, the last decision concerns how the inheritance reasoner draws inferences on frames. This decision is related to the foregoing ones, that is, it depends both on the inheritance network representation form and on the knowledge representation form in frames. This work may be classified as complex, since it is a system possessing the following characteristics: multiple inheritance, bipolarity, heterogeneity and nonmonotonicity, that is, it allows multiple ascending links, positive and negative links, strict and defeasible links, and exceptions to property inheritance.

The implemented inheritance reasoner for frames, in addition to meeting all these characteristics allows, being an encompassing reasoner, all other less complex inheritance forms, such as unipolar monotonic homogeneous inheritance, etc., to be carried out.

2.2 Algorithm

The paraconsistent inheritance reasoner is implemented in ParaLog_e on the basis of the algorithm shown in Figure 3. In this algorithm, regularization procedures are performed first. They are followed by the initialization of variables and the performance of the **ex_equal** procedure shown in Figure 4. At last, on the basis of the output values of such procedure, the **continue** procedure, shown in Figure 5, is performed, or an empty value is attributed to conclusion C.

Paraconsistent Inheritance Reasoner Algorithm

Input: an acyclic paraconsistent frame system Γ and a query $q(f, s, v):[q_f, q_c]$
Output: a set of conclusions C_n with its respective evidences $[C_{f_n}, C_{c_n}]$.

```
begin
   perform the Γ negation elimination procedure
   perform the Γ closure procedure
   Exceptions:= {}
   Equalₙ:= {elem(f:[e_f, e_c],Exceptions)}
   Parent:= {}
   Sol:= 0
   ex_equal(q(f, s, v):[q_f, q_c],Γ,C_n:[C_{f_n}, C_{c_n}],Equalₙ,Parent,Parent_ret,Sol,Sol_ret)
   if Sol_ret = 0
      then
         if Parent_ret ≠ {}
            then
               continue(q(f, s, v):[q_f, q_c],Γ,C_n:[C_{f_n}, C_{c_n}],Parent_ret)
            else
               C:[C_f, C_c] := {}:[0.0, 0.0]
         end if
   end if
end
```

Figure 3: Paraconsistent Inheritance Reasoner Algorithm

procedure ex_equal$(q(f, s, v):[q_f, q_c],\Gamma,C_n:[C_{f_n}, C_{c_n}],$ Equal$_n$,Parent,Parent_ret,Sol,Sol_ret)

```
begin
   if Equalₙ ≠ {}
      then
         find_sol(q(f, s, v):[q_f, q_c],Γ,C_n:[C_{f_n}, C_{c_n}],Equalₙ,Sol,Sol_ret)
         if Sol_ret ≠ 0
            then
               combine_sol(C:[C_f, C_c],C_n:[C_{f_n}, C_{c_n}])
            else
               find_parent(Γ,C:[C_f, C_c],Equalₙ,Parent,Parent_ret)
         end if
      else
         Sol_ret:= 0
   end if
end
```

Figure 4: The **ex_equal** procedure tries to find solutions to frames possessing equal inconsistency/under-determinedness degrees.

```
procedure continue(q(f,s,v):[q_f,q_c],Γ,C_n:[C_{f_n},C_{c_n}], Parent_new)

begin
  if Parent_new≠ 0
    then
      order_elem(Parent_new,Parent_ord)
      find_less_equal_parent(Parent_ord,Equal_n,Parent)
      Sol:=0
      ex_equal(q(f,s,v):[q_f,q_c],Γ,C_n:[C_{f_n},C_{c_n}],Equal_n,Parent,Parent_rt,Sol,Sol_rt)
      if Sol_rt=0
        then
          continue(q(f,s,v):[q_f,q_c],Γ,C_n:[C_{f_n},C_{c_n}],Parent_rt)
      end if
    else
      C_n:[C_{f_n},C_{c_n}] := {}:[0.0,0.0]
  end if
end
```

Figure 5: The **continue** procedure allows the search for ascendants in Γ.

The **ex_equal** procedure tries to find solutions to frames possessing equal inconsistency/under-determinedness degrees. When such solutions are found, a procedure combining the evidences of the solutions is performed. Otherwise, a procedure is performed trying to find parent frames for paths that did not produce solutions. This is necessary for the inheritance reasoner to search ascendants in Γ.

The **continue** procedure allows the search for ascendants in Γ, that it, since no solutions were found in the frames searched, the search must continue upward to the parent frames belonging to Γ. For this search to be performed, the parent frames found in the paths that did not produce solutions must first be ordered. Such ordering must be performed on the basis of the inconsistency/under-determinedness degree accrued from each path. Afterwards, the ordered elements are separated into two groups: elements with equal inconsistency/under-determinedness degrees and elements with higher inconsistency/under-determinedness degrees. Then, the **ex_equal** procedure is again performed based on these two groups of elements. This procedure must be repeated until no more parent frames exist or a solution is found.

3   Conclusions

The main objective of this work was to show that by using Paraconsistent Logic [7] [8], the frame systems may reason adequately in the presence of exceptions and inconsistent information. To achieve such objective, a paraconsistent inheritance reasoner was implemented in ParaLog_e.

The implemented inheritance reasoner represents knowledge through paraconsistent frames and performs inferences on tangled hierarchies, based on the inconsistency/under-determinedness degree. Furthermore, its main feature is not eliminating contradictions, *ab initio*. Thus, it may be concluded that the implemented paraconsistent inheritance reasoner allows better handling of the exceptions and inconsistent information in multiple inheritance frame systems.

Paraconsistent Logic, despite having been initially developed from the purely theoretical standpoint, found in recent years extremely fruitful applications in Computing Science [10] [2], thus solving the problem of justifying such logic from the practical standpoint.

References

[1] Abe, J.M., *Fundamentos da Lógica Anotada*, (Foundations of Annotated Logics), Ph.D. Thesis, Faculdade de Filosofia, Letras e Ciências Humanas, Universidade de São Paulo, São Paulo, Brazil, 1992. (in Portuguese)

[2] Angelotti, E.S.; Scalabrin, E.E.; Ávila, B.C.; Bortolozzi, F., *A Paraconsistent System of Autonomous Agents for Brazilian Bank Check Treatment*, XX International Conference of the Chilean Computer Science Society, IEEE Computer Society Press, Santiago, Chile, 2000, pp. 89-98.

[3] Ávila, B.C.; Abe, J.M.; Prado, J.P.A., *ParaLog_e: A Paraconsistent Evidential Logic Programming Language*, XVII International Conference of the Chilean Computer Science Society, IEEE Computer Society Press, Valparaiso, Chile, 1997, pp. 2-8.

[4] Blair, H.A.; Subrahmanian, V.S., *Paraconsistent Logic Programming*, Proc. 7th Conference on Foundations of Software Technology and Theoretical Computer Science, Lecture Notes in Computer Science, Springer-Verlag, Vol. 287, pp. 340-360, 1987.

[5] Brachman, R.J., *The Basics of Knowledge Representation and Reasoning*, AT&T Technical Journal, Vol. 67, pp. 7-24, 1988.

[6] Brewka, G., *The Logic of Inheritance in Frame Systems*, Proc. 10th IJCAI-87, pp. 483-488, Milan, 1987.

[7] da Costa, N.C.A., *Calculs Propositionnels pour les Systèmes Formels Inconsistants*, Compte Rendu Acad. des Sciences (Paris), 257, pp. 3790-3792, 1963.

[8] da Costa, N.C.A., *On the Theory of Inconsistent Formal Systems*, Notre Dame J. of Formal Logic 15 (1974), pp. 497-510.

[9] da Costa, N.C.A.; Prado, J.P.A.; Abe, J.M.; Ávila, B.C.; Rillo, M., *ParaLog: Um Prolog Paraconsistente Baseado em Lógica Anotada*, (ParaLog: An Annotated Logic-Based Paraconsistent Prolog), Coleção Documentos, Série: Lógica e Teoria da Ciência Nº18, Institute for Advanced Studies, University of São Paulo, São Paulo, Brazil, April, 1995, (in Portuguese)

[10] Enembreck, F.; Ávila, B.C.; Sabourin, R., *Decision Tree-Based Paraconsistent Learning*, XIX International Conference of the Chilean Computer Science Society, IEEE Computer Society Press, Talca, Chile, 1999, pp. 43-52.

[11] Krishnaprasad, T.; Kifer, M., *An Evidence-Based Framework for a Theory of Inheritance*, Proc. 11th IJCAI-89, pp. 1093-1098, Detroit, 1989.

[12] Minsky, M., *A Framework for Representing Knowledge*, MIT AI Memo 306, Cambridge, MA, June, 1974.

[13] Sandewall, E., *Nonmonotonic Inference Rules for Multiple Inheritance with Exceptions*, Proceedings of the IEEE, pp. 1345-1353, 1986.

[14] Selman, B.; Levesque, H.J., *The Tractability of Path-Based Inheritance*, Proc. 11th IJCAI-89, pp. 1140-1145, Detroit, 1989.

[15] Subrahmanian, V.S., *Towards a Theory of Evidential Reasoning in Logic Programming*, Logic Colloquium '87, The European Summer Meeting of the Association for Symbolic Logic, Granada, Spain, July, 1987.

[16] Touretzky, D.S., *The Mathematics of Inheritance Systems*, Research Notes in Artificial Intelligence, Pitman, London, 1986.

Some Model Theory of Ordered Structures and Undefinability Results

Ricardo Bianconi

Universidade de São Paulo
Instituto de Matemática e Estatística
Caixa Postal 66281, CEP 05315-970, São Paulo, SP, Brazil
e-mail: bianconi@ime.usp.br

Abstract: We survey some old, and present some new, results on the problem of characterizing definable sets, mainly in the case of expansions of the real closed field by adding analytic functions, in particular the exponential function.

1 Introduction

The study of the so called o-minimal structures started in the paper [7] and fully developped afterwards. For a survey see, for instance, [8] and the references therein. We will not give a survey of this rich theory but will concentrate on one aspect which is one line of research most alive nowadays in this subject. That is, definability of sets and functions in o-minimal expansions of the field of real numbers.

Let us first recall the definition of o-minimal structures. Let M be a linearly ordered structure (whose first order language is L). Then M is o-minimal if any definable set (in the first order language L) $A \subseteq M$ is a finite union of points and open intervals with endpoints in M or $\{\pm\infty\}$. This simple definition implies that this is a property of the theory of M, and that it admits a cell decomposition theorem (that is, any definable set $A \subseteq M^n$ can be decomposed in a finite union of definable cells, which are either definable open sets of M^n, or the graph of definable functions $f : D \subseteq M^k \to M^{n-k}$).

This property implies that we cannot intepret the theory of an infinite discrete linearly ordered structure in a densely ordered o-minimal structure (see [2]) and also it was an important ingredient to prove the model completeness of several theories expanding that of the field of real numbers (see [10]). This kind of result can be useful to prove the still unproved decidability of such theories (see [6]). Another ingredient wich imply the decidability of the expansion of the real field by the exponential function is the also still unproved Schanuell's conjecture (whose statement is: if $x_1, \ldots, x_n \in \mathbb{C}$ are linearly independent over $\mathbb{Q}$, then the transcendence degree of $\mathbb{Q}(x_1, \ldots, x_n, \exp(x_1), \ldots, \exp(x_n))$ over $\mathbb{Q}$ is at least n).

In this paper we prove some undefinability of certain functions in some expansions of the field of real numbers in section 2. In section 3 we sketch an application of model theory to Schanuell's conjecture and in section 4 we conclude the paper with some open problems.

2 Applications of Schanuel's conjecture to undefinability results

In [1] we have proved that the sine function is not definable from the (real) exponential function, and *vice versa*. Namely,

Theorem 1 *No restriction of the sine function to any interval is definable in $R_{\exp} = \langle \mathbf{R}, +, \cdot, \text{constants}, <, \exp \rangle$, and conversely, no restriction of the exponential function to any interval is definable in $\langle \mathbf{R}, +, \cdot, \text{constants}, <, \sin_0(x), \cos_0(x) \rangle$, where $\sin_0(x) = \sin(x)$, $\cos_0(x) = \cos(x)$, for $x \in [-\pi, \pi]$, and $\sin_0(x) = \cos_0(x) = 0$ for all $x \notin [-\pi, \pi]$.*

Usine the same techniques, we can prove the following result

Theorem 2 *Let $a \in \mathbb{R}$ and $f_a : \mathbb{R} \to \mathbb{R}$ be the function $f_a(x) = x^a$ if $x > 0$ and $f_a(x) = 0$ otherwise. For each $A \subseteq \mathbb{R}$, the function is definable in the structure $R_A = \langle \bar{\mathbb{R}}, f_a : a \in A \rangle$ if, and only if a is in the subfield of $\mathbb{R}$ generated by A.*

Proof: It is straightforward to show that if α is in the field $\mathbb{Q}(A)$, generated by A, then f_α is definable in R_A, so, without loss of generality, we can assume that A is a subfield of $\mathbb{R}$.

Now, if $a \notin A$ (a subfield) then (working with the model complete theory of $R_{\exp}$ if necessary), we can write an existential formula defining some restriction of x^a. Using the same kind of trick used in [1], we prove that x^a is not definable from x^b, $b \in A$.
$\square$

Now there is some research on applications of o-minimality to complex manifolds. In this line of research we have proved (with collaboration with Chris Miller from Ohio State University) the following result.

Theorem 3 *Let $u, v : D \to \mathbb{R}$ be two definable functions in $R_{\exp}$ such that $u + iv$ is holomorphic, $D \subset \mathbb{R}^{2n}$ is a definable polydisc. Then u and v are already definable in $\bar{\mathbb{R}}$.*

Proof: Let u and v be the real and imaginary parts of the holomorphic function definable in $R_{\exp}$ around the origin. Assume that they are not semialgebraic. Suppose that they are defined with $Z_1, ..., Z_p$ as witnesses, and variables $X_1, ..., X_n, Y_1, ..., Y_n$. Then, modulo desingularization tricks (as in [1]), we have polynomials P_j, $j = 1, ..., p + 2$ such that $F_j(\vec{X}, \vec{Y}, U, V, \vec{Z}) = P_j(\vec{X}, \vec{Y}, U, V, \vec{Z}, \exp \vec{X}, ..., \exp \vec{Y}, ..., \exp U, \exp V, \exp \vec{Z}) = 0$ and nonsingular Jacobian matrix $\partial(F_1, ..., F_{p+2})/\partial(U, V, \vec{Z})$. Imposing Cauchy Riemmann equations, we introduce $2n$ new equations, with polynomials P_j, $j = p + 3, ...p + 2 + 2n$. So we have $p + 2 + 2n$ polynomial relations on $\vec{X}, \vec{Y}, U, V, \vec{Z}, \exp X, ...$ totalling $2(p + 2 + 2n) = 2p + 4n + 4$ terms, giving at most $p + 2n + 2$ for the transcendence degree of $\vec{X}, \vec{Y}, U, V, \vec{Z}, \exp X, ...$

over Q. Now go to a nonstandard model (an ultrapower of R). Assume that all the functions $u, v, \vec{z}$ are zero in $\vec{0}$. We can choose infinitesimals $\vec{x}, \vec{y}$ such that $\vec{x}, \vec{y}, u(\vec{x}, \vec{y}), v(\vec{x}, \vec{y}), \vec{z}(\vec{x}, \vec{y})$ are linearly independent over R (and not just Q). Then Schanuel's conjecture applies in this setting, giving transcendence degree over R at least $p + 2n + 3$. So u, v, $\vec{z}$ must be semialgebraic. $\qquad\square$

Remark 1 *In this Theorem we have assumed that the real and imaginary parts of the complex analytic function are both definable in $R_{\exp}$. If only one of them is definable then the result is not true as for instance the complex logarithm, $\log(x + iy) = \log\sqrt{x^2 + y^2} + i\arctan(y/x)$, the real part is definable in $R_{\exp}$ but not definable in $\bar{\mathbb{R}}$.*

3 An application of model theory to Schanuell's conjecture

In the paper [4], we have turned our attention to how model theory can help in a possible proof of Schanuell's conjecture. Let us explain what we obtained.

Let $\mathbb{Q} \subseteq C \subseteq F$ be a tower of fields. We say that $x_1, \ldots, x_n \in F$ are linearly independent over $\mathbb{Q}$ modulo C, if no nontrivial $\mathbb{Q}$-linear combination of $x_1, \ldots, x_n$, belong to C.

Theorem 4 *If $x_1, \ldots, x_n \in \mathbb{C}$ are linearly independent over $\mathbb{Q}$ modulo $K\langle A\rangle$, then the transcendence degree of $K\langle A\rangle(x_1, \ldots, x_n, \exp x_1, \ldots, \exp x_n)$ over $K\langle A\rangle$ is at least $n + 1$.*

As a consequence we obtain

Theorem 5 *If Schanuell's conjecture is true in K_0 then it is true in $\mathbb{C}$, where K_0 is the prime model of the theory of the structure $R_{r.e.} = \langle \mathbf{R}, +, \cdot, 0, 1, <, \sin_0(x), \exp_0(x)\rangle$, where $\exp_0(x) = \exp x$ if $x \in [0, 1]$ and $\exp_0(x) = 0$ otherwise, $\sin_0(x) = \sin(x)$, for $x \in [-\pi, \pi]$, and $\sin_0(x) = 0$ otherwise (see [9]).*

S. Lang says in his book *Introduction to transcendental numbers*, [5] page 73, that Transcendental Number Theory determines which *classical numbers* are linearly independent or algebraically independent over $\mathbb{Q}$. What we prove here is if we define classical number as definable in such structure, then if Schanuell's conjecture is true for them, then it is true in $\mathbb{C}$.

4 Conclusion

Let us conclude this paper with some open problems.

We can prove that, for each $k > 0$ and any definable function f in an o-minimal expansion of the reals, f is piecewise k times continuosly differentiable. In all the known o-minimal such expansions the definable functions are piecewise C^∞. So it is natural to ask:

Problem 1 *Is there an o-minimal expansion or $\bar{\mathbb{R}}$ with a non piecewise C^∞ definable function?*

Referring to the notion of classical numbers in the end of last section, S. Lang defines in [5] classical numbers (in our parlance) as numbers obtained by a sequence of quantifier free definable elements of $\mathbb{C}$ using the numbers obtained in each step of this sequence as parameters for the definitions of the next step. We used the notion of definable by any formula. So it is natural to ask:

Problem 2 *Is any element of K_0 a classical number in the sense of Lang?*

We end this paper stating the problem left open in [2].

Problem 3 *A structure M is said to be a discrete o-minimal structure in the broad sense if it is an infinite linearly ordered structure such that the set of points which have no immediate successsor or predecessor is finite and the definable sets in M are finite unions of intervals with endpoints in $M \cup \{\pm\infty\}$. Does the theory of M interprets the theory of $(\mathbb{N}, <)$?*

References

[1] R. Bianconi, Nondefinability results for expansions of the field of real numbers by the exponential function and by the restricted sine function. J. Symbolic Logic 62 (1997), no. 4, 1173–1178.

[2] R. Bianconi, A Note on noninterpretability in o-minimal structures. Fundamenta Mathematicae, vol. 158 (1998), no. 1, pp. 19-22.

[3] R. Bianconi, Some remarks on Schanuel's conjecture. Annals of Pure and Applied Logic, 108 (2001), 15-18.

[4] R. Bianconi, Schanuel's conjecture and classical numbers, submitted.

[5] S. Lang, Introduction to transcendental numbers. Addison-Wesley Publishing Co., 1973.

[6] A. Macintyre, A. J. Wilkie, On the decidability of the real exponential field, pp. 441-467, in P. Odifreddi (ed.), Kreiseliana: About and around Georg Kreisel, A. K. Peters Ltd., Wellesley, MA, 1996.

[7] L. van den Dries, Tame topology and o-minimal structures. London Mathematical Society Lecture Note Series, 248. Cambridge University Press, Cambridge, 1998.

[8] L. van den Dries, Remarks on Tarski's problem concerning $(R, +, , \exp)$. Logic colloquium '82 (Florence, 1982), 97–121, Stud. Logic Found. Math., 112, North-Holland, Amsterdam, 1984.

[9] L. van den Dries, The elementary theory of restricted elementary functions. The Journal of Symbolic Logic, **53** (1988), 796-808.

[10] A. J. Wilkie, Model completeness results for expansions of the ordered field of real numbers by restricted Pfaffian functions and the exponential function. J. Amer. Math. Soc. **9** (1996), no. 4, 1051–1094.

Logic, Artificial Intelligence and Robotics
J.M. Abe & J.I. da Silva Filho (Eds.)
IOS Press, 2001

Evolutionary Approach to Design of Artificial Neural Networks

Lídio Mauro Lima de Campos, lidio@supridados.com.br
Mestrando em Ciência da Computação,
Universidade Federal de Santa Catarina-UFSC
Telefone: (091)223.3709

Mauro Roisenberg, mauro@inf.ufsc.br
Universidade Federal de Santa Catarina - UFSC
Centro Tecnológico
Telefone: (048) 331.7515

Gustavo Augusto Lima de Campos, gcampos@nautilus.com.br
CESUPA- Centro de Ensino Superior do Pará
Telefone: (091)249.6642

ABSTRACT

Nowadays several techniques of optimization of neural networks have being researched, one of them use evolutionary computation. The objective of this research is to introduce an Evolutionary system biologically plausible, as far as possible that can automatically generate Artificial Neural Networks (ANN) with good generalization capacity, smaller error and larger tolerance to noises. To this aid, three biological metaphors were used: Genetic algorithms (GA), Lindenmayer Systems (L-System) and ANN. First it was introduced the biological metaphors used in this research at the end the results of simulation for the parity problem are presented. The method is better than the other ones because it increases the level of implicit parallelism of genetic algorithm and for the aspects of biological plausibility. The system generates the minimum satisfactory architecture that solves a specific task, reducing the project costs and increasing the performance of the neural networks obtained.

Keywords

Artificial Neural Network, Genetic Algorithms, Evolutionary Computation, Lindenmayer Systems

1. Introduction

Nowadays the applications of Artificial Neural Networks (ANN) vary from pattern recognition, control systems, financial forecast, and medicine to agriculture. However, until the present moment, some of the first questions to be faced by a research when using the ANN technology to solve a given problem are: What kind of ANN should I use? How many neurons will be necessary to represent the problem solution with the desired precision?

The computer science area that worries about these questions is known as computational complexity theory. The complexity of ANN is a research field that attempt not only answer how big the neural network should be, but also, what should be the network structure.

The definition of the ANN architecture is an important parameter in its conception, because restricts the problem type that can be solved. ANN with one layer of neurons, for example, only solves linearly separable problems. A Recurrent Neural Network is more used to solve problems that involve voice-processing, prediction of temporary series. The parameters that make part of the definition of the architecture are: number of layers, number of nodes in each layer, connection type between the nodes and topology.

The determination of a satisfactory architecture, as well as of the parameters of its training algorithm (learning rate, momentum), improves its performance, that is, speed and accuracy of the learning, tolerance to noise and the generalization capacity. However that problem is not simple. There isn't a method to determine a satisfactory architecture for a given problem.

There are techniques that use empiric knowledge for the project. An approach very used in the practice consists in the construction of ANN with standardized architectures, already used in another applications. The same is tested for the function in study and the parameters (learning rate, momentum, number of epochs) are altered for the new application. This approach has a very high cost and it doesn't present reliable results. As well as it is not possible to guarantee the optimization of the solution.

The automatic project of ANN provides a more efficient form of search of architectures than the conventional methods. The Genetic Algorithms (GA) are an excellent tool for those types of problems, because it work with a population of individuals candidates to the solution of the problem and not only with one, moreover, it is possible to conduct the search of the (GA), in agreement with the aptitude function adopted, facilitating to obtain satisfactory ANN that solve specifics tasks.

Some works that use evolutionary algorithms for obtaining architectures of ANN are: [5] with a direct encoding method. [3] uses grammars for graphs generation. [1] and [4] looked for inspiration in the nature and used (GA) and Lindenmayer Systems, proposed by [6], to generate modular ANN. Those works differ basically for the encoding method that they use and for the degree of biological inspiration.

Some researchers do not share the concern with biological plausibility in the area of Artificial Intelligence. Others [1] and [4], incorporated biological inspirations in their methods. The researchers didn't get a consent on which is the best project methodology, because that depends on several factors as: flexibility, easiness of implementation, robustness, size of the chromosome, time of processing, performance obtained and degree of biological inspiration. Considering the biological plausibility the method that integrates GA, LS to generate ANN is the better than the others. Results of experiments with this method performed so far indicate this method to be very promising [1].

2. Biological Inspiration

Nowadays the heredity is described in terms of information, messages, and code. The reproduction of an organism became the reproduction of the molecules that constitute it. Not because each chemical species has the capacity to produce copies of them, but because the structure of the macromolecules is determined minutely by the sequence of four chemical radical contained in the genetic patrimony. Which is transmitted of a generation for another they are the " instructions " that specify the molecular structures. They are the architectural plans of the future organism. They are also the means to execute those plans and to coordinate the activities of the system. Therefore, each egg contains, in the parents' received chromosomes, all its future, the stages of its development, the form and the properties of being that will appear of him. The organism becomes like this the accomplishment of a program prescribed by the heredity.

The development of plants and animals is governed by the genetic information contained in each cell of the organism. Each cell contains the same genetic information (the genotype), which determines the way in which each cell behaves and, as a result of that, the final form and functioning of the organism (the phenotype). This genetic information is not a blueprint of that final form but can be seen as a recipe [7] that is followed not by the organism as a whole but by each cell individually. The shape and behavior of a cell depends on those genes that are expressed in its interior. Which genes actually are expressed depends on the context of the cell. Already at the very beginning of an organism's life, subtle intercellular interaction takes place, changing the set of genes that are expressed in each cell. This process of cell differentiation is responsible for the formation of all different organs.

3. Lindemayer Systems

Lindenmayer Systems are parallel string mechanisms. A grammar, the definition of what is called a language in formal language theory, consists of an alphabet, a starting string and a set of rewriting rules. The starting string, also known as the axiom, is rewritten by applying the rewriting rules: each rule describes how a certain symbol or string should be rewriting in another string of symbols. Whereas in most other grammars rewriting rules are applied sequentially, in L-systems all rewriting rules are applied in parallel to generate the next string in rewriting process [2].

The L-system grammar G of a Language L can be defined as $G=\{\Sigma, \Pi, \alpha\}$, where Σ is the finite set of symbols or alphabet of the language. O Π is the finite set of rewriting rules (also: production rules), and $\alpha \in \Sigma^*$ is the starting string (axiom) of **L** and $\Pi=\{\pi|\pi:\Sigma\rightarrow\Sigma^*\}$ is the set of rewriting rules. Each rewriting rule defines a unique rewriting of a symbol of Σ, the left side of rewriting rule, into a string s $\in \Sigma^*$, the right side of the rewriting rule. All symbols in Σ that do not appear as the left side of a rewriting rule are rewriting into themselves. As an example consider the L-system $G=\{\Sigma, \Pi, \alpha\}$ with $\Sigma=\{A,B,C\}$, $\Pi=\{A\rightarrow BA, B\rightarrow CB, C\rightarrow AC\}$ e $\alpha=ABC$, using the rewriting rules, the obtained language is **L={ABC, BACBAC, CBBAACCBBAAC}**.

An extension of L-systems is the context sensitive, which is used to generate conditional rules. In general the context sensitive rewriting rules will have the form **L<P>R→S** with $P \in \Sigma$ e $L,R,S \in \Sigma^+$. **P** is the predecessor e **S** o successor, and **L** and **R** are the left and right context respectively of the production rule. As an example, consider that a current string in the rewriting process of an L-system is **ABBACAADBAABBAC**, with the production rules: **CA<A→EG, A<A>B→BE** e **B→RT**. The string that results after one rewriting step is: **ARTRTACAEGDRTABERTRTAC**.

An interesting application of L-systems is a graphical interpretation of strings based on notion of a LOGO-style turtle. This interpretation was originally proposed by Szilard and Quinton [8]. A state of the turtle is defined as a triplet (x,y,α), where the Cartesian coordinates (x,y) represent the turtle's position and the angle α, called the turtle's heading, is interpreted as the direction in which the turtle is facing. Given the step size d and the angle increment δ, the turtle can respond to commands by the following symbols of the Table 1 a Figure 1.

Table 1 Commands for Turtle's Movement

	Action
F	Move forward a step of length d. The state of the turtle changes to (x',y',α), where $x'=x+d.\cos(\alpha)$ e $y'=y+d.\text{sen}(\alpha)$. A line segment between points (x,y) and (x',y') is draw.
f	Move forward a step of length d without drawing a line.
+	Turn right by an angle δ . The next state of the turtle is $(x,y, \alpha+\delta)$. It is assumed here that the positive orientation of angles is clockwise.
-	Turn left by an angle δ . The next state of the turtle is $(x,y, \alpha-\delta)$

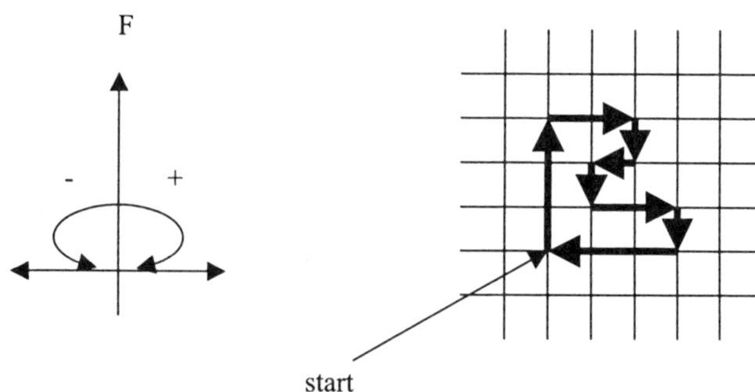

Figure 1 (a) the interpretation of symbols F, -, +. (b) The interpretation of a string FFF+FF+F+F-F-FF+F+FFF, for an angle of increment δ =90°.

Lindenmayer introduced in his paper in 1968 a notation for representing graph-theoretic trees using strings with brackets. The motivation was to formally describe branching structure found in many plants. An extension of turtle interpretation is described below. It was introduced two new symbols interpreted by turtle. Figure 2 shows examples of plant-like structures generated by bracketed L-systems.

'[' Push the current state of the turtle onto a pushdown stack. The information saved on the stack contains the turtle's position and orientation, as well as other attributes such as the color and width of lines being drawn.

']' Pop a state from the stack and make it the current state of the turtle. No line is drawn, although in general the position of turtles changes.

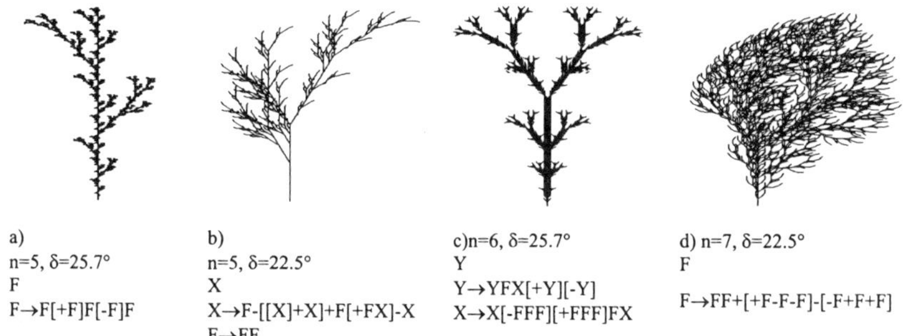

a)
n=5, δ=25.7°
F
F→F[+F]F[-F]F

b)
n=5, δ=22.5°
X
X→F-[[X]+X]+F[+FX]-X
F→FF

c)n=6, δ=25.7°
Y
Y→YFX[+Y][-Y]
X→X[-FFF][+FFF]FX

d) n=7, δ=22.5°
F

F→FF+[+F-F-F]-[-F+F+F]

Figure 2 Examples of plant-like structures generated by bracketed L-systems.

To imitate the mechanism of grown of structures, among them the neural networks. It was created a grammar based on Lindenmayer System, which can generate direct and recurrent ANN, the grammar was based on the model created by [2]. The grammar can be described as **G={Σ, Π, α}**, where **Σ={A;B;C;D;2; ,}**, the axiom is **α=A,B,C** and the production rules are **Π={A→ AB|ε,   B>C→ DB|ε,   B<C→ CD|ε,   D<D→ D2}**. The production rules are described in Table 2, where ε is empty chain.

Table 2 – Production Rules of the grammar

A→ AB|ε	A is replaced by **AB** or by ε .
B>C→ DB|ε	If **C** is the successor of **B**, B is replaced by **DB** or ε .
B<C→ CD|ε	If **B** is the predecessor of **B**, C is replaced by **CD** or ε .
D<D→ D2	If **D** is the predecessor of **D, D** is replaced by **D2**.

Supposing that starting with the axiom α= **A,B,C**, applying the first rule of Table 2 repeatedly a number of times equal to the number of inputs of the ANN, for example considering the number of inputs 2, the string that result is **ABB,B,C.** Considering the third rule applied to **ABB,B,C** a number of times equal to the number of outputs of the ANN, for example one output. The resulting string will be **ABB,B,CD**. Using the second rule of Table 2 applied to **ABB,B,CD** two times, the resulting string will be **ABB,DDB,CD** and finally considering the second option of the following rules **(B>C→ DB|ϵ)** , **(A→ AB|ϵ)** and **(B<C→ CD|ϵ)** the language generated will be **BB,DD,D**. That represents the ANN of Figure 3(a). : The comma represents another level. The definition of context used here it is not the same as in normal L-Systems. Usually the context of a symbol that is being rewriting is directly on the left and right of that symbol. Here it is considered in relation to another level. The Recurrent ANN of the Figure 3 (b) is obtained in a similar way, still considering the production rule **D<D→ D2** applied to **BB,DD,D** results in **BB,DD,D2**. The last **D2** represents the feedback connection . The ANN of the Figure 3(c) is obtained applying the rule **B>C→ DB|ϵ** three times instead of two how accomplished in the third stage previously presented.

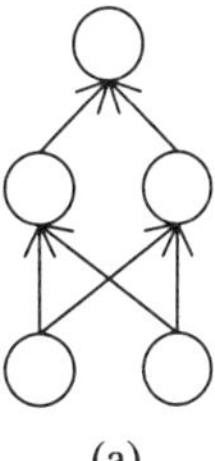 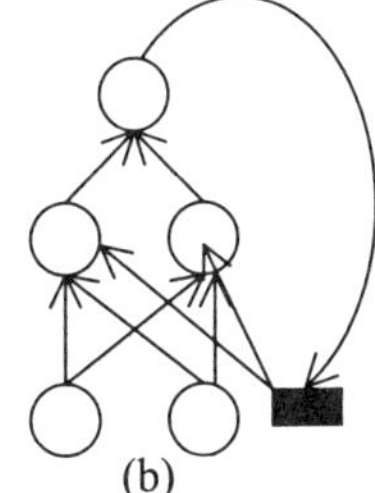 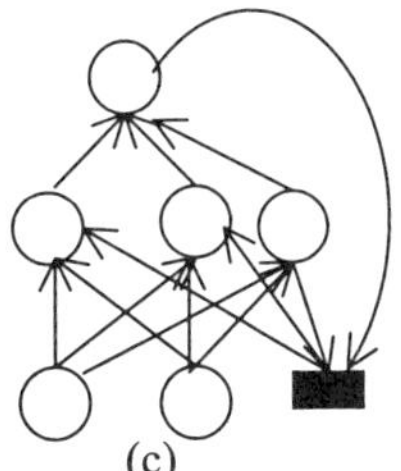

(a)   (b)   (c)

Figure 3 Artificial Neural Networks Generated

4. Genetic Algorithm

The evolution process attempt to create and select the individuals more adapted to the environment which are modeled with genetic algorithms, it is a simplification of what happens in the nature, because GA are often used with a well-defined fitness function to implement the evolution process. However, this does not give the needed adaptation to a dynamic environment, because the fitness function is static, i.e. designed to solve a well-defined task that may become obsolete in a dynamic environment. A population of string is manipulated, where each string can be seen as chromosome (the genotype), consisting of a number of genes. These genes are used to code the parameters for a problem. Each string can be assigned a fitness, which indicates the quality of the solution (the phenotype) it encodes. The strings used by the algorithm reproduce proportional to their fitness. A new generation is created by selecting and recombining existing strings based on their fitness, using genetic operators like selection, crossover, inversion and mutation. From generation to generation the mean fitness of the population should increase. The algorithm used for extraction of rules of the chromosome is shown below.

1. Read the chromosome, six bits at a time.
2. Each group of six bits is converted for a comma, or for a symbol of the alphabet defined in 3.1 by the grammar, in agreement with the Table 3, what will result in a string. The order of reading of the Table 3 is the following: to determine the string of bits that codes the character followed by the following steps: determine which of the four rows on the left corresponds to the first two bits of the string, then choose a column according to the middle two bits, and finally choose a row on the right using the last two bits. For example, the character corresponding to 001000 is A.
3. Find all the strings that code a production rule, following the representation adopted is shown in the Table 4.

ial neural network try, in principle, to imitate the computation as performed

4. Throw away all production rules that do not conform to the restrictions given in 3.1, leaving only valid production rules.
5. Repeat 1-4, by starting to read the bitstring not just at the first bit but also at bit 2,3,4,5,6 and at bit 512,511,510,509,508,507. All production rules are extracted in this way form the L-system for one network. Since the algorithm starts at all bit positions and reads in both directions, the chromosome of one member of the population is read   twelve times, which may eventually increase the level of implicit parallelism of genetic algorithm.

The genetic algorithm generates a population of initially random bitstrings. Each bitstring (Chromosome) is a member of the population. A solution is evaluated by having the decoded L-system produce the network architecture.

5. Artificial Neural Network

Artificial neural network try, in principle, to imitate the computation as performed by the neural system of humans and animals that have contact with the environment making control of their activities. A neuron and its axon dendrites and synapses, is usually modeled as a simple computational unit that implements a simple threshold function on a weighted sum of input values. The architectures of an ANN is generated by the L-System.

Table 3 – The Conversion Table used

	00	01	10	11	
00	B	2	A	B	00
	D	,	,	C	01
	D	B	,	,	10
	B	A	,	B	11
01	D	D	B	2	00
	D	A	D	,	01
	C	B	D	C	10
	B	,	C	D	11
10	A	B	2	C	00
	B	B	B	B	01
	,	D	D	D	10
	A	C	B	A	11
11	B	B	C	B	00
	B	A	D	C	01
	B	2	D	C	10
	,	B	,	B	11

Table 4 – Representation of Production Rules

Production Rules	Representation adopted
α= A,B,C	A,B,C
A→ AB|ε	AAB
B>C→ DB|ε	BCDB
B<C→ CD|ε	BCCD
D<D→ D2	DDD2

A Neural Network Simulator, trains the network for a specified problem. The fitness function adopted consider a minimum number of: neurons in the intermediary layer, production rules, recurrent connections and residual error, which can be represented by Eq.1 and Eq.2. The fitness is returned to the genetic algorithm, which produces a new solution from the population to be evaluated.

$$Fitness = \left[A1/ERM + A2/NCIH + A3/RP + A4/(REC+1)\right], \text{ if } ERM <= Av \qquad \text{(Eq.1)}$$

or

$$Fitness = \left[A1/ERM + A2/NCIH + A3/RP + A4/(REC+1)\right]*0.1, \text{ if } ERM > Av \qquad \text{(Eq.2)}$$

ERM= Average error of the patterns in the output the neural network. **NCIH**=Number of neurons in the intermediary layer. **RP**=Number of production rules. **REC**=Number of recurrent connections, **AV**= Acceptable value of error.

6. Experiments

The parity problem consists of presenting in the input of the neural network a string of bits and to obtain in the output the parity value of the string presented. The corresponding graph of the Mealy Machine, which implements the problem is shown in the Figure 4, where the second component of each state represents the output, that is the bit that indicates the parity of a string of bits. The Table 5, shows the training set for the problem studied.

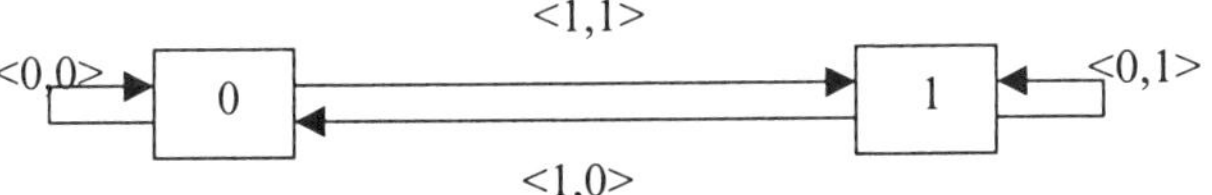

Figure 4- Mealy Machine ME={Σ,Q, δ, qo, F, Δ}=({0,1},{0,1},δ,0,{0,1},{0,1}}, for the parity problem.

Table 5- Training set for the Parity Problem

Input u(k)	State x(k)	State x(k+1)
0	0	0
1	0	1
0	1	1
1	1	0

The parameters of the fitness function used in the simulations were A1=0.000001, A2=100, A3=1 and A4=100, AV=0.01 the value of AV guarantees a good generalization capacity in the sense of the extrapolation, the values of A1,A2,A3 and A4 were chosen by heuristic. The number of epochs were 30000 and the number of generation was 100, the size of the chromosome used was of 512 bits. The minimum network obtained after the simulation is shown in Figure 5.

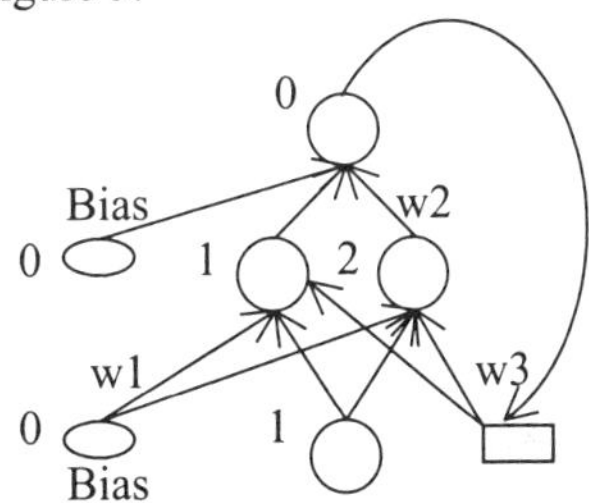

Figure 5 the minimum Artificial Neural Network obtained

The values of weights obtained are shown in Table 6, were w3 are the weights of recurrent connections. The Average Fitness and the Bests Fitness are shown in Figure 6 (a) and Figure 6 (b) respectively.

Table 6- Weights for the Network of Figure 5

W1[0][1]=-3.9739	w1[0][2]=-3.744871	w1[1][1]=-7.627204
W1[1][2]=7.220986	w2[0][0]=-6.726833	w2[1][0]=13.743343
W2[2][0]=13.678354	w3[0][1]=7.328455	w3[0][2]=-15.238744

In the second stage it was tested the learning of the ANN and it was checked the extrapolation capacity of the ANN. The bitstring 0,1,0,1,1,0,1 were presented in the input of the ANN, the outputs obtained and the respective error are shown below in the Table 7.

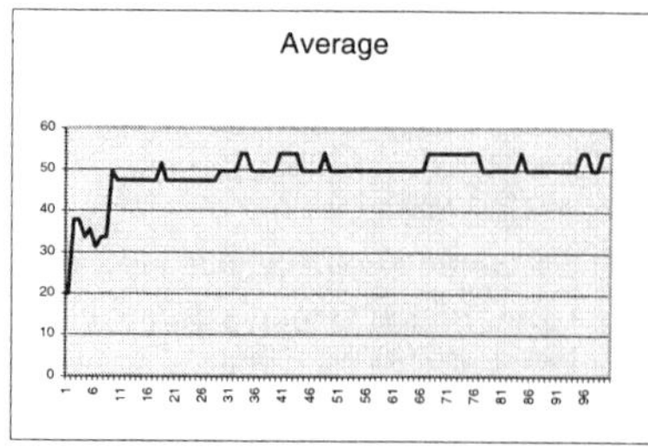

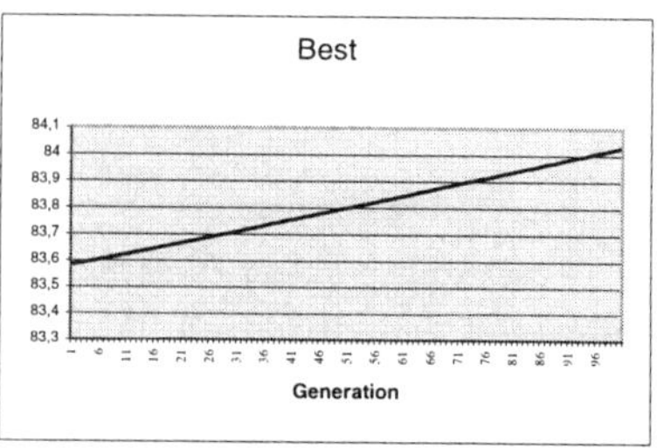

(a)   (b)

Figure 6   Best and Average Fitness versus generation for one run of GA

Table 7 – Obtained Outputs for the first experiment

Outputs	Obtained Outputs	Error
0	0.022707	0.000258
1	0.987723	0.000121
1	0.987723	7.5357E-05
0	0.016073	0.000129
1	0.984752	0.000116
1	0.987743	7.511157e-05
0	0.016074	0.000129

7. Conclusions

The task of modeling natural process through mathematical models is not easy, the system proposed in the present research is a simplification of the evolution process that happens in the nature. However it can be concluded that the automatic project of ANN facilitates the obtaining of satisfactory architectures with a smaller cost. The fitness function can address the search process. When it is adopted as fitness function the inverse of error, the generated ANN not always it is minimum.

References

[1]   BOERS E.J.W. AND KUIPER H. Biological metaphors and the design of modular artificial neural networks.Msc. dissertation, Leiden University, the Netherlands, Departments of Computer Science and Experimental and Theoretical Psychology, 1992.

[2]   BOERS E.J.W. (1995). Using L-Systems as Graph Grammar:G2L-Systems,Technical Report 95-30, 1995.

[3]   KITANO H."Designing neural networks using genetic algorithms with graph generation system, " Complex Systems, vol.4, no. 4, pp.461-476, 1990.

[4]   VAARIO J. An Emergent Modeling Method for Artificial Neural Networks. Doctor dissertation of engineering, Aeronautic an Astronautic Engineering Course, The University of Tokyo, Japan, 1993.

[5]   VICO F.J. AND SANDOVAL F. Use of Genetic algorithms in neural networks definition. In A. Prieto, editor Proc of the International Workshop on Artificial Neural Networks. IWANN´91, p.196-203, Granada, Spain, 17-19 Sep 1991. Springer-Verlag.

[6]   LINDENMAYER, A. "Mathematical models for cellular interaction in development, parts I and II". In Journal of theoretical biology, 18, 280-315, 1968.

[7]   DAWKINS R.; The Blind Watchmaker, Longman, Reprinted with appendix by Penguin, London, 1991.

[8]   SZILARD, A.L. and QUINTON, R.E. An Interpretation for DOL systems by computer graphics. The Science Terrapin, 4:8-13, 1979.

Logic, Artificial Intelligence and Robotics
J.M. Abe & J.I. da Silva Filho (Eds.)
IOS Press, 2001

Fuzzy Conceptual Graphs for the Semantic Web

T.H. Cao
Artificial Intelligence Group
Department of Engineering Mathematics, University of Bristol
United Kingdom BS8 1TR
Tru.Cao@bristol.ac.uk

Abstract. Despite its great usefulness in making information widely available to everyone, the current World Wide Web is showing its limitations with the explosion of information over the Internet. Its hypertext-based languages, like HTML, the information represented by which is mainly for human reading rather than machine processing, have hampered more advanced applications and better services on the Internet than ones currently available. For its next generation, the so-called Semantic Web, a logical formalism that can approach human expression and reasoning is a very good language candidate for Web documents. Conceptual graphs and fuzzy logic are two logical formalisms that emphasize the target of natural language, where conceptual graphs provide a structure of formulas close to that of natural language sentences while fuzzy logic provides a methodology for computing with words. This paper proposes fuzzy conceptual graphs, which combine the advantages of both the two formalisms, as a suitable Semantic Web language.

1. Introduction

We have seen how very useful the World Wide Web is to our daily life and work. It allows nearly instant access to an extremely large virtual pool of information from over the world and facilitates electronic services and businesses on the Internet. The current Web technology is however very simple. It represents information mainly in texts for human reading, using HTML (HyperText Markup Language)[1], for example, as a language to provide a structure for storing and displaying them. With the explosion of information on the Internet that we are witnessing, such a simple language without a machine-processable semantics becomes a bottleneck for finding information as well as for maintaining and presenting it. As an example, it restricts search engines to be just keyword-based and thus, when searching for information on the Web, one often receives many useless links. That is simply because, for instance, one cannot express the difference between the two queries to find Web pages about "Publications written by Tim Berners-Lee" and "Publications written about Tim Berners-Lee", which have the same keywords "book" and "Tim Berners-Lee" but different semantics.

To overcome this shortcoming, many researchers and practitioners have started making a great effort to extend the current Web towards the so-called Semantic Web ([1]), in which information is represented by languages with expressive but well-defined semantics to be processable by computers. In our view, this challenge is similar to the one of Artificial Intelligence, namely that of how to make computers approach human expression and reasoning. In particular, this requires a formal knowledge representation language that not only has the expressive power close to that of natural language, but also can be automatically processed by computers. The context of the Web adds one more difficult task to this, that is, the represented knowledge must also be comprehensible and interoperable widely over the Internet, and this requires accompanying ontologies.

---

[1] http://www.w3.org/MarkUp

How humans process information represented in natural language is still a challenge to science in general, and to Artificial Intelligence in particular. However, it is clear that for a computer with the conventional processing paradigm to process natural language, a formalism is required. For reasoning, it is desirable that such a formalism be a logical one. A logic for handling natural language should have not only a structure of formulas close to that of natural language sentences, but also a capability to deal with the semantics of vague linguistic terms pervasive in natural language expressions.

Currently, conceptual graphs (CGs) ([13]) and fuzzy logic ([18]) are two logical formalisms that emphasize the target of natural language, each of which is focused on one of the two mentioned desired features of a logic for handling natural language. Indeed, while a smooth mapping between logic and natural language has been regarded as the main motivation of conceptual graphs ([14], [15]), a methodology for computing with words has been regarded as the main contribution of fuzzy logic ([19], [20]). Conceptual graphs, based on semantic networks and Peirce's existential graphs, combine the visual advantage of graphical languages and the expressive power of logic. Meanwhile, fuzzy logic, based on fuzzy set theory, has been developed for approximate representation of, and reasoning with, imprecise and vague information.

This paper proposes fuzzy conceptual graphs (FCGs) ([2], [11], [16]), which combine the advantages of both conceptual graphs and fuzzy logic, as a suitable knowledge representation language for the Semantic Web. Firstly, Sections 2 and 3 briefly present the basic notions of conceptual graphs and fuzzy logic, respectively. Section 4 introduces fuzzy conceptual graphs with their recent development. Then Section 5 discusses and proposes fuzzy conceptual graphs as a suitable Semantic Web language. Finally, Section 6 concludes the paper and suggests future research.

2. Conceptual Graphs

A *simple CG* is a bipartite graph of *concept* vertices alternate with (conceptual) *relation* vertices, where edges connect relation vertices to concept vertices ([13]). Each concept vertex, drawn as a box and labelled by a pair of a *concept type* and a *concept referent*, represents an entity whose type and referent are respectively defined by the concept type and the concept referent in the pair. Each relation vertex, drawn as a circle and labelled by a *relation type*, represents a relation of the entities represented by the concept vertices connected to it. For brevity, we may call a concept or relation vertex a concept or relation, respectively. Concepts connected to a relation are called *neighbour concepts* of the relation. Each edge is labelled by a positive integer and, in practice, may be directed just for readability.

For example, the CG in Figure 1 says "John is a student. There is a subject. Computer Science is a field of study. The subject is in Computer Science. John studies the subject", or briefly, "John studies a subject in Computer Science". In a textual format, concepts and relations can be respectively written in square and round brackets as follows:

[STUDENT: John]->(STUDY)->[SUBJECT: *]->(IN)->[FIELD: Computer Science]

Here, for simplicity, the labels of the edges are not shown. In this example, [STUDENT: John], [SUBJECT: *], [FIELD: Computer Science] are concepts with STUDENT, SUBJECT and FIELD being concept types, whereas (STUDY) and (IN) are relations with STUDY and IN being relation types.

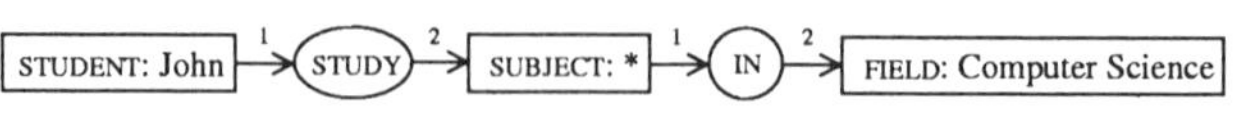

Figure 1: A simple CG

The referents John and Computer Science of the concepts [STUDENT: John] and [FIELD: Computer Science] are *individual markers*. The referent * of the concept [SUBJECT: *] is the *generic marker* referring to an unspecified entity. Two concepts with two different individual markers are assumed to refer to two different entities, while concepts with the same individual marker are assumed to refer to the same entity.

The first-order predicate logic semantics of conceptual graphs is defined through operator Φ ([13]) that maps a CG to a first-order predicate logic formula. Basically, Φ maps each vertex of a CG to an atomic formula of first-order predicate logic, and maps the whole CG to the *conjunction* of those atomic formulas with all variables being existentially quantified. Each individual marker is mapped to a constant, each generic marker is mapped to a variable, and each concept or relation type is mapped to a predicate symbol. For example, let G be the CG in Figure 1, then $\Phi(G)$ is:

$\exists x$ (student(John) $\wedge$ subject(x) $\wedge$ field(Computer Science) $\wedge$

study(John, x) $\wedge$ in(x, Computer Science)).

Conceptual graphs can also be nested. A *nested CG* is recursively defined as a simple CG extended by adding a *descriptor* field to each of its concepts, where a descriptor is either empty or a nested CG, which describes the referent of a concept. With this structure, *negation* of a proposition can be represented in conceptual graphs as a unary relation of type NEG whose connected concept is of type PROPOSITION and has the CG representing that proposition as its descriptor. This gives conceptual graphs the full expressive power of order-sorted predicate logic. For example, the following CG represents the negation of "John studies Literature":

(NEG)->[PROPOSITION: * [STUDENT: John]->(STUDY)->[SUBJECT: Literature]].

A fundamental operation on CGs is *CG projection*, which maps a CG to another more or equally specific one, by mapping each vertex of the former to a vertex of the latter that has a more or equally specific concept type and referent, or relation type. The mapping must also preserve the adjacency of the neighbour concepts of a relation. As such, if a CG has a projection to another one, then the latter logically implies the former. Figure 2 illustrates a projection from G to H, provided that STUDENT is a subtype of PERSON and STUDY is a subtype of ACT.

3. Fuzzy Logic

Often concepts encountered in the real world are vague in nature, like *young* or *old*, *short* or *tall*, *cheap* or *expensive*, as reflected in natural language. These concepts, or their expressions in natural language, are vague in the sense that, in most contexts, there is no clear-cut boundary between them and *not young* or *not old*, *not short* or *not tall*, *not cheap* or *not expensive*, respectively. In other words, the membership of an object in the extension of such a concept is not a matter of "to be or not to be", but rather a matter of degree. Classical set theory, in which the membership grade of an element in a set can only be either 0 or 1, is thus inadequate to deal with vague concepts. This was the main motivation of Zadeh founding fuzzy set theory ([17]), which generalizes classical set theory by defining membership grades to be real numbers in the interval [0, 1].

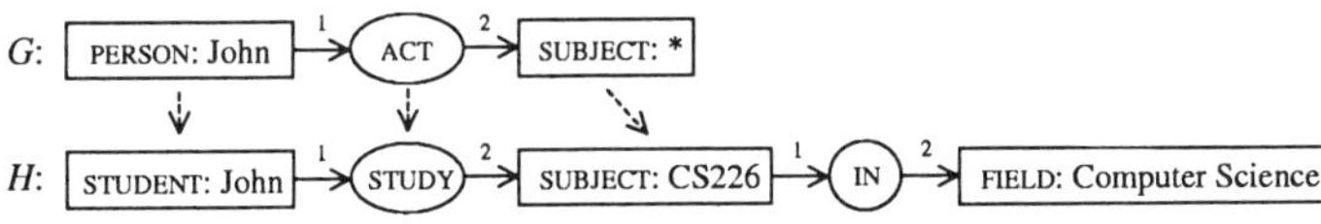

Figure 2: A CG projection

Fuzzy set theory then gave birth to fuzzy logic ([18]). In the literature, the term *fuzzy logic* has been used for different logic systems that have originated from the theory of fuzzy sets. However, they may have so different characteristics that they need to be distinguished to avoid confusion. We classify them into three main groups, namely, *partial truth-valued logic, possibilistic logic*, and *fuzzy set logic*. We call a fuzzy logic that deals with partial truth partial truth-valued logic, e.g. [12], which is a special multiple-valued logic, whose formulas are associated with real numbers in the interval [0, 1] interpreted as truth degrees of formulas. In contrast, in possibilistic logic ([7]), although formulas are also associated with real numbers in the interval [0, 1], they have the meaning of uncertainty degrees. Finally, in the broadest sense, we call a fuzzy logic whose formulas involve fuzzy set values fuzzy set logic, e.g. [3].

Fuzzy logic is supposed to be complementary to other approaches, in particular the long time and very well founded probability theory, for dealing with uncertainty and imprecision, which apparently exist in the real world and thus human expression and reasoning. The most controversial aspect of fuzzy set theory and fuzzy logic is also their cornerstone, the definition of fuzzy sets, that is, what a membership grade means. As answers, there have been so far two main schools of thoughts, one of which take membership grades as a primitive notion and the other links them with probability.

In the latter school, the voting model interpretation of fuzzy sets is as follows (cf. [9]). Given a fuzzy set A on a domain U, each voter has a subset of U as his/her own crisp definition of the concept that A represents. For example, a voter may have the interval [0, 35] representing human ages from 0 to 35 years as his/her definition of the concept *young*, while another voter may have [0, 25] instead. The membership function value $\mu_A(u)$ is then the proportion of voters whose crisp definitions include u. As such, A defines a probability distribution on the power set of U across the voters, and thus a fuzzy proposition "x is A" defines a family of probability distributions of the variable x on U.

However, we argue that the current lack of a unique and clearly defined semantics of membership grades of fuzzy sets should not discourage us from developing the theory. We, human beings, use many such vague concepts and degrees in [0, 1] in daily life, but it is still a challenge to science in general, and to Artificial Intelligence in particular, how we actually process these cognitively. It is fuzzy logic that takes this challenge, moving forward from the traditional probability theory, which is apparently not adequate to give the answer. We think theoretical research and practical experiments in both of the schools of fuzzy logic mentioned above should be encouraged. In practice, fuzzy logic has been successfully applied to several areas, such as expert systems, knowledge acquisition and fusion, decision making, and information retrieval, among others.

4. Fuzzy Conceptual Graphs

Firstly, *simple FCGs* extend simple CGs with linguistic labels defined by fuzzy sets as individual markers ([2], [11], [16]). For example, in Figure 3, the simple FCG G expresses "Peter, a Swede, is *tall*", where *tall* is the linguistic label of a fuzzy set. The contracted form of G is G^*, where the concepts [SWEDE: Peter] and [HEIGHT: *@*tall*] are called *fuzzy entity concept* and *fuzzy attribute concept*, respectively.

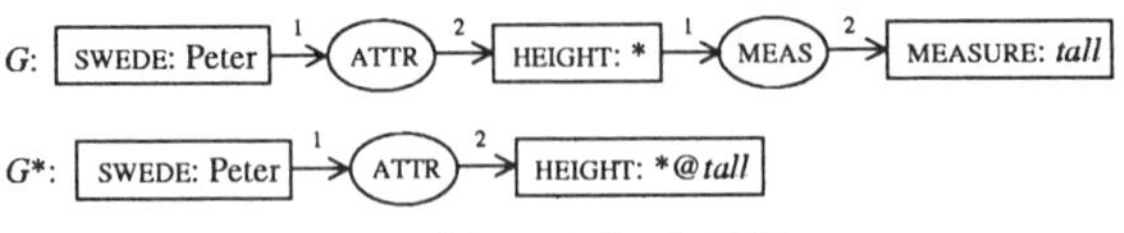

Figure 3: Simple FCGs

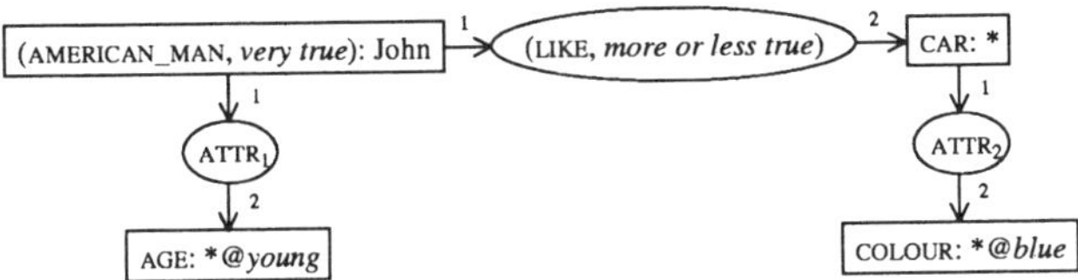

Figure 4: A simple FCG with fuzzy types

A concept type or a relation type can be a *fuzzy type* to represent uncertainty and/or partial truth about the type of the entity referred by a concept or the type of a relation, where a fuzzy type is defined as a pair of a *basic type* and an uncertainty and/or truth degree ([6]). An uncertainty and/or truth degree can be a *fuzzy truth value*, i.e., a fuzzy set on [0, 1] whose values are interpreted as truth degrees, or a *fuzzy probability value*, which is also defined by a fuzzy set on [0, 1] whose values are however interpreted as probability degrees. Examples of linguistic labels of a fuzzy truth value are *more or less true* and *not very false*, while ones of a fuzzy probability value are *quite likely* and *very unlikely*. A simple FCG with fuzzy types is shown in Figure 4, which says "It is *very true* that John is an American man, who is *young*, and it is *more or less true* that he likes a car, whose colour is *blue*", where *very true*, *more or less true*, *young* and *blue* are linguistic labels of fuzzy sets.

FCG projection, which matches an FCG to another one, extend CG projection with matching degrees between fuzzy sets and fuzzy types in the matched pairs of concepts and relations in the two FCGs. For representing complex information, we have introduced *nested FCGs* ([4]). As for nested CGs, a nested FCG is recursively defined as a simple FCG extended by adding a descriptor field to each of its concepts, where a descriptor is either empty or a nested FCG, which describes the referent of a concept.

For example, the nested FCG in Figure 5 expresses "Tom, a Scot, shows a picture of a monster which is *about 20 feet* in length, and tells about a *very unlikely* situation that it exists in Lake Loch Ness". Here, the dotted line is a *coreference link* denoting that the two concepts [MONSTER: *] refer to the same entity; in the textual format, coreference is represented by variables of the same name as concept referents. Meanwhile, representation of such nested pieces of information in linear notations, e.g. predicate logic, would be difficult to follow.

One aspect of natural language the logic for which has been the quest and focus of significant research effort is one of generalized quantifiers. In reasoning with quantifiers, *absolute quantifiers* and *relative quantifiers* on a set have to be distinguished, where the quantities expressed by the latter are relative to the cardinality of the set. Examples of absolute quantifiers are *only one*, *few*, or *several*, while ones of relative quantifiers are *about 9%*, *half*, or *most*.

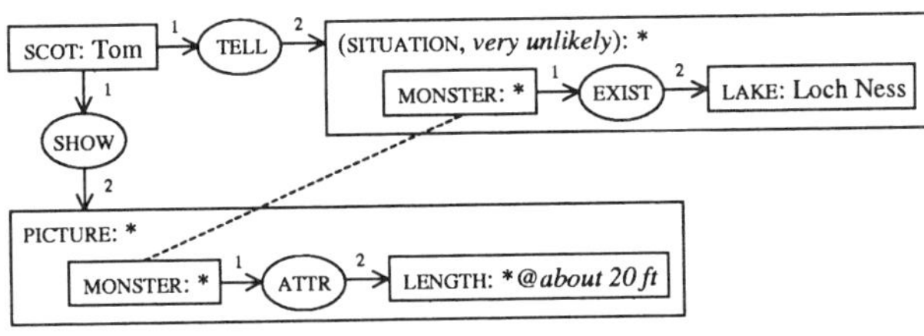

Figure 5: A nested FCG

In practice, there are quantifying words, e.g. *few* and *many*, that may be used with either meaning depending on the context. For instance, *few* in *"Few* people in this conference are from Asia" may mean a small number of people, while *few* in *"Few* people in the United Kingdom are from Asia" may mean a small percentage of population. Classical predicate logic with only the existential quantifier, equivalent to the absolute quantifier *at least 1*, and the universal quantifier, equivalent to the relative quantifier *all* or *every*, cannot deal with general quantification in natural language.

In the crisp case, absolute quantifiers can be defined by natural numbers, and relative quantifiers by non-negative rational numbers that are not greater than 1 measuring a proportion of a set, where 0 means 0% and 1 means 100%. Correspondingly, in the fuzzy case, absolute quantifiers can be defined by fuzzy sets on the set $\mathbf{N}$ of natural numbers, i.e., fuzzy numbers whose domain is restricted to $\mathbf{N}$, and relative quantifiers by fuzzy numbers whose domain is restricted to the set of rational numbers in [0, 1]. For simplicity without rounding of real numbers, however, we assume absolute quantifiers to be defined by fuzzy numbers on [0, +∞] and relative quantifiers by fuzzy numbers on [0, 1].

The existential quantifier in classical logic corresponds to *at least 1* in natural language, which is an absolute quantifier whose membership function is defined by $\mu_{at\ least\ 1}(x) = 1$ if $x \geq 1$, or $\mu_{at\ least\ 1}(x) = 0$ otherwise. Meanwhile, the universal quantifier, which corresponds to *all* or *every* in natural language, is a relative quantifier and its membership function is defined by $\mu_{all}(1) = 1$ and $\mu_{all}(x) = 0$ for every $0 \leq x < 1$.

An absolute quantifier Q on a type T whose denotation set in a universe of discourse has the cardinality $|T|$ corresponds to the relative quantifier $Q_T = Q/|T|$. A relative quantifier Q in a statement "Q A's are B's" can be interpreted as the proportion of objects of type A that belong to type B, i.e., $Q = |A \cap B|/|A|$, which is a fuzzy number. Equivalently, it can also be interpreted as the fuzzy conditional probability, which is a fuzzy number on [0, 1], of $B(x)$ being **true** given $A(x)$ being **true** for an object x picked at random uniformly.

We have applied this interpretation and the fuzzy set-theoretic semantics of generalized quantifiers to formally define the semantics of *generally quantified FCGs* as probabilistic logic rules comprising only simple FCGs ([5]). For example, Figure 6 shows a generally quantified FCG G and its defining expansion E, expressing *"Most* Swedes are *tall"*, where {*} denotes a *set referent, most* and *tall* are linguistic labels of fuzzy sets, and *most* = $Pr(tall(x)\ |\ Swede(x))$.

Furthermore, the type in a generally quantified concept can be represented by a simple FCG as a lambda expression defining that type. We call such a simple FCG a *lambda FCG*, which is like a simple FCG except that it has one concept, which we call a *lambda concept*, whose referent is denoted by λ to be distinguished from the generic and individual referents. For example, Figure 7 illustrates a generally quantified FCG G and its defining expansion E, expressing *"Most* people who are *tall* are *not fat"*. As such, a lambda FCG corresponds to a relative clause in natural language.

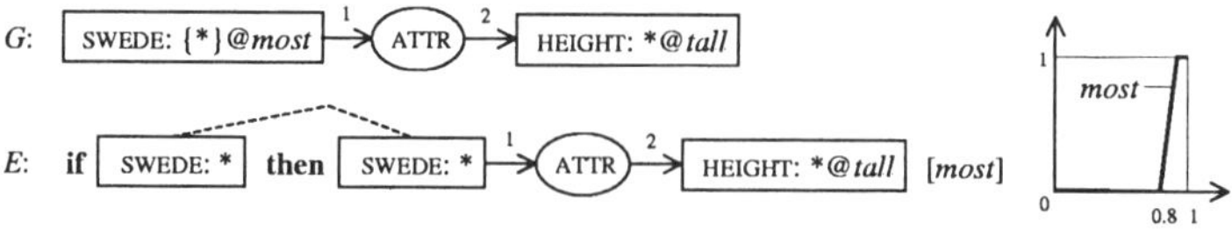

Figure 6: A generally quantified FCG and its defining expansion

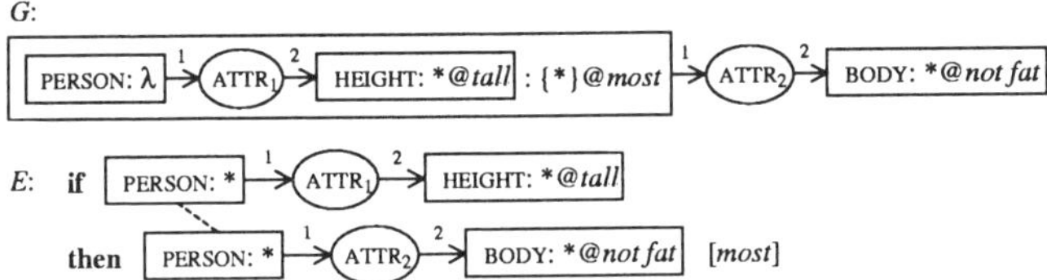

Figure 7: Quantification on a type defined by a lambda FCG

5. Towards the Semantic Web

As mentioned in the introduction section, despite its great usefulness in making information widely available, with the explosion of information on the Internet the simple hypertext-based Web is showing its limitations, so new knowledge representation languages are required for its next generation, the Semantic Web. As a first step, XML (eXtensible Markup Language)[2] has been introduced to allow everyone to add arbitrary structure to their documents and make use of that structure in sophisticated ways for domain- and task-specific purposes. Then higher-level languages based on XML, e.g., RDF (Resourse Description Framework), OIL (Ontology Inference Layer), or DAML (DARPA Agent Markup Language), have been developed with more logical and ontological modelling primitives ([8]).

Although, as one of its goals, XML was designed for Web documents written in it to be human-legible and reasonably clear, it turned out to be too verbose. That is why other more readable languages like the ones mentioned above have been introduced. Comparing Web languages to programming languages, we view XML as a low-level language, while others are high-level languages, which are closer to human expression and reasoning. As such, Web documents may be written in any high-level language and then translated to XML for machine processing or information exchange. As an alternative, Web documents written in a high-level language can be processed directly by a specialized language processor.

A formal language that has a smooth mapping to and from natural language, like conceptual graphs, and can deal with imprecision and vagueness, like fuzzy logic, is a very good candidate for the Semantic Web. Being a logical formalism itself is an advantage of a language for reasoning on knowledge represented by it. Therefore we propose fuzzy conceptual graphs as a suitable Semantic Web language, which combines the mentioned advantages of both conceptual graphs and fuzzy logic as summarized below:

- The language is more human-readable than XML-based languages and other logical formalisms, like predicate logic. A query in natural language can be smoothly translated to an FCG to be matched to FCGs in Web documents, and an answer FCG can also be smoothly translated to natural language for human reading.

- The inherent graphical notation of the language is useful for knowledge visualization, in particular for presenting nested information whose representation in linear notations is difficult to follow. The structure for abstracting knowledge into nested concepts is helpful for knowledge organization and storage.

- The language itself is a logical formalism, which is desirable for knowledge reasoning. With its order-sorted feature, knowledge can be represented more concisely and reasoning can be performed more efficiently via inheritance. Its fuzzy logic core is significant for representation and reasoning with uncertain and imprecise information.

---

[2] http://www.w3.org/XML

For example, Web pages about "Books written by Tim Berners-Lee" and "Books written about Tim Berners-Lee" as mentioned in the introduction section can be indexed by the following CGs

[PERSON: Tim Berners-Lee]<-(AGNT)<-[WRITE: *]->(OBJ)->[BOOK: *]

and

[PERSON: *]<-(AGNT)<-[WRITE: *]->(OBJ)->[BOOK: *]->(CONT)->[PERSON:Tim Berners-Lee]

where (AGNT) and (OBJ) are relations between the action concept [WRITE: *] and its agent and object concepts, and (CONT) is a relation between an entity concept and its content as another entity concept.

Then the query for "Publications written by Tim Berners-Lee" can be translated into the CG

[PERSON: Tim Berners-Lee]<-(AGNT)<-[WRITE: *]->(OBJ)->[PUBLICATION: *]

which can then be matched to the first indexing CG above by CG projection and the corresponding Web pages are retrieved, provided that BOOK is a subtype of PUBLICATION. We note that this query CG cannot be projected to the second indexing CG above because "Tim Berners-Lee", referring to a specific entity, is more specific than "*", referring to an unspecified entity.

For the fuzzy case, let us take an example adapted from [10], which discusses the advantages of conceptual graphs over XML-based languages and uses them for embedding knowledge to Web documents, in particular for indexing document elements. For instance, the following CG can be used to index an image, among others, about a red vehicle:

[VEHICLE: *]->(ATTR)->[COLOR: red]

where "red" is a non-interpreted individual marker.

The implemented knowledge processor, called WebKB, is equipped with ontologies of concept types and relation types, in order to access knowledge represented in conceptual graphs. A query to retrieve an image about something that is red can be translated in to the following CG

[SOMETHING: *]->(ATTR)->[COLOR: red]

which can then be projected to the indexing CG above and the corresponding image is retrieved.

That work however did not consider vague linguistic terms, of which *red* is an example. Therefore, a query to find an image about something that is *very red* would fail, as its corresponding CG could not match to the indexing CG. Meanwhile, using fuzzy conceptual graphs, such an image can be indexed by the FCG

[VEHICLE: *]->(ATTR)->[COLOR: *@*red*]

where *red* is a linguistic term defined by a fuzzy set.

Then the query about something *very red* represented by the following FCG

[VEHICLE: *]->(ATTR)->[COLOR: *@*very red*]

can be answered to a certain degree defined by the matching degree of the fuzzy set defining *very red* to that defining *red*, using FCG projection from the query FCG to the indexing FCG. Here, for dealing with fuzziness, besides the ontologies of concept and relation types, one needs also an ontology of vague concepts with their defining fuzzy sets.

6. Conclusion

We have presented the basic notions of conceptual graphs and fuzzy logic, showing that these two logical formalisms actually have the common target of natural language. We then have introduced fuzzy conceptual graphs, which combine the advantages of both conceptual graphs and fuzzy logic, as a knowledge representation language for Artificial Intelligence approaching human expression and reasoning. In particular, we have proposed fuzzy conceptual graphs as a suitable knowledge representation language for the Semantic Web, in which information is not only for human reading but also for machine processing.

There are still many aspects to be investigated for this proposal. A specific subset of fuzzy conceptual graphs with a balance between the expressive power and the computational tractability in the Web context is to be identified. The management of an ontology of vague concepts is a problem to be studied. Followed up is the implementation of tools for processing fuzzy conceptual graphs in Web documents. These are some of the topics that we suggest for future research.

References

[1]   T. Berners-Lee and J. Hendler and O. Lassila, The Semantic Web, *Scientific American*, May 2001.

[2]   T.H. Cao, Foundations of Order-Sorted Fuzzy Set Logic Programming in Predicate Logic and Conceptual Graphs. PhD Thesis, University of Queensland, 1999.

[3]   T.H. Cao, Annotated Fuzzy Logic Programs, *International Journal for Fuzzy Sets and Systems* **113** (2000) 277-298.

[4]   T.H. Cao, Fuzzy Conceptual Graphs: A Language for Computational Intelligence Approaching Human Expression and Reasoning. In: Sincak, P. et al. (eds), The State of the Art in Computational Intelligence. Physica-Verlag, 2000, pp. 114-120.

[5]   T.H. Cao, Generalized Quantifiers and Conceptual Graphs. In: H.S. Delugach and G. Stumme (eds), Conceptual Structures: Broadening the Base, Lecture Notes in Artificial Intelligence, Vol. 2120. Springer-Verlag, 2001, pp. 87-100.

[6]   T.H. Cao and P.N. Creasy, Fuzzy Types: A Framework for Handling Uncertainty about Types of Objects, *International Journal of Approximate Reasoning* **25** (2000) 217-253.

[7]   D. Dubois and J. Lang and H. Prade, Possibilistic Logic. In: Dov M. Gabbay et al. (eds), Handbook of Logic in Artificial Intelligence and Logic Programming, Vol. 3. Oxford University Press, 1994, pp. 439-514.

[8]   D. Fensel (ed.), The Semantic Web and Its Languages, *IEEE Intelligent Systems*, November/December 2000.

[9]   B.R. Gaines, Fuzzy and Probability Uncertainty Logics, *Journal of Information and Control* **38** (1978) 154-169.

[10] P. Martin and P. Eklund, Embedding Knowledge in Web Documents. In: Proceedings of the 8th International World Wide Web Conference, 1999.

[11] S.K. Morton, Conceptual Graphs and Fuzziness in Artificial Intelligence. PhD Thesis, University of Bristol, 1987.

[12] J. Pavelka, On Fuzzy Logic, *Zeitschrift fur Mathematik Logik und Grundlagen der Mathematic* **25** (1979), Part I: 45-72, Part II: 119-134, Part III: 447-464.

[13] J.F. Sowa, Conceptual Structures - Information Processing in Mind and Machine. Addison-Wesley Publishing Company, 1984.

[14] J.F. Sowa, Towards the Expressive Power of Natural Language. In: J.F. Sowa (ed.), Principles of Semantic Networks - Explorations in the Representation of Knowledge. Morgan Kaufmann Publishers, 1991, pp. 157-189.

[15] J.F. Sowa, Matching Logical Structure to Linguistic Structure. In: N. Houser and D.D. Roberts and J. Van Evra (eds), Studies in the Logic of Charles Sanders Peirce. Indiana University Press, 1997, pp. 418-444.

[16] V. Wuwongse and M. Manzano, Fuzzy Conceptual Graphs. In: G.W. Mineau and B. Moulin and J.F. Sowa (eds), Conceptual Graphs for Knowledge Representation, Lecture Notes in Artificial Intelligence, Vol. 699. Springer-Verlag, 1993, pp. 430-449.

[17] L.A. Zadeh, Fuzzy Sets, *Journal of Information and Control* **8** (1965) 338-353.

[18] L.A. Zadeh, Fuzzy Logic and Approximate Reasoning, *Synthese* **30** (1975) 407-428.

[19] L.A. Zadeh, PRUF - A Meaning Representation Language for Natural Languages, *International Journal of Man-Machine Studies* **10** (1978) 395-460.

[20] L.A. Zadeh, Fuzzy Logic = Computing with Words, *IEEE Transactions on Fuzzy Systems* **4** (1996) 103-111.

Logic, Artificial Intelligence and Robotics
J.M. Abe & J.I. da Silva Filho (Eds.)
IOS Press, 2001

Logic and Valuations

Newton C. A. DA COSTA

Paulista University – UNIP
Rua Dr. Bacelar, 1212 – São Paulo – SP – Brazil.

Abstract. We introduce the notion of valuation of a logical system with the aim of systematize the various classical and non-classical logics. We show the relationships between logics and mathematical structures. We also show how to build a generalized Galois theory encompassing the various logics.

Logic, Artificial Intelligence and Robotics
J.M. Abe & J.I. da Silva Filho (Eds.)
IOS Press, 2001

Emmy: A Paraconsistent Autonomous Mobile Robot

J. I. DA SILVA FILHO and J. M. ABE
SENAC-College of Computer Science and Technology
and
Institute for Advanced Studies, University of São Paulo, São Paulo, Brazil

Abstract. Contradictions or inconsistencies are common when we describe parts of the real world. The control systems used in Automation and Robotics work in general based on classical logic. As it is well known, these binary systems are inadequate to treat the contradictory situations properly. In some cases the systems of classic control are projected to ignore these situations, losing information that could be important for the increasing of the control efficiency. The treatment of ambiguous information by classical control in Robotics may take a long time. This delay reduces the robot's reaction capability and may cause control to stop.

It is difficult to get efficiency in control systems for autonomous mobile robot navigation. The difficulties are many. This is because in the project of a control of autonomous mobile robot, several expected situations causing inconsistencies are involved. These inconsistencies can happen when the robot moves in an unknown environment. For instance: when passing by close to walls, in narrow roads, curved and in uneven lands or still in the presence of traffic, it is obtained a lot of contradiction.

In this work we present the specification of a prototype of an autonomous mobile robot consisting mainly by a mobile platform of aluminium of circular format of 30 cm of diameter and 60 cm of height projected for researches in application of the paraconsistent logics in Robotics. The control system of the robot's navigation is made by the Para-Control [13] built in Hardware and by Para-sonic [12]. The autonomous mobile robot is destined to traffic in a non-structured environment avoiding collision and obstacles consisting by objects that appear suddenly in its way and its main characteristic is allowing programming the robot's actions when it faces with inconsistent or vague situations captured by the sensors.

The robot was named "Emmy" after the great mathematician Emmy Nöether. Taking into account the robot's technical limitations, we think that results were satisfactory showing the possibility to do treatment of uncertain knowledge through the paraconsistent logics.

1. The Control System Based On $E\tau$

The paraconsistent annotated logics are a class of non-classic logics that allow in their structure the treatment of contradictory signals. They are quite appropriate when the knowledge of evidences is used to help to solve conflicts.

The paraconsistent logic of annotation with two values - $E\tau$ - is a class of logic that allows evidential reasoning where two values are associated to an annotation: (μ_1, μ_2). Thus, a proposition $p_{(\mu1, \mu2)}$ of $E\tau$ is read intuitively: it is believed that p's truth value is at least (μ_1, μ_2). μ_1 is interpreted here as the belief degree of p while μ_2 is interpreted as the disbelief degree of p. μ_1, μ_2 belong to the real unitary interval $[0, 1]$. There is also a natural operator (epistemic negation) defined among annotations: $\sim : [0, 1] \times [0, 1] \rightarrow [0, 1] \times [0, 1]$ defined by $\sim(\mu_1, \mu_2) = (\mu_2, \mu_1)$. Also, we associate each annotation (μ_1, μ_2) the two values: Dce = degree of certainty = $\mu_1 - \mu_2$ and Dct = degree of contradiction = $\mu_1 + \mu_2 - 1$

With the values of Dce and Dct it is made the description of areas limited in the lattice that are representative of states logical exit resultants. We have 4 extreme states that are those corresponding to the extreme regions of the lattice and 8 non-extreme states that are those corresponding to the remaining areas.

The extreme logical states with its corresponding symbols are:

T $\Rightarrow$ Inconsistent[1]; F $\Rightarrow$ False; $\perp$ $\Rightarrow$ Paracomplete[2]; V $\Rightarrow$ True

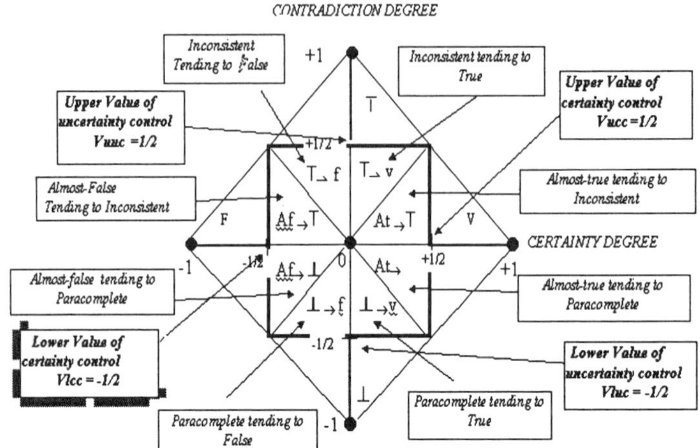

Figure 1 – Certainty and contradiction degrees with control value limit: $V_{uuc} = V_{ucc} = 1/2$ e $V_{lcc} = V_{luc} = -1/2$.

And the non-extreme logical states:

$\perp \to$ f $\Rightarrow$ Paracomplete tending to False ; $\perp \to$ v $\Rightarrow$ Paracomplete tending to True;

T $\to$ f $\Rightarrow$ Inconsistent tending to False; T$\to$ v $\Rightarrow$ Inconsistent tending to True;

At $\to$ T $\Rightarrow$ Almost - True tending to Inconsistent ; Af $\to$ T $\Rightarrow$ Almost - False tending to Inconsistent; Af $\to \perp$ $\Rightarrow$ Almost - False tending to Paracomplete and At $\to \perp$ $\Rightarrow$ Almost - True tending to Paracomplete. In the paraconsistent logic controller Para-control, the values μ_1 and μ_2 are considered as imputs and by the equations of the definition of certainty and contradiction values we get the analogic values of Dce and Dct besides of a binary word composed by 12 digits. Each active digit corresponds to the resultant output logical state.

2. The Autonomous Mobile Robot Emmy

The autonomous mobile robot Emmy consists of a mobile platform of aluminum of circular format of 30 cm of diameter and 60 cm of height projected for researches of application of the paraconsistent annotated logics in Robotics. The robot was projected in separated functional modules of the control system. This allows an easy visualization of the action of each module in the control of the robot's movement.

In Emmy's movement in non-structured environment, the information about the existence or not of an obstacle in its way are obtained through the device denominated Para-sonic. The Para-sonic is capable to capture obstacles in an autonomous mobile robot's path transforming proportional the distance measures between the robot and the obstacle in electric signals, in the form of a continuous electric tension that can vary within the range

---

[1] Intuitively, a value corresponding to a proposition and its negation both true at same time.
[2] Intuitively, a value corresponding to a proposition and its negation both false at same time.

of 0 to 5 volts. The Para-sonic is basically composed by two ultrasonic sensors of type POLAROID 6500 [17] controlled by a 8051 Microprocessor. The 8051 Microprocessor is programmed properly to do the synchronization between the measurements of the sensor two and the transformation of the distance in electric tension.

The system of control of the robot's navigation is made by a Logical Controller paraconsistent - Para-Control [13] built in Hardware and that receives and makes the treatment of electric signals based on the theoretical concepts of the paraconsistent logic $E\tau$.

The Figure 2 (b) shows the robot's main parts Emmy where the modules are considered in first plane that compose the whole control system.

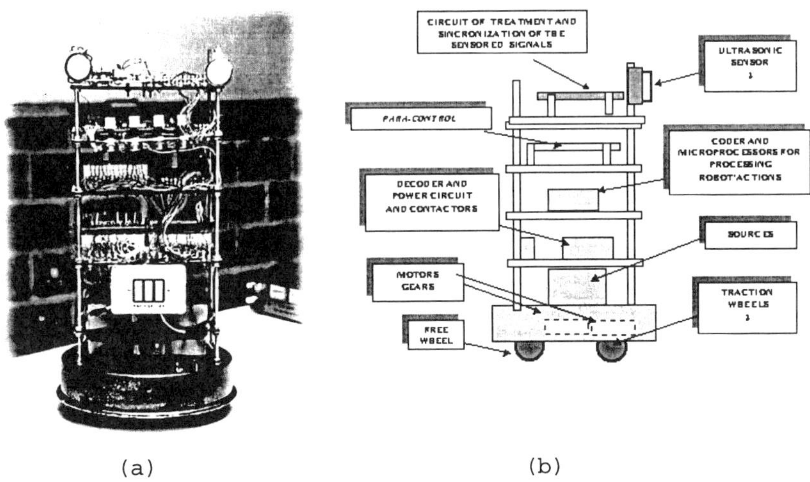

(a)　　　　　(b)

Figure 2 – (a) The robot Emmy – (b) The robot's main parts.

The Para-Control receives information in the form of belief degrees and disbelief degrees, making a paraconsistent analysis that outputs in representative signals of logical states and certainty and uncertainty degrees. The two exit forms can be used in the control, depending on the project. The Para-Control still offers alternative of adjusts external control through potentiometer and all its functions can be programmed in software with the use of a certain algorithm called "Para-analyzer" [10, 11]. With the information of the resulting logical state, a programmed microprocessor with certain sequences of routines will address the actions that the robot will take in each situation.

3. Basic Operation

The robot Emmy uses the para-control system to traffic in non-structured environments avoiding collisions with human beings, objects, walls, tables, etc. The form of reception of information on the obstacles [14] is named *non-contact* which is the method to obtain and to treat signals from ultra-sonic or optical sensors in order to avoid collisions.

The system of the robot's control is composed by the Para-sonic, Para-control and supporting circuits. See Figure 3.

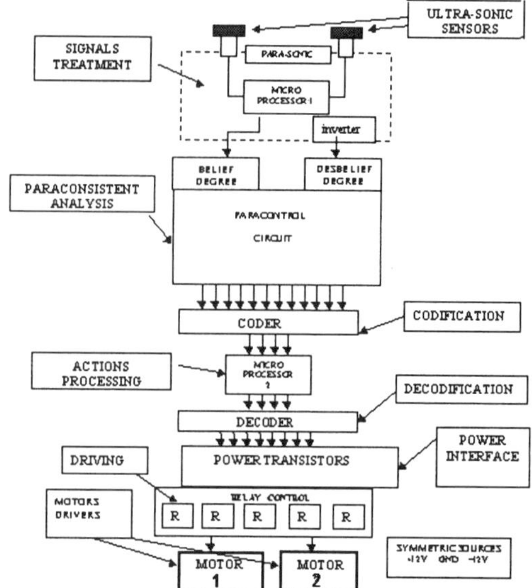

Figure 3 - Scheme of the system of the robot's control Emmy.

- **Ultra-Sonic Sensors** - the two sensors, of ultra-sonic sound waves accomplish the detection of the distance between the robot and the object through the emission of a pulse train in ultra-sonic sound waves frequency and the return reception of the signal (echo).
- **Signals Treatment** - The treatment of the captured signals is made through the Para-sonic. The microprocessor is programmed to transform the time elapsed between the emission of the signal and the reception of the echo in an electric signal of the to 5 volts for degree of belief, and from 5 to 0 volts for disbelief degree. The width of each voltage is proportional at the time elapsed between the emission of a pulse train and its receivement by the sensor ones.
- **Paraconsistent Analysis** - The paraconsistent logical circuit control makes the logical analysis of the signals according to the logic $E\tau$.
- **Codification** - The coding circuit change the binary word of 12 digits in a code of 4 digits to be processed by the personal computer.
- **Actions Processing** - The microprocessor is programmed properly to work with the relay in sequences that establish actions for the robot.
- **Decodification** - The circuit decoder change the binary word of 4 digits in signals to charge the relay in the programmed paths.
- **Power Interface** - The power interface circuit potency interface is composed by transistors that amplify the signals making possible the relay's control by digital signals.
- **Driving** - The relays ON and OFF the motors M_1 and M_2 according to the decoded binary word.
- **Motor Drives** - The motors M_1 and M_2 move the robot according to the sequence of the relays control.
- **Sources** - The robot Emmy is supplied by two batteries forming a symmetrical source of tension of $\pm$ 12 Volts DC.

As project is built in hardware besides the para-control it was necessary the installation of components for supporting circuits allowing the resulting signals of the paraconsistent analysis to be addressed and indirectly transformed in action.

In this first prototype of the robot Emmy it was necessary to have a coder and a decoder such that the referring signals to the logical states resultants of the paraconsistent analysis had its processing made by a microprocessor of 4 inputs and 4 outputs.

4. Programming the Robot's Behavior

In this work, the objective is to make Emmy move from a point *A* to a point *B* detecting and avoiding the obstacles in its path. For that they were programmed sequences of routines for each logical resultant state of the paraconsistent analysis.

The change of the sequences of routines of the microprocessor can modify the programming of the actions of the robot's movement.

The analyzed proposition "there is an obstacle ahead" for each resulting logical state regarding this proposition has a programmed routine in the microprocessor that will make possible an action to the robot. In this control system, the analysis of that proposition supplies us 12 different resulting logical states.

Taking advantage of this paraconsistent analysis, the microprocessor programming was built such that all actions taken by the robot are similar to the attitude taken by a human being. The situations, the human attitudes, and the robot's behavior are as follows:

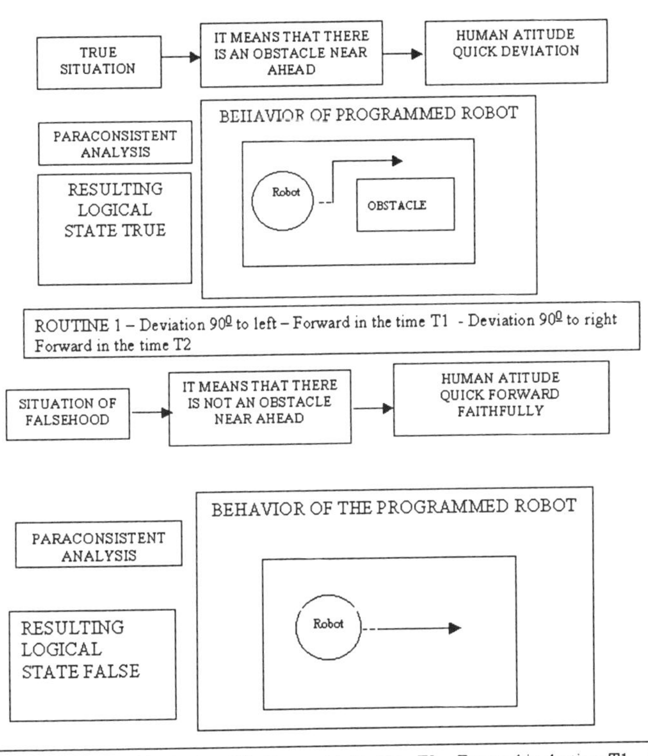

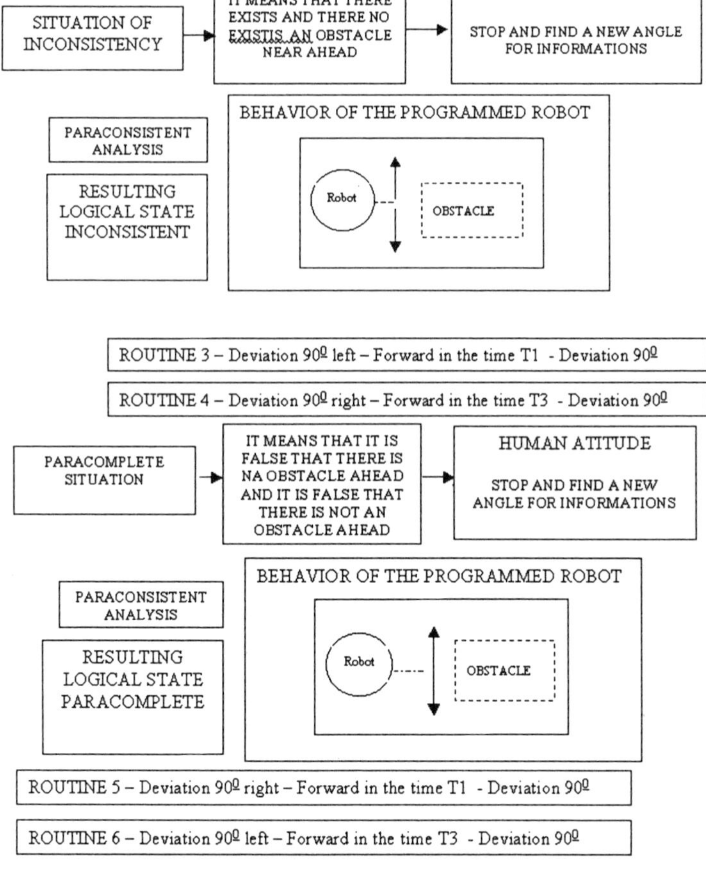

For the referring situations to the non-extreme logical states the routine sequences are related with human attitude in the following way:

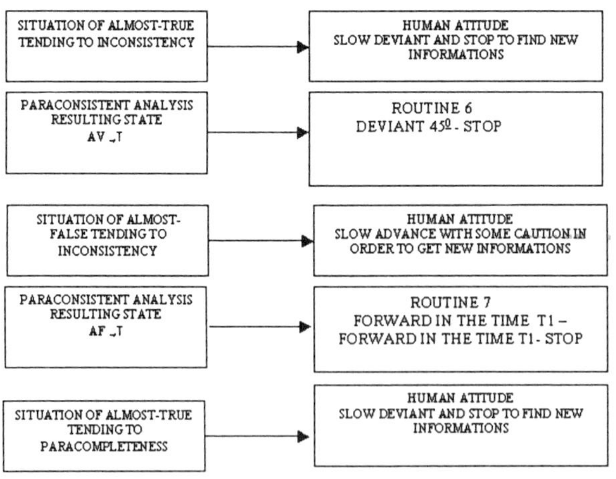

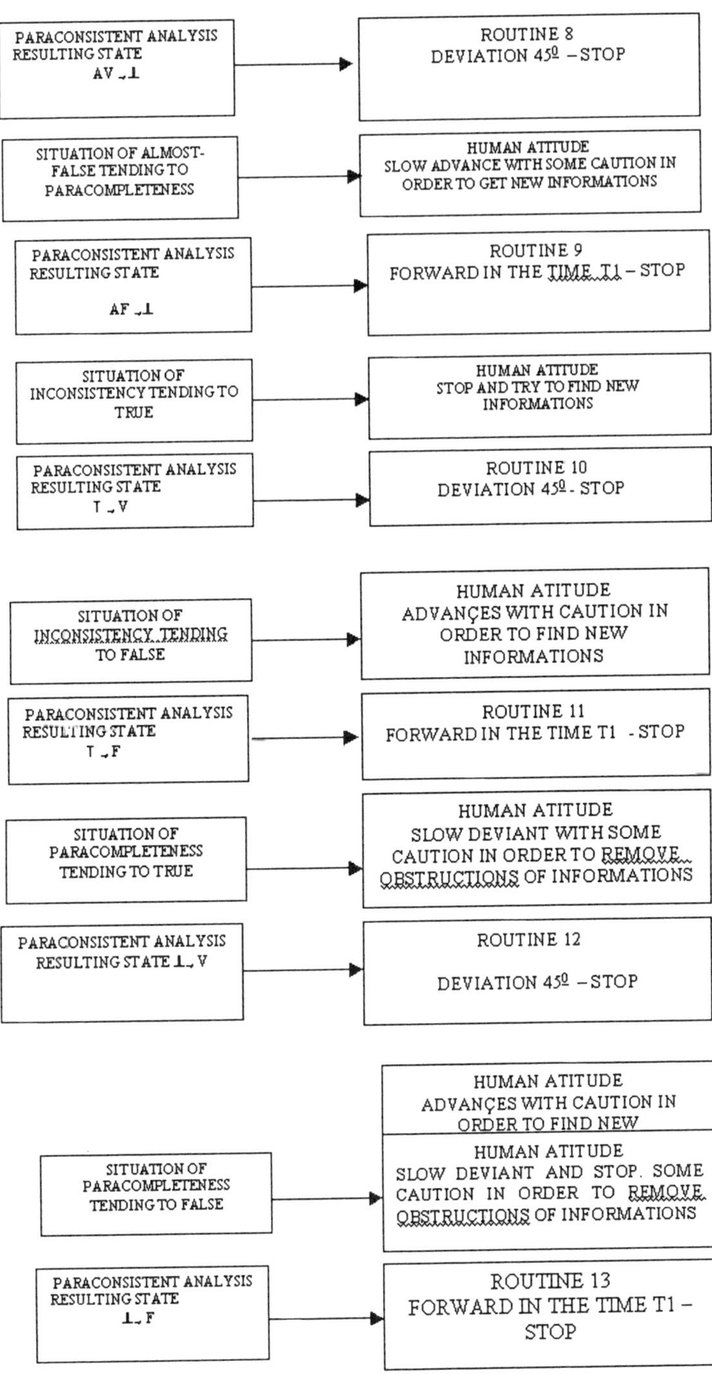

The times T_1, T_2 and T_3 that appear in the routines are certain starting from the behavior presented by the robot in the made tests. In this work the used times were:

$T_1 = 0,75s$, $T_2 = 1,5s$ and $T_3 = 2s$.

To obtain the deviation of the obstacles in a certain angle for the routine a code of 4 digits applied to the potency circuits it just works one of the motors in an enough time for the robot to rotate 900 or 450.

5. Conclusions

In the researches related to autonomous mobile robots' traditional classic control techniques of planning are described and of navigation in known and/or unknown routines. And many of the results present large computational time or a very heavy hardware what disables the effective application of the control. The major difficulty is the recognition of environment changes in real situations, because it generates a lot of indefinitions and inconsistencies in the analyses of the sensored signals, increasing in a big way the hardware of the control system and the time of computation.

In many studies, the logic $E\tau$ presents results that make possible to consider the inconsistencies in its structure in a non-trivial way and for that, they are shown more favorable in dealing problems caused by situations of contradictions that appear when we worked with the real world.

It is known that due to complexity in the construction of sensor in order to capture the several forms of information, the imprecision and noises in the signals, the compulsory nature of having an embarked hardware and the countless situations that should be modeled, when in the movement for non-structured environments, the development of systems of autonomous mobile robots' control is one of the largest questions in the field of the Engineering. For a better treatment of all these problems it is needed new researches seeking a technological improvement in this area. It is possible to say that with all these technical barriers is difficult to obtain a good result in the autonomous mobile robot's behavior using the classic method, therefore this work contributes with the application of the paraconsistent logic by presenting a new control system that shows more adequate to do this type of control.

In this work it was obtained good results in spite of the technical limitations presented by the characteristics of the robot's mechanical structure as: absence of multiple speeds, different types of sensors, movements of mechanical arms, synchronization of speed among the motors, shock absorber, brakes, etc. The tests made showed us that the Para-control can be applied to help to solve linked problems the navigation and treatment of representative signals of information on the environment.

The control system using the Para-control presents good capacity to modify the robot's behavior imitating the human instinctive reaction when in an unexpected modification of the environmental conditions.

Due to the low capacity of memory of the microprocessor that elaborates the actions in this first prototype it was programmed some sequences of routines that present few differences of actions among some logical states. This technical limitation makes poor the objective of the control of giving similarity of behavior between the robot and a human being. All these problems of technical order will be analyzed in the projects of the next prototypes. In spite of that it is important the form presented by the Para-control considering in real time the many factors of generation of inconsistencies presenting means of being adjusted. This offers to Emmy an adaptive behavior to the environment in a similar way to the human behavior easily. In this work it is shown that the Para-control presents means for control fittings, flexibility in an optimization, correction and modification of the characteristics of analysis and of control. Surely the autonomous robot Emmy is a contribution to the development of researches that treat of uncertainty knowledge and it brings new and promising control forms in this area.

References

[1] J. M. Abe, *Fundamentos da lógica anotada*, (Foundations of annotated logics), in Portuguese, Ph.D. thesis, University of São Paulo, São Paulo, 135p.

[2] J. M. Abe and N. Papavero, Teoria Intuitiva dos Conjuntos, *MAKRON Books do Brasil* - São Paulo, 1992.

[3] J. M. Abe and J. I. Da Silva Filho, Implementação de circuitos eletrônicos de funções lógicas paraconsistentes Radix N, *Coleção Documentos, Série Lógica e Teoria da Ciência*, IEA-USP, n_o 22, 37 p., 1996.

[4] J. M. Abe and J. I. Da Silva Filho, Inconsistency and Electronic Circuits, *Proceedings of The International ICSC Symposium on Engineering of Intelligent Systems* (EIS'98), Volume 3, Artificial Intelligence, Editor: E. Alpaydin, ICSC Academic Press International Computer Science Conventions Canada/Switzerland, ISBN 3-906454-12-6, 191-197, 1998

[5] R. Anand, and V.S. Subrahmanian, A Logic Programming System Based on a Six-Valued Logic, *AAAI/Xerox Second Intl. Symp. on Knowledge Eng.* -Madri-Espanha, 1987.

[6] J. Borestein and Y. Koren, Obstacle avoidance with ultrasonic sensors" *IEEE Journal of Robotcs and automation,*N.Y, 1988 .

[7] N.C.A. Da Costa, J.M. Abe, and V.S. Subrahmanian, Remarks on Annotated Logic *Zeitschrift fur Mathematische Logik und Grundlagen der Mathematik*, Vol.37, pp.561-570, 1991.

[8] J. I. Da Silva Filho, *Implementação de circuitos Lógicos fundamentados em uma classe de Lógicas Paraconsistents Anotada*, MSc. Dissertation, EPUSP, São Paulo, 1997.

[9] J.I. Da Silva Filho and J.M.Abe, Paraconsistent Electronic Circuits, presented in the Fourth International Conference on Computing Anticipatory Systems, CASYS'2000, Liege, Belgium, Aug. 7-12, 2000.

[10] J.I. Da Silva Filho and J.M.Abe, Para-Analyser and Inconsistencies in Control Systems, Proceedings of the IASTED *International Conference on Artitficial Intelligence and Soft Computing* (ASC'99), August 9-12, Honolulu, Hawaii, USA, 78-85, 1999.

[11] J.I. Da Silva Filho and J. M Abe, Paraconsistent Analyser Module, presented in the Fourth International Conference on Computing Anticipatory Systems, CASYS'2000, Liege, Belgium, Aug. 7-12, 2000.

[12] J. I. Da Silva Filho, A.M. Santos, J. M. Abe, and D. M. Salles, Para-Sônico: Um Sistema de Sensoramento Ultra Sônico para ser Utilizado em Controle de robôs Móveis autônomos , to appear.

[13] J. I. Da Silva Filho, J. M. Abe, C. R. Torres, A. M. Cézar, and I. J. Cancino, "PARA-CONTROL – Circuito Controlador Lógico Paraconsistente", to appear.

[14] A. Elfes, Sonar based real-world mapping and navegation" *IEEE Journal of Robotcs and automation*, N.Y, 1987.

[15] P. Mckerrow, *Introduction to Robotcs*, Addison-Wesley Publishing Company, New York, 1992.

[16] K-Team *KHEPERA-User's manual*, K-Team- Laboratorie de microinformatique- Lausane, 1994.

[17] POLAROID 500 system ultra-sonic sensor manual, 1996.

[18] V.S Subrahmanian, On the semantics of quantitative Lógic programs, *Proc. 4 th. IEEE Symposium on Logic Programming, Computer Society press,*Washington D.C, 1987.

[19] E. Rich and K. Knigth, "*Artificial Intelligence*", Mc Graw Hill, NY, 1983.

Logic, Artificial Intelligence and Robotics
J.M. Abe & J.I. da Silva Filho (Eds.)
IOS Press, 2001

A Treatment of Limit Figures of Quadratic Transformations as One-way Functions in Authentication for Pictures or Documents of Digital Contents

Tsutomu Da-te*, Takeshi Yoshikawa*, Hiroyuki Shioya* and Hidetoshi Noanaka*

**Laboratory of Intelligence-Oriented Computer Engineering,
Division of Systems and Information Engineering,
Hokkaido University, JAPAN
Kita 13, Nishi 8, Kita-ku, Sapporo, 060-8628, JAPAN
E-mail: date@main.eng.hokudai.ac.jp*

Abstract. We propose, in this paper, an authentication method for pictures or documents of digital contents using a divergence-convergence boundary (DCB) of a quadratic transformation. The DCB has a property of so-called one-way function. It is not difficult to draw a DCB figure from the parameters of given transformation by using a dedicated software on any platform. On the other hand, it is difficult to know the values of parameters from the illustrated figure, because of its complexity. Moreover, the reliability can be raised by repeating the requests and answers between the trust center and the claimers. Each request specifies the area and the number of pixels to redraw, and each answer is the specified part of enlarged DCB.

We describe the details of the authentication system using the DCB, introduce the properties of the DCB in the itemized form, and discuss the   effectiveness of the proposed method

1. Introduction

This paper deals with an authentication method for pictures or documents of digital contents.   It makes use of a divergence - convergence boundary (DCB) of a quadratic transformation, and shows that it is advantageous in distinguishing a man impersonating the copyright holder, or an unjustified claimer. The advantage comes from the following properties of DCB's.

The DCB of a quadratic transformation has, generally, a complex figure even in two-dimensional homogeneous case. It is not difficult to draw a DCB from the parameters of any given quadratic transformation by using an ordinary computer. Whereas it is difficult to know the values of parameters in this transformation from the illustrated figure because of its complexity and the restrictions on the number of pixels in display images. This relation can be considered a property of so-called one-way functions, and hence we use this property to authenticate a digital content like computer graphics or digital documents.   Of course, we are not confined to a DCB of quadratic transformation and we can use general algebraic polynomials or rational functions as an extended version of the quadratic ones for this purpose if necessary.

Any user in this system is assumed to be able to access the list of published digital contents of display images or documents, and to be able to get an algorithm of drawing a display image of DCB for any given parameters of quadratic transformations.

A copyright holder "A" of certain digital contents of his/her published works draws a DCB on the work as his/her signature and send the transformation parameters to a trust center to claim his/her copyright..

A user "B" who wants to use an A's digital content has to pay a certain expense to the copyright holder "A". But it is necessary for "B" to find out a true right holder if several persons come forward and claim to be the right holder.   Then "B" asks to the trust center who is the true right holder. The center requests, to all the persons claiming to be the copyright holder, to offer certain kind of information about the DCB in question. Then one who knows the right transformation parameters can offer the requested information in the form of DCB correctly. However, the impersonator not knowing the right parameters of this transformation cannot respond correctly to all requested questions. Thus the center distinguishes the true right holder.

A DCB figure marked on electronic pictures corresponds to a public key in a public key system, and the transformation parameters correspond to a hidden key. The difference of our system from the ordinary public key system exists in the fact that the trust center can request a DCB figure in any enlargement by specifying the drawing area and number of pixels for checking any number of times, and that the degree of reliability can be raised as much as the center wants.

Any user in this system is assumed to get an algorithm of drawing DCB for any given parameters of quadratic or algebraic transformations.   A copyright holder "A" of certain digital contents of his or her published works draws a DCB on the work as his or her signature and send the transformation parameters to a trust center to claim his or her copyright..

A user "B" who wants to use the digital contents has to pay a certain expense to the copyright holder "A". But it is necessary for "B" to find out whether a person coming forward and claiming to be the right holder tells the truth or not.   And "B" asks to the trust center who is the true right holder. The center requests, to all the persons claiming to be the copyright holder, to offer certain kind of information about the DCB in question. Then one who knows the right transformation parameters can offer the requested information in the form of DCB correctly, however, the impersonator not knowing the right parameters of this transformation cannot respond correctly to all requested questions. Thus the center distinguishes the true right holder.

A DCB figure marked on electronic pictures corresponds to a public key in a public key system, and the transformation parameters correspond to a hidden key. The difference of our system from the ordinary public key system exists in the fact that the trust center can easily find the figures to request for checking any number of times, and the degree of reliability can be raised as much as the center wants.

2. Divergence-Convergence Boundary of a quadratic transformation

A real p-ic transformation of an n-dimensional Euclidean space R^n into itself is determined in terms of a tensor $P^k{}_{r1...rp}$ of contravariant valence 1 and covariant valence p, symmetric in the covariant indices, by the formula:

$$x'^k = \sum_{s=1}^{p} P^k_{r_1...r_s} x^{r_1}...x^{r_s} \quad (k=1,...,n; \; r_1,...,r_s = 1,...,n), \qquad (1.1)$$

with reference to a rectilinear coordinate system, where Einstein's convention is used in the right side term of Σ notation and the point $\mathbf{x}'$ with co-ordinates x'^k is the image of the point $\mathbf{x}$ with coordinates x^k $(k=1,...,n)$ under the transformation.

We deal with the case for $p = 2$ in this paper, then an n-dimensional quadratic transformation represented by the formula

$$x'^k = f^k(x_1,...,x_n) = P^k_{r_1 r_2} x^{r_1} x^{r_2} + P^k_r x^r \quad (k=1,...,n), \qquad (1.2)$$

where we consider a mapping $f: R^n \to R^n$ such that the m-th iterate of $\mathbf{x}$ by $\mathbf{x}^{(m)} = f^m(\mathbf{x})$.

Then we can introduce a convergence region C $(\in R^n)$ as a set of initial values $\mathbf{x}$ satisfying

$$\lim \; \| \mathbf{x}^{(m)} \| = 0 \text{ for } m \to \infty \qquad (1.3)$$

The origin is always included in C. C is not always bounded in finite region.
We show, in the next section, several figures of convergence regions illustrated for explaining how divergence-convergence boundaries in 2-dimensional case are used for authentication.

A fixed point of a quadratic transformation satisfies for $\Lambda \in R$,

$$\sum_{i,j=1}^{n} (P^k_{ij} x^i x^j + P^k_i x^i) = \Lambda x^k \quad (k = 1,...,n). \qquad (1.4)$$

We call the above $\mathbf{x}$ a finite fixed point if $\Lambda \neq 0$, and an infinite fixed point, if $\Lambda = 0$ holds.

Next we confine ourselves to a homogeneous case, i.e., when the second term $P^k_r x^r$ in the right side of (1.2) vanishes strictly. Thus we treat hereafter

$$x'^k = f^k(x_1,...,x_n) = P^k_{r_1 r_2} x^{r_1} x^{r_2} \quad (k=1,...,n). \qquad (1.5)$$

Then complex figures of the convergence regions are classified in relatively simple types and we can see in [1] many properties of such transformations, and several of them are used in the following without proof..
1. The convergence region C is an open set and forms a single connected region.
2. A point in the small neighborhood of the origin is included in C.
3. There exists in each direction only one point which neither converges to the origin nor diverges to infinity under the transformation processes, when the convergence region is bounded finitely.
This property holds for homogeneous algebraic transformations but does not hold in case of general inhomogeneous ones.
4. A point in the direction of an infinite fixed point is transformed into the origin by a single transformation.
5. A point in the direction of a finite fixed point is transformed into a point in the same direction, that is, the direction is kept unchanged.

Next we denote a mapping from the unit sphere $S = \{x| \, \| x \| = 1\}$ into $R \cup \{\infty\}$ by $r(\Theta) = \sup \{\alpha \, | \, \alpha \Theta \subset C\}$ for $\Theta \in S$, then a divergence-convergence boundary (DCB) as
$$B = \{r(\Theta)\Theta \, | \, r(\Theta) < \infty\}. \qquad (1.6)$$
Then we can easily see that $B \cup C$ coincides with a Julia set in the case of homogeneous transformations.

It is easily seen that the convergence region exists if an integral form $\mathbf{f}$ does not include a linear term but includes only the terms of higher orders.

The DCB remains finite in most case, but it happens to include infinite point.

It is shown in [1] that there is a series of invariants K_1, K_2, K_3,..., in terms of $p^k{}_{ij}$ representing typical properties of DCB.

6. By using K_2, one of those invariants, we can show that there exists an infinite point if and only if $K_2 = 0$.

  We say simply a DCB for the divergence-convergence boundary of real homogeneous quadratic transformation.

7. A DCB is lower semicontinuous.

8. There are directions in which DCB is unbounded, if there exists an infinite fixed direction. A point of DCB in a certain direction, which is transformed into an infinite fixed direction within the finite number of transformations, is unbounded.

9. A DCB is bounded and continuous if the infinite fixed direction does not exist.

  These properties of DCB with respect to a homogeneous quadratic transformation assure that they enable the trust center to distinguish the true right holder from the impersonators by enlarging a part of DCB and clearing the fine structure of DCB, and that it is difficult to know the transformation parameters from the DCB due to its complexity

3. Illustrations of DCB of quadratic transformations

  We show several examples of the DCB figures for two-dimensional quadratic transformation in this section. They are composed of curves or regions in two-dimensional plane.

  Figs.1-3 are the examples of approximated images of a certain DCB having a self-similar structure for a two-dimensional real homogeneous quadratic transformation.   As shown in this figure, a DCB of homogeneous quadratic transformation has a center of symmetry. Fig.1 shows its DCB with $-4 \leqq x \leqq 4$, $-2 \leqq y \leqq 2$ in a window of 401 x 201 pixels.   Fig.2 is an enlargement of the square part in Fig.1 with $1 \leqq x \leqq 3$, $0 \leqq y \leqq 2$ in 101 x 101 pixels, that is, a simple zoomed up image of Fig.1.

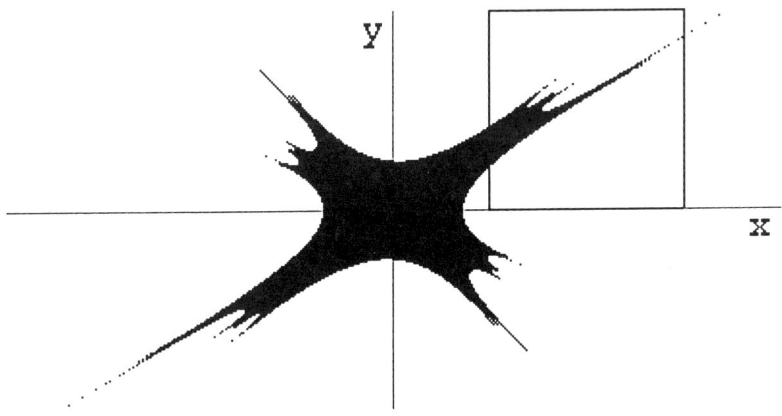

Fig.1 A DCB with self-similar structure in homogeneous quadratic transformation

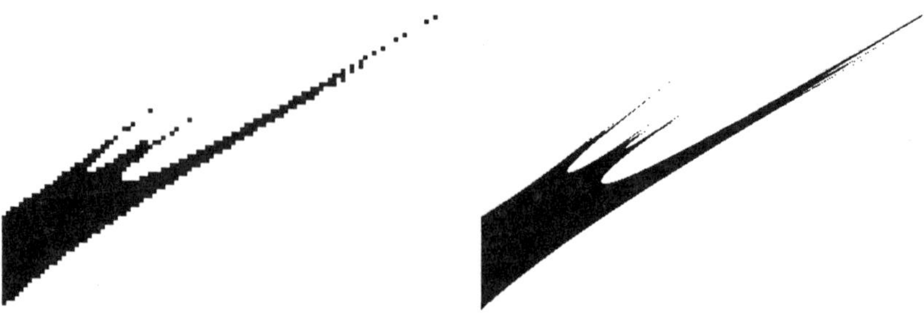

Fig.2 Enlarged DCB of the square in Fig.1   Fig.3 Enlarged DCB of the square in Fig.1

Fig.3 is an enlargement of the square part in Fig.1 with $1\leq x\leq 3$, $0\leq y\leq 2$ in 401 x 401 pixels. An impersonator draws such a figure as Fig.2 since he/she does not know the parameters of the transformation and can not draw the DCB in higher resolution than Fig.1, whereas the true right holder, knowing the parameters, can produce Fig.3. We can easily see that Fig.2 is apparently different from Fig.3, thus the authentication center distinguishes a true right holder.

Figs.4,5 show the examples of approximated images of DCBs for two-dimensional real inhomogeneous quadratic transformations. In the case of inhomogeneous transformations, a point in the neighborhood of the origin is not generally included in the convergence region C, but, for the sake of simplicity, we confine ourselves to the case where the small neighborhood of the origin is included in C, in this paper,.

Fig.4 is an image of 401 x 401 pixels. The gray region in this figure represents a set of initial points converging to the fixed points other than the origin. Fig.5 is an enlarged image of the second quadrant of Fig.4 in the higher resolution with 801 x 801 pixels. We can observe that the DCB has many dotted curves in Fig.4, but these dots form stratified curves in Fig.5. Hence we easily distinguish the Fig.4 from Fig.5.

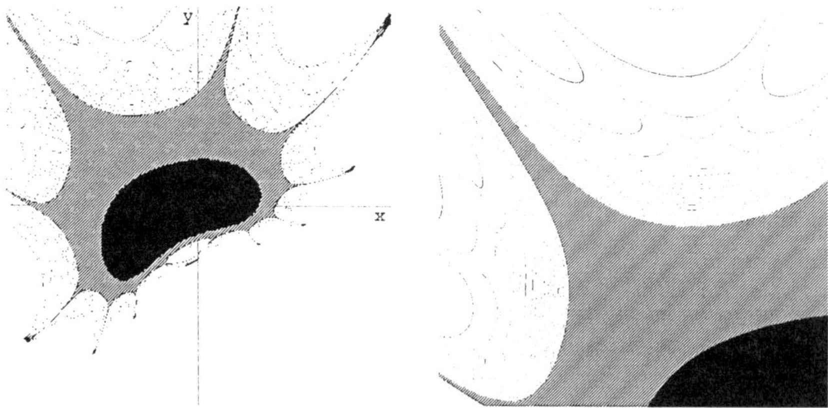

Fig.4 A DCB of an inhomogeneous case   Fig.5 An enlarged DCB of Fig.4

4. Discussions

An authentication system requires that the methods must have the property of one-way function in making calculations or illustrations. Moreover, it requires the other securities in communication systems, the shortness of computing time or sufficiency for user's convenience, for example, it doesn't need much prior knowledge for using the system.

In this proposition to use the DCB of quadratic transformations in authentication systems, the property of one-way functions is sufficiently satisfied, but the user's conveniences will be a topic for further discussion.

Our system will be more easily understandable if we compare it with the digital signature of the well-known public key system. An illustration of DCB corresponds to a public key, and the parameters of a quadratic transformation correspond to a secret key. In connection with the figures of DCB, the number of pixels in each figure is different for each machine but it's upper limit is bounded to a certain extent. Thus it is impossible for anyone to estimate the parameters of the transformation exactly through the DCB figures. Moreover the verifier can raise the scale of enlargement on the specific portions of DCB as much as possible by repeating the questions and answers if necessary. Those impersonators of the right holder, not knowing the true parameters of the transformation, shall not be able to go through the trials. Computing time of illustrating each DCB is around 100ms for a homogeneous quadratic transformation, but around 500 ms for an illustration of 400 x 400 pixels in an inhomogeneous case.

The number of times the verifier requests the claimers to illustrate a DCB will be at most one or several in an authentication procedure. A user has to input a window size, the scale of enlargement and the several parameters of quadratic transformations, in using a dedicated software for illustraing a DCB. This much will be tolerated in the sense of saving time in an authentication procedure.

5. Conclusion

This paper deals with a new authentication system for a display image of pictures or documents of digital contents with the DCB of quadratic transformation attached.

The DCB is a limit figure of the transformations and has a property of one-way functions. By using this property, one can extract its detailed structure as fine as possible, if necessary, by raising the resolution and distinguish the true right holder from impersonators.

Theoretical effectivity of the system can be considered to be verified sufficiently, however, it has not been examined through practical uses and requires a thorough going over.

References

[1] Tsutomu Da-te and Takeshi Yoshikawa, Hidetoshi Nonaka and Mayuka Kawaguchi, On Divergence-Convergence Boundary of Homogeneous Quadratic Transformations in Two Dimensions, Proc. of Third European Congress on Intelligent Techniques and Soft Computing in Aachen, Germany, (1995), pp.141-144.
[2] Tsutomu Date and Masao Iri, Canonical Forms of Real Homogeneous Quadratic Transformations. J. Mathematical Analysis and Applications, 56, 3(1976), pp.650-682.
[3] Tsutomu Date, Classification and Analysis of Two-Dimensional Real Homogeneous Quadratic Differential Equation Systems, J. Diff. Equations 32, 3(1979), pp.311-334.
[4] Takeshi Yoshikawa and Tsutomu Da-te, On Self-Similarity in Homogeneous Quadratic Transformations. American Institute of Physics Conference Proceedings, 517(2000), pp.574-579.
[5] Takeshi Yoshikawa and Tsutomu Da-te, Invariants for Nesting in the Divergence-Convergence Boundary of Two-Dimensional Real Homogeneous Quadratic Transformations (in Japanese). Trans. of the Japan Society for Industrial and Applied Mathematics, 10, 1(2000), pp.283-294.

Logic, Artificial Intelligence and Robotics
J.M. Abe & J.I. da Silva Filho (Eds.)
IOS Press, 2001

Toward a Domain-Theoretic Modelling of Measuring Processes

Graçaliz Pereira Dimuro
Antônio Carlos da Rocha Costa
Escola de Informática, Universidade Católica de Pelotas
Felix da Cunha 412, 96010-000, Pelotas, RS, Brazil

Abstract. This work[1] provides for a general domain-theoretic framework for modelling measuring instruments and processes, with a geometric interpretation based on simplicial complexes. An analogy between the notions of approximation and computation in domain theory and the notion of measuring process performed with a measuring instrument is established. A qualitative domain of coherent simplexes of instrument readings is defined. The histories of the measuring processes performed in that domain are modelled in an adjoint qualitative domain of coherent simplicial complexes.

1 Introduction

Scott and Strachey [1] introduced *Domain Theory* as a mathematical framework for the semantics of programming languages. The main idea is that programming language semantics can be formally specified in terms of constructions on domains (partial orders of information that model the notions of approximation and of computation steps). Girard [2] introduced the *qualitative domains* – a subcategory of Scott domains having a strictly finitary structure – to model polymorphism in type theoretic languages. Dimuro [3] represented real intervals and real numbers with *bi-structured* coherence spaces (Girard binary qualitative domains), in her study of computation models for classical spaces used in Interval Mathematics and Scientific Computing.

In this paper, we introduce *qualitative domains of coherent simplexes* to model measuring processes and their uncertainty. Section 2 gives the intuition behind the work. Section 3 offers some background in qualitative domains. Simplexes and simplicial complexes are reviewed in section 4. Section 5 introduces qualitative domains of coherent simplexes and their adjoint qualitative domains of coherent simplicial complexes. The domain of simplicial complexes of instruments readings and the geometric interpretation of measuring processes in terms of simplicial complexes are introduced in section 6. The conclusion is in section 7.

2 Measurements as Computations

It is possible to establish an analogy between the notions of approximation and computation in domain theory and the notion of measuring process performed with a measuring in-

---

[1] This work has financial support from CNPq and FAPERGS.

strument. This provides for a general domain-theoretic framework for modelling measuring instruments and processes, with a geometric interpretation based on simplicial complexes.

In general, considering a measuring instrument MI of an arbitrary accuracy, it is common to obtain many different possible readings for the same actual value v. Each possible reading has an interpretation, which should also consider the inherent error of MI. A reading process using MI consists of repeated reading experiments, executed in discrete steps. At each step (a reading experiment), more information about v is obtained and partial coherent sets of readings are progressively constructed. Maximal sets are achieved when no new information is possible, that is, no different reading is possible for v with MI. Maximal sets give the best approximation in MI for the values they are measuring.

If one considers a measuring instrument MI of an infinite accuracy (taken as a limiting possibility), then certainly one has also to consider a denumerable set of possible readings for v. In such case, for each ideal denumerable set of possible readings one should also have an associated interpretation - the infinitary approximation of v given by MI.

The basic idea of the approach, thus, is to model measuring processes as approximation processes in domains, conceiving measuring processes as composed of two sub-processes: first, getting a set of readings from an instrument and modelling then in a domain of readings; second, interpreting those readings in a space of values, also modelled as a domain, where increasing information means decreasing uncertainty.

Formally, sets of instrument readings constitute so-called *coherent simplexes*, whose collections are construed as qualitative domains - the *domains of coherent simplexes*. Sets of coherent simplexes, downward closed for the inclusion relation and satisfying a coherence condition, constitute so-called *coherent simplicial complexes*, collected together in *qualitative domains of coherent simplicial complexes*. A domain-theoretic operator is thus established, giving for each domain of coherent simplexes an adjoint qualitative domain of coherent complexes, so that each coherent simplex has a corresponding coherent complex representing the history of the measuring process that it models.

3 Qualitative Domains

A *complete partial order of information objects* (*cpo*), is a partial order $(D, \sqsubseteq)$ having a least element $\perp$ and all lubs of directed subsets of D. A *compact* (or *finite*) object of a cpo $(D, \sqsubseteq)$ is an $a \in D$ such that for any directed subset $X \subseteq D$, if $a \sqsubseteq \bigsqcup X$ then there exists an $x \in X$ such that $a \sqsubseteq x$. The set of compact elements of D is denoted by D^0. A cpo $(D, \sqsubseteq)$ is *algebraic* if, for all $x \in D$, the set of finite approximants of x, $\{d \in D^0 | d \sqsubseteq x\}$, is directed and $x = \bigsqcup \{d \in D^0 | d \sqsubseteq x\}$. It is *consistently complete* if every consistent subset $X \in D$ has a lub in D. An algebraic and consistently complete cpo $(D, \sqsubseteq)$ is a *Scott domain* [1].

Definition 3.1. *A* **qualitative domain** *[2]* $(qD, \subseteq)$ *is a collection of sets* qD, *ordered under the inclusion relation* $\subseteq$, *such that: (i)* qD *is non-empty:* $\emptyset \in qD$; *(ii)* qD *is closed under directed unions: for every directed subset* $S \subseteq qD$, $\bigcup S \in qD$; *(iii)* qD *is downward closed: if* $a \in qD$ *and* $b \subseteq a$, *then* $b \in qD$.

Qualitative domains are algebraic cpo's. They have the interesting property that any compact element dominates only finitely many elements, which is not generally true in domains. The *set of tokens* of a qualitative domain $(qD, \subseteq)$, denoted by $|qD|$, is given by $|qD| = \{\alpha | \{\alpha\} \in qD\} = \bigcup qD$. Therefore, a qualitative domain qD appears as a subset of the powerset of the set $|qD|$ of its tokens. A subset $x \subseteq |qD|$ is said to be *coherent* if $x \in qD$, that is, x is an object of $(qD, \subseteq)$. $a, b \in qD$ are said to be *consistent* if $a \cup b \in qD$; by 3.1 *(iii)* this is equivalent to the existence of $c \in qD$ such that $a, b \subseteq c$.

4 Simplexes and Complexes

In this section some extensions to the standard theory [4] about *simplicial complexes* are made to consider not only finite simplexes of dimension p (p-simplexes), but also denumerable simplexes (ω-simplexes) and the non-dimensional simplex (ϵ-simplex).

Definition 4.1. *A **simplex** σ is a countable (i.e., finite or denumerable) set. A simplex of dimension p (p-simplex) is a finite set of $p+1$ elements, denoted by σ^p. An ω-simplex is a simplex of infinite dimension σ^ω. The ϵ-simplex is the empty non-dimensional simplex, denoted by σ^ϵ.*

The elements of σ are called *vertices*. A subset σ^q of σ is called a *q-face* of σ. For a geometric interpretation, a 0-simplex σ^0 is merely a point. A 1-simplex σ^1 is an arc (open 1-cell) which has two distinct end-points which are not considered, however, as part of σ^1. A 2-simplex σ^2 is an open rectilinear or curvilinear triangle (perimeter excluded).

Definition 4.2. *A **simplicial complex** is a countable collection K of simplexes such that if σ is in K so are all its faces. The **dimension** of K is the largest of the dimensions of the simplexes of K. In particular, the empty complex is non dimensional.*

5 Domains and Simplicial Complexes

In this section, we show how interpretations can be given to the elements of simplexes, like the indices of tokens introduced in [3]. Such interpretations induce the notion of coherent simplexes and the corresponding kind of qualitative domains – *the qualitative domains of coherent simplexes*. In such domains, the computation of a coherent simplex is construed as a simplicial complex, inducing adjoint qualitative domains of coherent simplicial complexes.

5.1 Qualitative Domains of Coherent Simplexes

We consider a countable set B of elements having an associated interpretation in a poset $IV \equiv (IV, \sqsubseteq, \bot, \top)$ of *interpretation values*, where $\bot$ is the least element and $\top$ is the greatest element. The elements of B are called *tokens*. An *interpretation of the tokens* is given by an interpretation function $i : B \to IV$ associating each token $\beta \in B$ to an interpretation $i(\beta) \in IV$. The tokens $\beta_1, \beta_2, \ldots, \beta_n \in B$ are said to be *coherent* if and only if there exists $\bigsqcup \{ i(\beta_1), i(\beta_2), \ldots, i(\beta_n) \} \neq \top$.

Definition 5.1. *A **simplex** $\sigma \subseteq B$ is said to be a **coherent simplex** in B if and only if there exists $\bigsqcup \{ i(\beta) \in IV \mid \beta \in \sigma' \} \neq \top$, for every finite face $\sigma' \subseteq \sigma$. The **interpretation** of σ, if exists, is given by $i(\sigma) = \bigsqcup \{ i(\beta) \in IV \mid \beta \in \sigma \}$. In particular, the empty simplex is coherent and $i(\sigma^\epsilon) = \bot$.*

The collection of coherent simplexes induced on B by the interpretation i is denoted by $CohSimp(B, i)$. The following result is immediate:

Proposition 5.1. *For all coherent simplexes* $\sigma, \sigma' \in CohSimp(B, i)$, $\sigma \subseteq \sigma'$ *if and only if* $i(\sigma) \sqsubseteq i(\sigma')$.

Proposition 5.2. $(CohSimp(B, i), \subseteq)$ *is a qualitative domain –* the **Domain of Coherent Simplexes** *induced on* B *by the interpretation* i.

Proof: Let $S \subseteq CohSimp(B, i)$ be a directed set and $\bigcup S \notin CohSimp(B, i)$. Then there exists a finite face $\sigma^p \subseteq \bigcup S$ such that $\bigsqcup\{i(\beta) \in IV \mid \beta \in \sigma^p\} = \top$, and a finite $\sigma^q, \sigma^r \in S$, with $q, r \leq p$, such that $\bigsqcup\{i(\beta) \in IV \mid \beta \in \sigma^q \cup \sigma^r\} = \top$. Then there exists finite $\sigma \in S$, such that $\sigma^q \cup \sigma^r \subseteq \sigma$, which leads to a contradiction.

5.2 Qualitative Domains of Coherent Complexes

Let $(CohSimp(B, i), \subseteq)$ be a qualitative domain of coherent simplexes induced on a countable set B by the interpretation function $i : B \rightarrow IV$. Denote by $\mathbf{Fin}CohSimp(B, i)$ the subfamily of finite coherent simplexes.

Definition 5.2. *A simplicial complex* $K \subseteq \mathbf{Fin}CohSimp(B, i)$ *is said to be a* **coherent simplicial complex** *if and only if whenever* $\{\beta\}, \{\beta'\} \in K$ *then* $i(\beta) \sqcup i(\beta') \neq \top$. *In particular, the empty complex is coherent.*

The set of coherent complexes associated with $(CohSimp(B, i), \subseteq)$ is denoted by
$$CohComp(B, i) = \{K \subseteq \mathbf{Fin}CohSimp(B, i) \mid K \text{ is a simplicial complex s.t. } i(\beta) \sqcup i(\beta') \neq \top, \forall \{\beta\}, \{\beta'\} \in K\}.$$

Proposition 5.3. $(CohComp(B, i), \subseteq)$ *is a qualitative domain –* the **Domain of Coherent Simplicial Complexes** *– adjoint to the domain* $(CohSimp(B, i), \subseteq)$.

Proof: Let $\kappa \subseteq CohComp(B, i)$ be a directed set and $\bigcup \kappa \notin CohComp(B, i)$. There are $\{\beta\}, \{\beta'\} \in \bigcup \kappa$ such that $i(\beta) \sqcup i(\beta') = \top$, and $K, K' \in \kappa$ such that $\{\beta\} \in K$ and $\{\beta'\} \in K'$. Then, there exists $K'' \in \kappa$ such that $K \cup K' \subseteq K''$, which leads to a contradiction.

Abstracting from these ideas, it is possible to define a domain-theoretic *exponential operator* $\Downarrow$ associating each domain $(CohSimp(B, i), \subseteq)$ of coherent simplexes to its respective adjoint domain of coherent complexes $(CohComp(B, i), \subseteq)$.

Definition 5.3. $\Downarrow (CohSimp(B, i), \subseteq) = (CohComp(B, i), \subseteq)$.

As in [5], any coherent simplicial complex of $CohSimp(B, i)$ is seen as the *history of the computation* of its maximal elements in the domain $(CohSimp(B, i), \subseteq)$. The domain $(CohComp(B, i), \subseteq)$ thus represents the set of the possible histories of computation in the domain $(CohSimp(B, i), \subseteq)$. The total objects in $(CohComp(B, i), \subseteq)$ are coherent complexes representing complete histories of computations; partial objects are coherent complexes indicating computations partially performed.

6 The Domain-Theoretic Modelling of Measuring Processes

In this section we apply the theory stated in the previous sections to model measuring instruments and measuring processes.

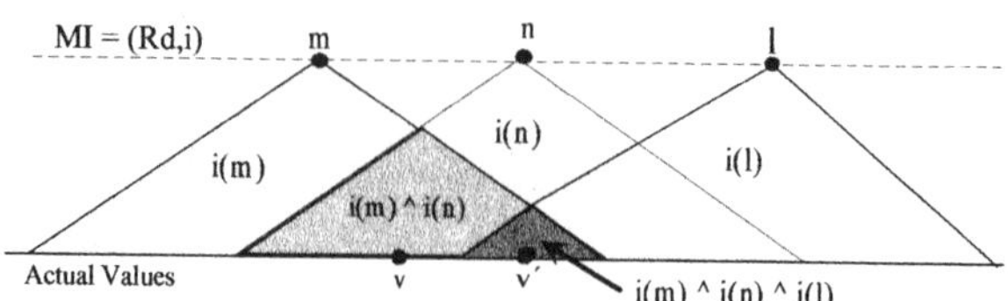

Figure 1: Measuring the actual values v and v' using a measuring instrument MI.

6.1   Measuring Instruments

Consider the lattice of interpretation values $\mathbb{IV} = (\wp(\mathbb{V}), \supseteq, \mathbb{V}, \emptyset)$, where $\mathbb{V}$ is the set of actual values being measured. A *measuring instrument* MI can be identified with an interpretation function i from the set Rd of all *readings* one can obtain when using MI to the poset $\mathbb{IV}$. The *accuracy* of MI is given by the cardinality of Rd. We shall consider only measuring instruments with countable sets of readings.

Definition 6.1. *A* **measuring instrument** *is defined as a structure* MI $\equiv (Rd, i, \mathbb{IV})$, *where Rd is a countable set of readings and $i : Rd \rightarrow \mathbb{IV}$ is a reading interpretation function.*

We write simply MI $\equiv (Rd, i)$ whenever one can identify the poset of reading interpretation values from the context.

Example 6.1. *In the measuring process of the actual value v using the measuring instrument* MI *of Figure 1, there are two possible readings for v: m and n, uniquely interpreted by* MI. *The interpretations $i(m)$ and $i(n)$ result in bounded sets of actual values, which consider the inherent error of* MI. *To obtain the readings m and n, reading experiments have to be done at least twice to complete the reading process. $\{m\}$, $\{n\}$ and $\{m, n\}$ are coherent simplexes of readings in* MI. *$\{m\}$ and $\{n\}$ are partial readings and $\{m, n\}$ is a total reading, in the sense that it represents maximal coherent information in that instrument.*

Generalizing these ideas, for a measuring instrument MI, readings $m, n, \ldots, s$ are *coherent readings* if and only if $v \in i(m) \cap i(n) \cap \cdots \cap i(s) \neq \emptyset$. In this situation, by definition 5.1, $\{m, n, \ldots, s\}$ is said to be a *coherent simplex of readings* in MI, whose interpretation is given by $i(m) \cap i(n) \cap \cdots \cap i(s)$. In particular, the empty set is a coherent set of readings, the initial one in every reading process. The collection of all coherent simplexes of readings of MI is denoted by $CohSimp(Rd, i)$.

Definition 6.2. *The* **interpretation** *of a coherent simplex of readings σ in a measuring instrument* MI $= (Rd, i)$, *denoted by $i(\sigma)$, is given by $i(\sigma) = \bigcap_{x \in \sigma} i(x)$. In particular, for the empty simplex, $i(\sigma^\epsilon) = \mathbb{V}$.*

From proposition 5.1 it follows that:

Proposition 6.1. *$\sigma \subseteq \sigma'$ if and only if $i(\sigma') \subseteq i(\sigma)$, for all coherent simplexes of readings $\sigma, \sigma' \in CohSimp(Rd, i)$.*

Let $CohSimp(Rd, i)$ be the set of coherent simplexes of readings of a measuring instrument MI. From proposition 5.2, it follows that:

Proposition 6.2. *$(CohSimp(Rd, i), \subseteq)$ is a qualitative domain – the* **Domain of Coherent Simplexes of Readings** *induced by the measuring instrument* MI $= (Rd, i)$.

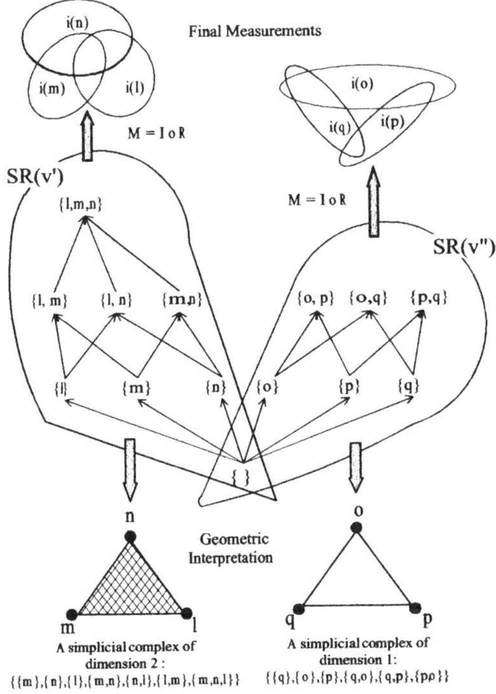

Figure 2: Diagram of a Domain of Coherent Simplexes of Readings.

6.2 Reading Processes

In this section, reading processes in measuring instruments are formally modelled.

Definition 6.3. *The* **scope of a reading process** *of* $v \in \mathbb{V}$ *with* MI $= (Rd, i)$ *is given by a function* $S\mathcal{R} : \mathbb{V} \to \wp(CohSimp(Rd, i))$, *defined by* $S\mathcal{R}(v) = \{\sigma \in CohSimp(Rd, i) \mid v \in i(\sigma)\}$.

For $v \in \mathbb{V}$, the scope $S\mathcal{R}(v)$ in MI is a maximal coherent complex in $CohComp(Rd, i)$. Its dimension is the maximum of the set of dimensions of the maximal simplexes of the scope. The collection of such maximal simplicial complexes is denoted by $\mathrm{Max}CohComp(Rd, i)$. Analogously, the *scope of interpretation values (measurements)* adjoint to the scope $S\mathcal{R}$ of a reading process in the measuring instrument MI is defined:

Definition 6.4. *The* **scope of interpretation values** *adjoint to the scope of a reading process in* MI $= (Rd, i)$ *is given by the function* $S\mathcal{I} : \mathrm{Max}CohComp(Rd, i) \to \wp(\mathbb{IV})$, *defined by* $S\mathcal{I}(K) = \{i(\sigma) \in \mathbb{IV} \mid \sigma \in K\}$.

Example 6.2. *Figure 2 shows the Domain of Coherent Simplexes of Readings induced by a measuring instrument* MI. *The empty set* {} *represents the initial step of a reading process. The levels of the diagram where the coherent simplexes of readings are located correspond to*

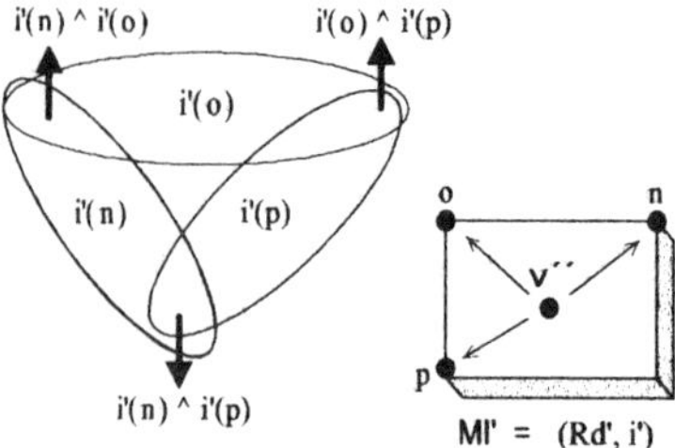

Figure 3: Measuring the actual value v''

their dimensions. The scope of the reading process of v (Figure 1) is the simplicial complex $S\mathcal{R}(v) = \{\emptyset, \{m\}, \{n\}, \{m, n\}\}$. The scope of the respective interpretations (measurements) is $S\mathcal{I}(v) = \{\mathbb{V}, i(m), i(n), i(m) \cap i(n)\}$. $\mathbb{V}, i(m)$ and $i(n)$ are partial measures of v and $i(m) \cap i(n)$ is a maximal measure attainable in this scope. The scope of the reading process of v' is $S\mathcal{R}(v') = \{\emptyset, \{m\}, \{n\}, \{l\}, \{m, n\}, \{m, l\}, \{n, l\}, \{m, n, l\}\}$. Its maximal coherent simplex is $\sigma^2 = \{m, n, l\}$. Comparing $S\mathcal{R}(v')$ with $S\mathcal{R}(v)$ one can surely affirm that v' is not v, because when measuring v the reading l shall never be obtained. The scope of interpretations for v' is $S\mathcal{I}(v') = \{\mathbb{V}, i(m), i(n), i(l), i(m) \cap i(n), i(m) \cap i(l), i(n) \cap i(l), i(m) \cap i(n) \cap i(l)\}$, whose maximal measurement is $i(m) \cap i(n) \cap i(l)$. The minimal uncertainty attainable when measuring v' is certainly less than the one attainable when measuring v.

Example 6.3. *In the measuring instrument* MI$' = (Rd', i')$ *of Figure 3, the reading process of the actual value v'' has the scope of readings $S\mathcal{R}(v'') = \{\emptyset, \{q\}, \{o\}, \{p\}, \{q, o\}, \{q, p\}, \{o, p\}\}$, shown in Figure 2. This complex presents three maximal coherent simplexes of readings, namely $\sigma_a{}^1 = \{q, o\}$, $\sigma_b{}^1 = \{q, p\}$ and $\sigma_c{}^1 = \{o, p\}$.*

6.3   Measuring Function

Any function $R : \mathbb{V} \rightarrow \wp(CohSimp(Rd, i))$, assigning a set of coherent simplexes of readings for an actual value, is a reading function in the measuring instrument MI. Reading functions are generally not unique for a given MI, so measurements can not be characterized by reading functions of measuring instruments. We thus define the concept of maximal reading function. Let $S\mathcal{R}$ be the function that gives the scope of the reading processes in the measuring instrument MI $= (Rd, i)$. The *maximal reading function* of MI selects the maximal coherent simplexes of readings in the scope of the reading processes.

Definition 6.5. *The* **maximal reading function** *associated to* MI *is the function* $\mathcal{R} : \mathbb{V} \rightarrow \wp(CohSimp(Rd, i))$, *defined as* $\mathcal{R}(v) = \{\sigma \in S\mathcal{R}(v) \mid$ *whenever there exists* $\sigma' \in S\mathcal{R}(v)$ *s.t.* $\sigma \subseteq \sigma'$ *then* $\sigma = \sigma'\}$.

The final result of a measuring process of an actual value v in a measuring instrument MI may then be defined as the interpretation of the maximal coherent simplexes of readings selected by the maximal reading function $\mathcal{R}$. Thus, we can extend the interpretation function:

Definition 6.6. *The* **extended interpretation function** *of a measuring instrument* MI $= (Rd, i)$ *is the function* $\mathcal{I} : \wp(CohSimp(Rd, i)) \rightarrow \wp(\mathbb{IV})$, *defined by* $\mathcal{I}(S) = \{i(\sigma) \in \mathbb{IV} \mid \sigma \in S\}$.

Definition 6.7. *The* **measuring function** *associated to* MI $= (Rd, i)$ *is the function* $\mathcal{M}$: $\mathbb{V} \to \wp(\mathbb{IV})$ *such that* $\mathcal{I} \circ \mathcal{R} = \mathcal{M}$.

Proposition 6.3. *For a measuring instrument* MI *with infinite precision, the measuring function* $\mathcal{M}$ *is the identity, that is,* $\mathcal{M}(v) = \{v\} \equiv v$, *for all* $v \in \mathbb{V}$.

Proof: A sketch of the proof is given. We observe that if the accuracy of MI is infinite, than the *lub* of the set of diameters of the interpretations of the scope of interpretations (measurements) for an actual value v, $\mathcal{SI}(v)$, is zero. In other words, the *lub* of $\mathcal{SI}(v)$ ordered by reverse inclusion is a degenerated interval, that is, $\bigsqcup(\mathcal{SI}(v), \supseteq) = \bigcap \mathcal{SI}(v) = w \in \mathbb{V}$. Since $v \in i(\sigma)$, for all $\sigma \in \mathcal{SR}(v)$, then $w = v$. Therefore, $\mathcal{M}(v) = \mathcal{I} \circ \mathcal{R}(v) = \{v\} \equiv v$.

6.4 Measuring Processes

Example 6.4. *In Figure 2, the scopes of reading processes* $\mathcal{SR}(v')$ *and* $\mathcal{SR}(v'')$ *are given geometric representations as simplicial complexes. The maximal simplexes are obtained by function* $\mathcal{R}$, *whose interpretations are given by function* $\mathcal{I}$. *The measuring function* $\mathcal{M} = \mathcal{I} \circ \mathcal{R}$ *determines the final measurements and their uncertainties.*

Let $\mathbf{Fin}CohSimp(Rd, i)$ be the set of finite coherent simplexes and $CohComp(Rd, i) = \{K \subseteq \mathbf{Fin}CohSimp(Rd, i) \mid K$ is a simplicial complex s.t. $i(\beta) \cap i(\beta') \neq \emptyset, \forall \{\beta\}, \{\beta'\} \in K\}$. Applying the exponential operator $\Downarrow$ (def. 5.3), we obtain a domain modelling total and partial histories of measuring processes. From the proposition 5.3, it follows that:

Proposition 6.4. $\Downarrow (CohSimp(Rd, i), \subseteq) = (CohComp(Rd, i), \subseteq)$ *is a qualitative domain of coherent simplicial complexes, called the* **Domain of Measuring Processes** *adjoint to the Domain of Coherent Simplexes of Readings induced by a measuring instrument* MI $= (Rd, i)$.

7 Conclusion and Final Remarks

This paper introduced a domain-theoretic framework to model measuring instruments and processes, interpreting geometrically the related uncertainty with the help of simplicial complexes that trace the history of the reading processes. The domain-theoretic approach, based on qualitative domains, allowed us to deal with infinite measuring processes in a simply natural way. On the other hand, the geometric interpretation of such processes, given in terms of simplicial complexes, suggests an interesting representation for computational processes performed in qualitative domains.

References

[1] D. S. Scott and C. Strachey, Towards a Mathematical Semantics of Computer Languages, Oxford University Computing Lab. (1971).

[2] J-Y Girard, The System F of Variable Types - Fifteen Years Later, Theor. Comp. Sc. **45** (1986) 159–192.

[3] G. P. Dimuro, A. C. R. Costa and D. M.Claudio, A coherence space of rational intervals for a construction of IR, Reliable Computing **6** (2) (2000) 139–178.

[4] S. Lefschetz, Introduction to Topology, Princeton University Press (1949).

[5] R. G. S. Sellanes and A. C. R. Costa, Sequencial and Parallel Computation Strategies on Coherence Spaces, proceedings: XXII Conferencia Latinoamericana de Informatica, Colombia (1996).

Logic, Artificial Intelligence and Robotics
J.M. Abe & J.I. da Silva Filho (Eds.)
IOS Press, 2001

Logic of Incursive Synchronization Applied to the Anticipation of a Chaotic Epidemic

Daniel M. DUBOIS
Centre for Hyperincursion and Anticipation in Ordered Systems,
CHAOS asbl, Institute of Mathematics, B37, University of Liège
Grande Traverse 12, B–4000 LIEGE 1, Belgium
Fax: + 32 (0) 4 366 94 89 – Email: Daniel.Dubois@ulg.ac.be
http://www.ulg.ac.be/mathgen/CHAOS/CASYS.html

Abstract. This paper deals with a general theory of synchronization of systems coupled by an incursive connection. For systems with a time shift, the slave or driven system anticipates the values of the master or driver system by a future time period giving rise to an anticipatory synchronization. Some extensions show the possibility to enhance the anticipatory synchronization, what we call meta-anticipatory synchronization. An application is shown in the case of an epidemic system modelled by a chaotic delayed Pearl-Verhulst map. A slave model is incursively synchronized to the master system, the simulation of which an anticipation of the epidemic evolution by the synchronized model. The application to chaos epidemic is just an example to show the power of such an approach. This anticipatory synchronization logic can be applied to many natural systems (biology, ecology, economy, sociology) and engineering systems (artificial intelligence, neural networks, automation, robotics, computer science, information systems).

1   Introduction

Some years ago, Dubois and Resconi [1] proposed an incursive system for synchronizing systems with an application to two chaotic Pearl-Verhulst maps.
It was proposed that two disjoint recursions

$$x(n + 1) = \mu x(n)[1 - x(n)] \tag{1a}$$

$$y(n + 1) = \mu y(n)[1 - y(n)] \tag{1b}$$

can be connected by the following incursion

$$x(n + 1) + D_1(n)[x(n + 1) - y(n + 1)] = \mu x(n)[1 - x(n)] \tag{2a}$$

$$y(n + 1) + D_2(n)[y(n + 1) - x(n + 1)] = \mu y(n)[1 - y(n)] \tag{2b}$$

by which the recursive system is obtained

$$x(n + 1) = [(1 + D_2)\mu x(n)[1 - x(n)] + D_1\mu y(n)[1 - y(n)]]/(1 + D_1 + D_2) \tag{3a}$$

$$y(n + 1) = [(1 + D_1)\mu y(n)[1 - y(n)] + D_2\mu x(n)[1 - x(n)]]/(1 + D_1 + D_2) \tag{3b}$$

Because the two independent recursions assume values in the interval between zero and one, the two recursions joined by the incursion can be synchronized for adequate values of D_1 and D_2.

An extension of this incursive synchronization for a master (driver) system and a slave (driven) system is developed in our recent paper [3].

This paper presents the main logical aspects.

2 Theory of the Logic of Incursive Synchronization

Let us consider the two general disjoint recursions

$$x(t + \Delta t) = x(t) + \Delta t f(x(t - \tau)) - \Delta t b x(t) \tag{4a}$$

$$y(t + \Delta t) = y(t) + \Delta t f(y(t - \tau)) - \Delta t b y(t) \tag{4b}$$

that are the discrete system of the differential equations

$$dx(t)/dt = f(x(t - \tau)) - bx(t) \tag{5a}$$

$$dy(t)/dt = f(y(t - \tau)) - by(t) \tag{5b}$$

which are retarded differential equations by the time shift τ. When this system is chaotic, the evolution of the two independent variables $x(t)$ and $y(t)$ are not synchronized due to the sensitivity to initial conditions.

The purpose of this paper is to propose an incursive connection of these two systems in view of synchronizing the second equation of $y(t)$, considered as the slave or driven system, with the first equation of $x(t)$, considered as the master or driver system.

Let us generalize the incursive connection, given by Dubois and Resconi [1], with $D_1 = 0$, and $D_2 = D \geq 0$, in the following way

$$x(t + \Delta t) - x(t) + \Delta t b x(t) = \Delta t f(x(t - \tau)) \tag{6a}$$

$$y(t + \Delta t) - y(t) + \Delta t b y(t) + D[(y(t + \Delta t) - y(t)$$
$$+ \Delta t b y(t)) - (x(t + \tau + \Delta t) - x(t + \tau) + \Delta t x(t + \tau)] = \Delta t f(y(t - \tau)) \tag{6b}$$

This incursive system corresponds to the following differential equations

$$dx(t)/dt + bx(t) = f(x(t - \tau)) \tag{7a}$$

$$dy(t)/dt + by(t) + D[(dy(t)/dt + by(t)) - (dx(t + \tau)/dt + bx(t + \tau))] = f(y(t - \tau)) \tag{7b}$$

The connection of the slave equation of $x(t)$ to the master equation of $y(t)$, is incursive because, in the factor depending of D, the future value of $y(t + \Delta t)$ depends of itself at the future time $t + \Delta t$ and of $x(t + \tau + \Delta t)$ at the future time $t + \tau + \Delta t$. Thus the connection is anticipatory.

The second differential equation 7b can be transformed in the following way:

$$dy(t)/dt + Ddy(t)/dt = f(y(t - \tau)) - by(t) + D[- by(t)) + (dx(t + \tau)/dt + bx(t + \tau))]$$

and with eq. 7a, $dx(t)/dt + bx(t) = f(x(t - \tau)$, we obtain

$$dy(t)/dt = [f(y(t - \tau)) - by(t) + D[- by(t)) + f(x(t))]]/(1+D) \text{ or}$$

$$dy(t)/dt = [(1 + D)f(y(t - \tau)) - (1 + D)by(t) + D[f(x(t)) - f(y(t - \tau))]]/(1+D)$$

so the two equations system is

$$dx(t)/dt = f(x(t - \tau)) - bx(t) \tag{8a}$$

$$dy(t)/dt = f(y(t - \tau)) - by(t) + K[f(x(t)) - f(y(t - \tau))] \tag{8b}$$

where $K = D/(1 + D)$, is the coupling factor.

When $\tau = 0$, this synchronization is similar to the weak ($K = 1/3$) and strong ($K = 1/2$) synchronization presented by Pyragas [4] for the chaos map $f(x) = 4x(1 - x)$.

As $D \geq 0$, we have the interval of values for K, $0 \geq K \geq 1$, so

$$dx(t)/dt = f(x(t - \tau)) - bx(t) \tag{9a}$$

$$dy(t)/dt = (1 - K)f(y(t - \tau)) - by(t) + Kf(x(t)) \tag{9b}$$

If the values of $x(t)$ and $y(t)$ are in the interval $]0,1[$ in the disjoint systems, the resulting values of $y(t)$ in the coupling will remain in the same interval, due to the fact that K plays the role of a weighting of $f(y(t - \tau)$ and $f(x(t))$.

Let us notice that when $D = 0$, $K = 0$, the original disjoint system 5ab is obtained.

When D is very large, $D \gg 1$, K tends to $K = 1$. In this limit case, one obtains

$$dx(t)/dt = f(x(t - \tau) - bx(t) \tag{10a}$$

$$dy(t)/dt = - by(t) + f(x(t)) \tag{10b}$$

This limit case is similar to the anticipating synchronization presented by Voss [5].

Let us show that this system is stable by introducing a difference between x and y as

$$z(t) = x(t + \tau) - y(t) \tag{11}$$

The differential equation of $z(t)$ is then

$$dz(t)/dt = f(x(t)) - bx(t + \tau) + by(t) - f(x(t)) = - bz(t) \tag{12}$$

and it is clear that $z(t)$ will tend to zero for $b > 0$. In this case, we obtain the anticipatory relation

$$z(t) = x(t + \tau) - y(t) = 0 \tag{13}$$

or

$$y(t) = x(t + \tau) \tag{14}$$

This remarkable result means that the slave system $y(t)$ is synchronized to the master system $x(t)$ at the future potential value of $x(t + \tau)$, where τ is the time shift of the master equation of $x(t)$.

More the time shift τ is large, more the slave system anticipates the values of the master system, even if it is chaotic, as we will show in the next section.

When the master system is without time shift, $\tau = 0$, the two systems are synchronized in a normal way.

A first extension to synchronize the two system at any time shift consists in making the following modification of the synchronization system 8ab:

$$dx(t)/dt = f(x(t - \tau)) - bx(t) \tag{15a}$$

$$dy(t)/dt = f(y(t - \tau)) - by(t) + K[f(x(t - \tau_0)) - f(y(t - \tau)] \tag{15b}$$

where τ_0 is a time shift the interval of which is given by $\tau \geq \tau_0 \geq 0$.

When $\tau_0 = 0$, we obtain an anticipatory synchronization.

When $\tau_0 = \tau$, the synchronization is normal, similarly to the time-delayed dissipative coupling presented by Voss [5].

So our general anticipatory synchronization is governed by the anticipatory time shift τ_a

$$\tau_a = \tau - \tau_0 \tag{16}$$

The discrete equation system of this differential synchronization is written as

$$x(t + \Delta t) = x(t) + \Delta t f(x(t - \tau)) - \Delta t bx(t) \tag{17a}$$

$$y(t + \Delta t) = y(t) + \Delta t f(y(t - \tau)) - \Delta t by(t) + \Delta t K[f(x(t - \tau_0)) - f(y(t - \tau))] \tag{17b}$$

A second extension to the system 15ab is given by

$$dx(t)/dt = f(x(t - \tau)) - bx(t) \tag{18a}$$

$$dy(t)/dt = f(y(t - \tau)) - by(t) + K[f(x(t - \tau_0)) - f(y(t - \tau - \tau_1)] \tag{18b}$$

where τ_1 is an additional anticipatory time of synchronization for which an anticipatory synchronization is possible even if $\tau = \tau_0 = 0$.

The discrete equations of this differential synchronization are written as

$$x(t + \Delta t) = x(t) + \Delta t f(x(t - \tau)) - \Delta t bx(t) \tag{18c}$$

$$y(t + \Delta t) = y(t) + \Delta t f(y(t - \tau)) - \Delta t by(t) + \Delta t K[f(x(t - \tau_0)) - f(y(t - \tau - \tau_1))] \tag{18d}$$

When the two systems are synchronized, $f(x(t - \tau_0)) = f(y(t - \tau - \tau_1))$, which means that

$$y(t) = x(t + \tau + \tau_1 - \tau_0) = x(t + \tau_a) \tag{19}$$

with

$$\tau_a = \tau + \tau_1 - \tau_0 \tag{16a}$$

A third extension consists is adding a second slave synchronization system which will synchronize with the slave system in the following way

$$dx(t)/dt = f(x(t - \tau)) - bx(t) \tag{20a}$$

$$dy(t)/dt = f(y(t - \tau)) - by(t) + K[f(x(t - \tau_0)) - f(y(t - \tau)] \tag{20b}$$

$$dy_1(t)/dt = f(y_1(t - \tau)) - by_1(t) + K_1[f(y(t - \tau_{01})) - f(y_1(t - \tau)] \tag{20b}$$

When the three systems are synchronized,

$$y_1(t) = y(t + \tau - \tau_{01}) = x(t + 2\tau - \tau_0 - \tau_{01}) = x(t + \tau_a) \tag{19a}$$

with

$$\tau_a = 2\tau - \tau_0 - \tau_{01} \tag{16b}$$

so the anticipation is extended to two times the delay τ. I will call this second order anticipatory synchronization, a meta-anticipatory synchronization.

The discrete equation system of this differential synchronization is written as

$$x(t + \Delta t) = x(t) + \Delta tf(x(t - \tau)) - \Delta tbx(t) \tag{21a}$$

$$y(t + \Delta t) = y(t) + \Delta tf(y(t - \tau)) - \Delta tby(t) + \Delta tK[f(x(t - \tau_0)) - f(y(t - \tau))] \tag{21b}$$

$$y_1(t + \Delta t) = y_1(t) + \Delta tf(y_1(t - \tau)) - \Delta tby_1(t) + \Delta tK_1[f(y(t - \tau_{01})) - f(y_1(t - \tau))] \tag{21c}$$

Further extensions can be made in adding more slave systems $y_2(t)$, $y_3(t)$, ... $y_n(t)$ in cascade, giving a meta-anticipation of $\tau_a = (n + 1)\tau$.

3   Modelling and Simulation of the Anticipation of a Chaotic Epidemic

An application of the incursive synchronization is now considered for a delayed Pearl-Verhulst system.

A rather simple model of epidemic is given by

$$dS(t)/dt = -aS(t)I(t) + bI(t) \tag{22a}$$

$$dI(t)/dt = +aS(t)I(t) - bI(t) \tag{22b}$$

where $S(t)$ is the susceptible population and $I(t)$ is the infectious population. The parameter a is the contact rate between the susceptible and infectious populations and b is the rate of decrease of the infected population which recovers as susceptible. In this model, the total population $S(t) + I(t)$ is constant. Indeed

$$d(S(t)/dt + dI(t)/dt = 0 \tag{23}$$

so

$$S(t) + I(t) = C \tag{24}$$

and the two equations reduce to the following equation

$$dI(t)/dt = +aI(t)[C - I(t)] - bI(t) \tag{25}$$

This equation is similar to the Pearl-Verhulst equation. The discrete form of this equation is similar to the well-known chaos map given by May [6].

As pointed out by Dubois and Sabatier [2], a time delay of the susceptible population to become an infectious population is more adequate: it is not instantaneously that the susceptibles become infected.

In general, susceptibles become infected after a certain time period τ of incubation. In taking into account such a time shift, the equation system 22ab can be generalized as

$$dS(t)/dt = -aS(t - \tau)I(t - \tau) + bI(t) \tag{26a}$$

$$dI(t)/dt = +aS(t - \tau)I(t - \tau) - bI(t) \tag{26b}$$

Similarly to the original model, there is also the conservation of susceptible and infected populations

$$dS(t)/dt + dI(t)/dt = 0 \tag{27}$$

$$S(t) + I(t) = C \tag{28}$$

and the two eqs. 26ab are reduced to the following equation

$$dI(t)/dt = +aI(t - \tau)[C - I(t - \tau)] - bI(t) \tag{29}$$

which is a time delayed Pearl-Verhulst equation.

For simulating such a delayed Pearl-Verhulst system, eq. 29 is written in a discrete equation

$$I(t + \Delta t) = I(t) + \Delta t \, aI(t - \tau)[C - I(t - \tau)] - \Delta t bI(t) \tag{30}$$

Figure 1 is the simulation of eq. 30, with $\Delta t = 0.1$, $b = 1$, $\tau = 50$, $C = 1$, and the parameter a varying from 0 to 4. This is a strange attractor, as shown in fig. 1a.

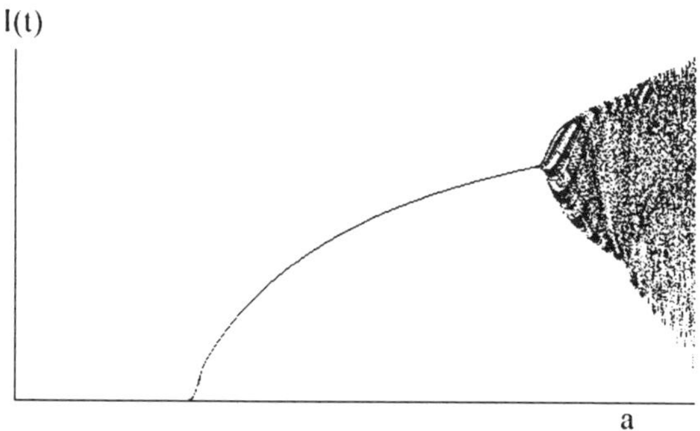

Figure 1: Bifurcation diagram of the delayed Pearl-Verhulst map, with a time shift of $\tau = 50$, and $\Delta t = 0.1$.

Figure 1a: Diagram of $I(t + \tau)$ versus $I(t)$.

Let us now apply the incursive synchronization theory to the delayed Pearl-Verhulst system in view of anticipating the evolution of the such a system.

Let us consider the equation 30 as the master or driver system and let us construct a slave or driven system.

In applying eqs. 17ab to the eq. 30 system, the synchronization system is given by

$$I(t + \Delta t) = I(t) + \Delta t a I(t - \tau)[C - I(t - \tau)] - \Delta t b I(t) \tag{31a}$$

$$I^*(t + \Delta t) = I^*(t) + \Delta t a I^*(t - \tau)[C - I^*(t - \tau)] - \Delta t b I^*(t)$$
$$+ \Delta t[D/(1 + D)][a I(t - \tau_0)[C - I(t - \tau_0)] - I^*(t - \tau)[C - I^*(t - \tau)]] \tag{31b}$$

where eq. 31a is the master system $I(t)$ and eq. 31b the slave system $I^*(t)$.

Figures 2ab are the simulation of eqs. 31ab for $\Delta t = 0.1$, $a = 4$, $C = 1$, $\tau_0 = 0$, $\tau = 50$, and $b = 1$. This gives the time evolution of $I(t)$ and $I^*(t)$ as a function of time $t = 0$ to 1000. It is well-seen that $I^*(t)$ is synchronized to $I(t)$ with an anticipation $\tau_a = \tau = 50$.

Figures 3ab are the simulation of eqs. 31ab for $\Delta t = 0.1$, $a = 4$, $C = 1$, $\tau_0 = 25$, $\tau = 50$, and $b = 1$. This gives the time evolution of $I(t)$ and $I^*(t)$ as a function of time $t = 0$ to 1000. So, $I^*(t)$ is synchronized to $I(t)$ with an anticipation $\tau_a = \tau - \tau_0 = 50$.

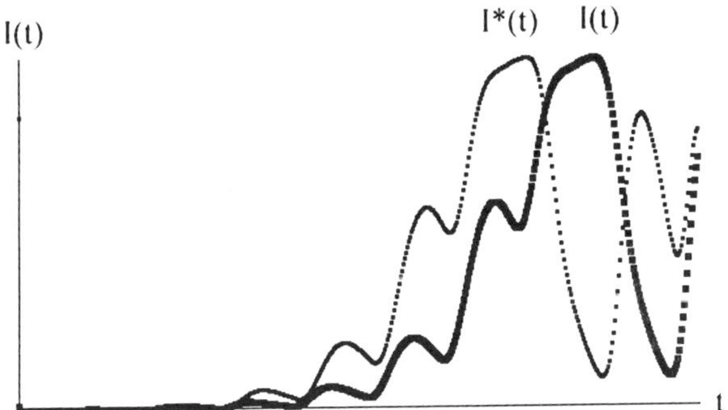

Figure 2a: Anticipatory synchronization of $I^*(t)$ and $I(t)$ with $\tau_a = 50$, versus time t.

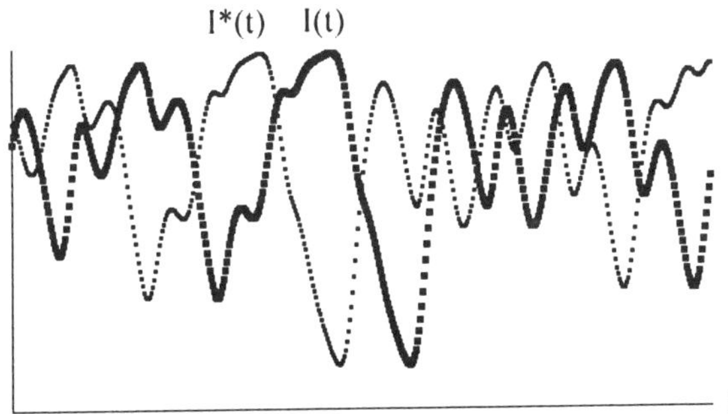

Figure 2b: Continuation of Fig. 2a.

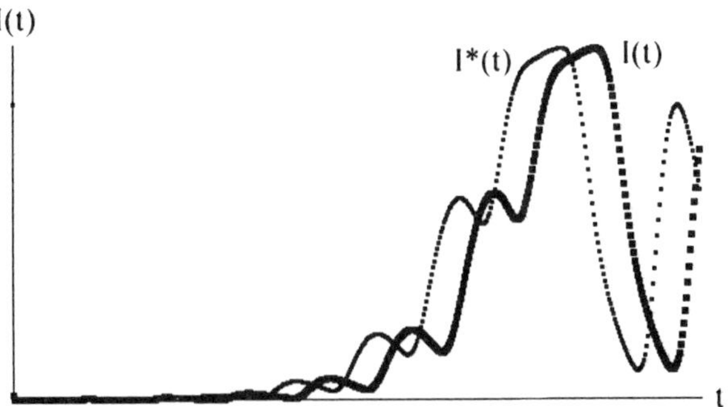

Figure 3a: Anticipatory synchronization of I*(t) and I(t) with $\tau_a = 25$, versus time t.

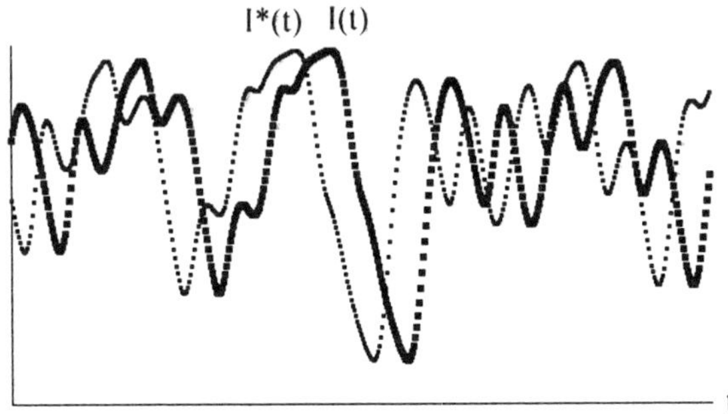

Figure 3b: Continuation of Fig. 3a.

In applying eqs. 18cd to the eq. 30, the synchronization system is given by

$$I(t + \Delta t) = I(t) + \Delta t a I(t - \tau)[C - I(t - \tau)] - \Delta t b I(t) \qquad (32a)$$

$$I*(t + \Delta t) = I*(t) + \Delta t a I*(t - \tau)[C - I*(t - \tau)] - \Delta t b I*(t)$$

$$+ \Delta t[D/(1 + D)][aI(t - \tau_0)[C - I(t - \tau_0)] - I*(t - \tau - \tau_1)[C - I*(t - \tau - \tau_1)]] \qquad (32b)$$

where τ_1 is an additional anticipatory time. Figures 4ab are the simulation of eqs. 32ab for $\Delta t = 0.1$, $a = 4$, $C = 1$, $\tau_0 = 0$, $\tau_1 = 10$, $\tau = 50$, and $b = 1$.

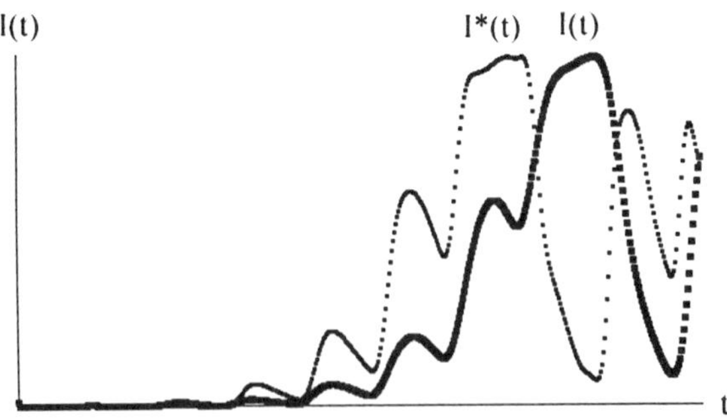

Figure 4a: Anticipatory synchronization of I*(t) and I(t) with $\tau_a = 60$, versus time t.

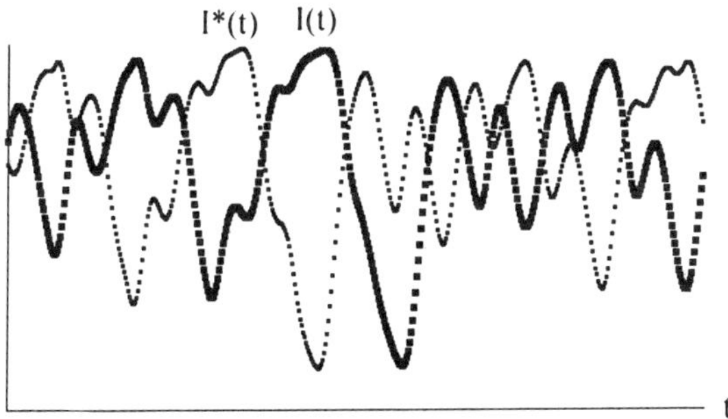

Figure 4b: Continuation of Fig. 4a.

This gives the time evolution of I(t) and I*(t) as a function of time t = 0 to 1000. So, I*(t) is synchronized to I(t) with an anticipation $\tau_a = \tau + \tau_1 - \tau_0 = 60$.

For a master equation without delay, the anticipatory synchronization is $\tau_a = \tau_1 - \tau_0$.

4   Conclusion

This paper presents a theoretical logic of anticipatory synchronization of slave or driven systems on master or driver systems. The application to chaos epidemic is just an example to show the power of such an approach. This anticipatory synchronization logic can be applied to many natural systems (biology, ecology, economy, sociology) and engineering systems (artificial intelligence, neural networks, automation, robotics, computer science, information systems). As an example of a very exciting new application, which will be developed in a forthcoming paper, let us mention the anticipatory synchronization of delayed chaotic neurons. It is well-known that, at one hand, delay exists in chaotic neural networks, and at the other hand, anticipatory processes were experimentally verified in the neural dynamics of the cortex in the brain. Finally, this anticipatory logic will be enlarged, in forthcoming papers, to a theoretical basis of computing anticipatory systems based on differential retarded-advanced difference equation systems. This will demonstrate how and why anticipatory logic is present in natural systems and will represent a new very important avenue for technological applications.

References

[1]   Dubois D. M., G. Resconi (1993), Introduction to Hyperincursion with Applications to Computer Science, Quantum Mechanics and Fractal Processes, CC-AI, vol. 10, NOS 1-2, pp. 109-148.

[2]   Dubois D. M. and Sabatier Ph. (1998), Morphogenesis by Diffusive Chaos in Epidemiological Systems, CP437, Computing Anticipatory Systems, CASYS - First International Conference, edited by Daniel M. Dubois, published by The American Institute of Physics, pp. 295-308.

[3]   Dubois Daniel M. (2001), Theory of Incursive Synchronization and application to the Anticipation of a Chaotic Epidemic, International Journal of Computing Anticipatory Systems, volume 10, pp. 3-18.

[4]   Pyragas K. (1995), Weak and Strong Synchronization of Chaos, Applied Nonlinear Dynamics and Stochastic Systems near the Millennium, AIP Conference Proceedings 411, edited by J. B. Kadke and A. Bulsara, pp. 63-68.

[5]   Voss Henning U. (2000), Anticipating Chaotic Synchronization, Physical Review E, volume 61, number 5, pp. 5115-5119.

[6]   May R. M. (1976) Simple mathematical models with very complicated dynamics. Nature 261, 459–467

Logic, Artificial Intelligence and Robotics
J.M. Abe & J.I. da Silva Filho (Eds.)
IOS Press, 2001

Rules from Supervised Neural Networks in Data Mining Tasks

Nelson F. F. Ebecken
COPPE / Federal University of Rio de Janeiro
nelson@ntt.ufrj.br

Eduardo R. Hruschka
COPPE / Federal University of Rio de Janeiro
erh@onda.com.br

Abstract

The main challenge to the use of supervised neural networks in data mining applications is to get explicit knowledge from these models. For this purpose, a clustering genetic algorithm for rule extraction from artificial neural networks is developed. The methodology is based on the clustering of the hidden units activation values. The developed algorithm is experimentally evaluated in two databases that are benchmarks for data mining applications. The clustering genetic algorithm is employed both in the original space of attributes and in the hidden units activation space. The experimental results show that the clustering genetic algorithm provides better results when it is applied to the hidden units activation space.

1. Introduction

A primary goal of data mining in practice is *description*, which focuses on finding human-interpretable patterns describing the data. Multilayer perceptrons (MP) adjust their internal parameters performing vector mappings from the input to the output space. Although they may achieve high accuracy of classification, the knowledge acquired by such neural networks is usually incomprehensible for humans [1]. This fact can be a major obstacle in domains such as data mining, where it is important to have symbolic rules or other forms of knowledge structure [2]. Therefore, many methods have been developed to get explicit knowledge from these models. One observes that, fundamentally, the knowledge acquired by a neural network is codified on the connection weights [3]. Thus, the knowledge acquisition process from supervised neural networks implies the use of algorithms based either on the connection weight values or on the hidden units activation values, since they are determined by the connection weights. The algorithms designed to perform this task are usually called *algorithms for rule extraction from neural networks*. A survey of the main approaches developed for rule extraction from neural networks until 1995 can be found in [4] and other recent approaches can be found in [5-28].

This paper describes the application of a Clustering Genetic Algorithm (CGA) for rule extraction from multilayer perceptrons (MP). The methodology is based on the activation values of the hidden units, i.e. the CGA is applied in the hidden units activation space. The rule extraction algorithm basically consists of two steps. First, we apply a clustering algorithm to find clusters of hidden unit activation values. Second, we generate a set of rules that describes the hidden unit discretized values in relation to the inputs. This methodology provides a way of relating domain regions to the classes, through the clustering of the activation values. The rules are obtained for each class separately, i.e. the

CGA is applied independently in the examples of each different class. The motivation to extract rules for each class separately comes from the fact that the neural network model is supposed to get different groups of activation values for each class. Therefore, it is possible to find rules that describe the involved sub-classes. On top of that, this methodology eliminates the extra effort to establish what class the clusters describe. Basically, the *CGA* provides a set of rules in the following form:

$$\text{If } \{ v^1_{min} \leq a_1 \leq v^1_{max} \text{ and } ... \text{ and } v^n_{min} \leq a_n \leq v^n_{max} \} \text{ then class } C_j ; \qquad (1)$$

where a_is are the hidden unit activation expressions (as a function of the input units), v^i's are the maximum or minimum values from the obtained clusters for each class C_j and n is equal to the number of hidden units.

One observes that this method can provide satisfactory results when the neural network uses linear transformation functions in the input units, since in this way the hidden unit activation expressions are also linear functions of the original attributes. Besides, hundreds of logical rules produced by some algorithms provide opaque description of the data and therefore are not more comprehensible than any black-box classification system [1]. Therefore, complex neural networks are unlikely to provide suitable classification rule sets when our methodology is used. Several authors noticed the need for simplification of neural networks to facilitate the rule extraction process and are in favor of using specialized training schemes and architectures. We argue that it is also important to try to develop algorithms for *general application*, since their application is twofold, because they would allow the use of supervised neural networks in new data mining classification problems as well as they would provide an approach for clarifying the knowledge encoded in previously defined neural network models. The next section describes the CGA applied to extract rules from MP. The third section presents the experimental results obtained in the Wisconsin Breast Cancer Database and in the Australian Credit Approval Database. In the fourth section we conclude our work.

2. Clustering Genetic Algorithm

Clustering is a task where one seeks to identify a finite set of categories or clusters to describe the data. A generic description of the clustering objective is to maximize homogeneity within each cluster while maximizing heterogeneity among clusters [30]. Basically, there are three kinds of clustering techniques: overlapping, hierarchical and partitioning. This work deals with the partitioning approach, which assigns each object to exactly one cluster. Thus, this work considers that clustering involves the partitioning of a set X of objects into a collection of mutually disjoint subsets C_i of X. Formally, let us consider a set of N objects $X = \{X_1, X_2, ..., X_N\}$ to be clustered, where each $X_i \in \Re^p$ is an attribute vector consisting of "p" real measurements. The objects must be clustered into non-overlapping groups $C = \{C_1, C_2, ..., C_k\}$ where k is the number of clusters, such that [31]:

$$C_1 \cup C_2 \cup ... \cup C_k = X, \quad C_i \neq \varnothing, \quad \text{and} \quad C_i \cap C_j = \varnothing \text{ for } i \neq j. \qquad (2)$$

The problem with finding an optimal solution to the partition of N data into C classes is NP-complete [32] and because the number of distinct partitions of *n* objects into *m* clusters increases approximately as $m^n/m!$, attempting to find a globally optimum solution is usually not computationally feasible [30].

Genetic algorithms are widely believed to be effective on NP-complete global optimization problems and they can provide good sub-optimal solutions in reasonable time [33]. Thus, we believe that a clustering genetic algorithm can provide a way of finding the right clustering. Some recent approaches that describe the application of genetic algorithms to clustering problems can be found in [31-34].

There are several complications concerning the application of traditional genetic algorithms for clustering problems. The use of the simple encoding scheme that yields to constant-length chromosomes usually causes the problems of redundant codification and context insensitivity [32]. We have developed a Clustering Genetic Algorithm (CGA) that allows the utilization of such schemes and avoids all of their problems. The CGA performance has already been evaluated in four different domains, considering initial populations randomly generated, and the results show that the method is very promising [35].

2.1 Individual Representation

Considering that there are N objects to be clustered, a phenotype is represented as a one dimensional integer array with (N+1) elements. As each data unit can be numbered from 1 to N, the *ith* element (*ith* gene) of a genotype represents the *ith* data unit, whereas the last gene represents the number of clusters of the genotype. Therefore, each gene of an individual has a value over the alphabet {1,2,3,4,...,k}, where *k* is the maximum number of clusters. For example, considering a dataset with 20 objects one can get the following cluster:

Genotype1: 22345123453321454552 5

This means that the cluster whose label is *2* is formed by 5 objects, corresponding to the set {1,2,7,13,20}, and the cluster whose label is *1* has 2 objects (6 and 14). The last gene corresponds to the number of clusters encoded by the solution, which is five. The first problem with the straightforward encoding scheme above is that it is highly redundant (there are 5! chromosomes encoding the same solution to the original problem), i.e. the encoding scheme is *one-to-many*.

Thus, the size of the space the CGA has to search is much larger than the original space of solutions. Indeed, the cost function of the clustering problem depends exclusively on the grouping of the objects, rather than on the group label, that is, what matters is not the cluster label but the objects it contains. In addition, there is another problem caused by this encoding scheme when applied with the traditional genetic operators. It is the undesirable effect of casting context-dependent information *out of context* under the standard crossover, i.e. the context insensitivity [32]. These problems are explained in [28,35] and happen when one is dealing with the classic genetic operators of recombination and mutation. Thus, we have developed genetic operators that are group-oriented, i.e. context-sensitive. Basically, the developed CGA is as described in Figure 1.

> 1) Initialize a population of genotypes;
> 2) Evaluate each genotype in the population;
> 3) Apply a linear normalization;
> 4) Select genotypes by proportional selection;
> 5) Apply crossover and mutation;
> 6) Replace the old genotypes by the ones formed by 6);
> 7) If the convergence is attained, stop; if not, go to step 2).

Figure 1. Clustering Genetic Algorithm (CGA).

2.2 Objective Function

Determining the optimal number of clusters in a dataset is one of the most difficult aspects in the clustering process [31]. Most of the approaches found in the clustering literature determine the right number of clusters according to criteria based on given clusterings. However, the CGA is suppose to optimize not only the clusterings for a given number of clusters but also the number of clusters, i.e. the objective function must maximize the homogeneity within each cluster, the heterogeneity among clusters, and also must be the highest one when the solution represents the *right* number of clusters.

The CGA employs an objective function based on the Average Silhouette Width developed by Kaufman and Rousseeuw [36]. In principle, this concept was developed in order to choose the right number of clusters considering that there are different clusterings to be evaluated. However, we believe that this approach can be used as the objective function for the CGA. As genetic algorithms work in parallel with a certain number of solutions, it is possible to work with genotypes representing different clusterings both in terms of the number of clusters and of the clustering structure itself.

Basically, let us consider an object "i" belonging to cluster A. So the average dissimilarity of "i" to all other objects of A is denoted by $a(i)$. Now consider a different cluster C and let us calculate the average dissimilarity of "i" to all objects of C, which will be here denoted by $d(i,C)$. After computing the $d(i,C)$ for all clusters $C \neq A$ we select the smallest of those, which will be here called $b(i)$, i.e.:

$$b(i) = \min d(i,C) , C \neq A . \quad (3)$$

This number represents the dissimilarity of "i" to its neighbor cluster. Now one defines the silhouette $s(i)$ like follows:

$$s(i) = \frac{b(i) - a(i)}{\max\{a(i), b(i)\}} . \quad (4)$$

When cluster A contains only one object we consider that $s(i)=0$, which is the most neutral choice [36]. In addition, it is easy to see that: $-1 \leq s(i) \leq 1$. The objective function is the average of $s(i)$ for $i=1,2,...,N$. It implies that the best value of "k" happens when the objective function value is as high as possible. In order to provide a mutual comparison among the chromosomes, a linear normalization [37,38] is applied to the objective function values.

2.3 Selection

The genotypes that make part of each generation are selected accordingly to the roulette wheel selection strategy. As this strategy does not allow negative objective function values, a constant number equal to one is summed up to each objective function value. Doing so, one observes that the average silhouette range changes to $0 \leq s(i) \leq 2$. Besides, an elitist strategy is used, that is, the best individual of each generation is always copied into the succeeding generation. However, before the objective function values are shown by the developed program, they are summed up with a constant equal to -1.

2.4 Operators

Classic genetic operators are not suitable for clustering problems [32] because they usually work with a set of unrelated genes. In clustering problems the genes are related, i.e. equal allele genes mean that the objects they represent belong to the same cluster. In other words, the groups are the meaningful building blocks of a solution. Thus, it is necessary to use operators that work with groups of genes, instead of with the genes themselves. These operators are referred to as group oriented [32].

2.4.1 Crossover

The crossover operator is carefully described in [28,35], where examples of how it works are given. Let us consider two genotypes - A and B. Thus, the crossover operator basically works in the following way:
1) Select two genotypes (A and B);
2) Considering that A represents $k1$ clusters, choose randomly n ($1 \leq n \leq k1$) clusters to copy in B. The unchanged groups of B are maintained and the changed ones have their object allocated to the cluster that has the nearest centroid. This way the child C is obtained.
3) To get child D, we copy all the clusters whose labels were changed by the n clusters copied in B to D. Then the unchanged clusters of A are computed and they are copied in D. The objects which do not belong to any cluster are allocated to the cluster that has the nearest centroid.

2.4.2 Mutation

There are two operators for mutation. The operator 1 (which only works in clustering formed by more than two groups) eliminates a randomly chosen group and places all their objects to the remaining cluster that has the nearest centroid. The operator 2 divides a randomly selected group into two new ones. The first group is formed by the objects closer to the centroid, whereas the other group is formed by those objects nearer to the farthest object to the centroid. These operators are suitable for grouping problems because they just change the genotypes in the smallest possible way, i.e. dividing groups and eliminating others in the hope of finding better solutions to the problem being solved. Besides, 50% of the genotypes are crossed-over, 25% are mutated by operator 1 and 25% are mutated by operator 2.

2.5 Initial Population

Basically, we have also employed the methodology developed by Kaufman and Rousseeuw[36] in order to set up the initial population. The initial clustering process is based on the selection of representative objects. The first selected object is the most centrally located in the set of objects. Subsequently, at each step another object is selected by means of the following steps:

1. Consider an object i which has not been selected.
2. Consider a nonselected object j and calculate the difference between its dissimilarity D_j with the most similar previously selected object, and its dissimilarity $d(j,i)$ with object i.
3. If the difference is positive, object j will contribute to the decision to select object i. Therefore we calculate $C_{ji} = \max \{D_j - d(j,i), 0\}$.
4. Calculate the total gain obtained by selecting object i: $\sum_j C_{ji}$.
5. Choose the not yet selected object i that maximizes the total gain calculated in step 4.

This process is continued until k objects have been found. In fact, our algorithm stops when the total gain is not positive anymore, because in this case the selection of object i is not appropriate. Besides, the algorithm developed in [36] has a second phase, where it is attempted to improve the set of representative objects. This is done by considering all pairs of objects (i,h) for which object i has been selected and object h has not. It is determined what effect is obtained on the value of the clustering when a swap is carried out, i.e., when object i is no longer selected as a representative object but object h is. As all potential swaps are considered, we believe that this process is not efficient enough in data mining applications. In addition, we are not looking for representative objects, but for good clusterings that could represent good classification rules. Thus, our method aims to provide a way of looking for the best clustering in a space where it is likely to be found. This process is carried out by the CGA.

After selecting the representative objects, which are the *cluster seeds*, the initial population is formed considering that the objects nonselected must be clustered according to their proximity with the representative objects. Considering that we have selected k representative objects, the first genotype represents two clusters, the second genotype represents three clusters, the third genotype represents four clusters and so on, and the last one represents k clusters. Thus, we have employed populations formed by (k-1) genotypes, where each genotype represents a different clustering.

3. Experimental Results

We have evaluated the proposed methodology in two databases that are well known to the data mining community: Wisconsin Breast Cancer Database and Australian Credit Approval Database. In both cases, we have employed the CGA in the original space of attributes and in the hidden units activation space of the best neural network model.

We have employed the Euclidean distance in order to calculate the dissimilarities among the objects and we have stopped the CGA after 2,000 generations. Besides, we just run the CGA one time, but we believe that the obtained results are likely the best ones, because the CGA has always converged before the final generation. In fact, we think that running the CGA several times would not provide a different final solution, but indeed the CGA could converge - which here means to get the best solution - after a different number of generations. Besides, we have set the maximum number of clusters equals to the number of examples.

3.1 Wisconsin Breast Cancer Database

This database is available at the UCI Machine Learning Repository [29]. Each object has 9 attributes and an associated class label (benign or malignant). The attributes are: clump thickness (A), uniformity of cell size (B), uniformity of cell shape (C), marginal adhesion (D), single epithelial cell size (E), bare nuclei (F), bland chromatin (G), normal nucleoli (H) and mitoses (I). The two classes are known to be linearly inseparable. The total number of objects is 699 (458 benign and 241 malignant), of which 16 have a single missing feature. We have removed those 16 objects and used the remaining ones not considering the class label. First we applied the CGA to the original space of attributes [35]. Second, we trained some MP and applied the CGA in the hidden units activation space of the best MP model.

3.1.1 Applying the CGA to the original space of attributes

This section describes the application of the CGA to the original space of attributes. Our first experiment describes the application of the CGA in order to find the right groups, i.e. those clusters formed by the objects that represent each different class in the original database. To do so, we did not employ any kind of cross-validation method. Thus, the CGA has found two clusters after 26 generations with an objective function value equals to 0.59. A silhouette coefficient between [0.51,0.70] indicates that a reasonable structure has been found [36]. However, a 95.9% accuracy was obtained, which is slightly better than the results described by [39]. However, when one is dealing with data mining applications, it is necessary to find rules that appropriately describe the database in hand. Besides, the rule set usually should have the generalization property, which means that the rule set will accurately predict or classify examples that were not used to get the knowledge structure model.

This work is supposed to provide a way of extracting knowledge from MP trained in data mining tasks. Therefore, we employed the cross-validation method suggested by Haykin [42]. Thus, we used a training set and a validation set to train the MP model. The neural network training is stopped when the error on the validation sample goes up. The idea is stopping training before overfitting occurs, as it is well described in [40]. Besides, the neural network performance is also evaluated in an independent sample – the test set. As we are interested in comparing the results obtained in the original space of attributes with those obtained in the hidden units activation space, we employed the same datasets in both experiments. The three samples were randomly extracted from the original database: 60% to define the model, which can be either the rule set or the neural network model; 20% for the validation set and 20% for the test set. One can observe that it would not be necessary to apply these three datasets when dealing with the task of extracting classification rules by means of the CGA in the original space of attributes, but we have employed this methodology to provide a comparison between the two approaches.

The training set is formed by 409 examples (266 examples of benign cases and 143 of malignant cases). Both the validation and the test set are formed by 137 examples (89 of the benign cases and 48 of malignant cases). First, we separated, considering the training set, the examples of each class in order to get rules that describe the sub-classes involved. These rules were evaluated both in the validation set and in the test set. Considering the benign class, the CGA found 71 seeds, so a population formed by 70 genotypes was employed. The best solution was obtained after 126 generations and the fitness function value was equal to 0.69. This solution represents two clusters, c1 and c2, formed by 260 and 6 objects respectively. Considering the malignant class, the CGA found 61 representative objects, so a population formed by 62 genotypes was employed. The CGA converged after 296 generations and the fitness function value was equal to 0.25. This solution represents two clusters, c3 and c4, formed by 141 and 2 objects respectively. These clusters can be translated into the following rules:

If $\{1 \leq A \leq 8 \wedge 1 \leq B \leq 4 \wedge 1 \leq C \leq 6 \wedge 1 \leq D \leq 6 \wedge 1 \leq E \leq 10 \wedge 1 \leq F \leq 10 \wedge 1 \leq G \leq 7 \wedge 1 \leq H \leq 8 \wedge 1 \leq I \leq 7\}$ then c1;
If $\{5 \leq A \leq 8 \wedge 3 \leq B \leq 9 \wedge 3 \leq C \leq 8 \wedge 1 \leq D \leq 5 \wedge 3 \leq E \leq 7 \wedge 4 \leq F \leq 10 \wedge 3 \leq G \leq 7 \wedge 2 \leq H \leq 8 \wedge 1 \leq I \leq 3\}$ then c2;
If $\{1 \leq A \leq 10 \wedge 1 \leq B \leq 10 \wedge 1 \leq C \leq 10 \wedge 1 \leq D \leq 10 \wedge 1 \leq E \leq 10 \wedge 1 \leq F \leq 10 \wedge 1 \leq G \leq 10 \wedge 1 \leq H \leq 10 \wedge 1 \leq I \leq 10\}$ then c3;
If $\{A=10 \wedge 8 \leq B \leq 10 \wedge C=10 \wedge D=10 \wedge 6 \leq E \leq 10 \wedge F=1 \wedge 3 \leq G \leq 8 \wedge 1 \leq H \leq 8 \wedge 8 \leq I \leq 10\}$ then c4;

This rule set only classifies 30.65% of the examples of the validation set correctly, as well as only 32.11% of the examples of the test set are correctly classified. The problem is that the third rule, corresponding to cluster c3, does not have any classification power. However, considering only the rules that describe the benign class (c1 and c2), the following rule set can be obtained:

If $\{1 \leq A \leq 8 \wedge 1 \leq B \leq 4 \wedge 1 \leq C \leq 6 \wedge 1 \leq D \leq 6 \wedge 1 \leq E \leq 10 \wedge 1 \leq F \leq 10 \wedge 1 \leq G \leq 7 \wedge 1 \leq H \leq 8 \wedge 1 \leq I \leq 7\}$ then benign;
If $\{5 \leq A \leq 8 \wedge 3 \leq B \leq 9 \wedge 3 \leq C \leq 8 \wedge 1 \leq D \leq 5 \wedge 3 \leq E \leq 7 \wedge 4 \leq F \leq 10 \wedge 3 \leq G \leq 7 \wedge 2 \leq H \leq 8 \wedge 1 \leq I \leq 3\}$ then benign;
Else malignant.

This rule set classifies 94.89% of the validation set correctly as well as 97.81% of the examples of the test set are correctly classified. However, these rules do not allow us to know when a determined example represents a malignant condition, except by means of not being a benign condition. This kind of association rules are not well suited for data mining tasks, because it is not possible to know the characteristics that describe the *default class*.

3.1.2 Applying the CGA to the hidden units activation space

This section describes the application of a MP to get explicit rules that describe appropriately the classes. First, 40 multilayer perceptrons were trained in order to try to get the best possible results in terms of the average classification rate in the validation set. Our experiments just consider the application of three layer networks, with standard connections. We employed 5 hidden units at most. Considering the activation functions, we used linear transformations in the input units, the logistic and the hyperbolic tangent in the hidden and in the output units. Besides, we also tried to apply a linear activation function in the output units. We employed a backpropagation learning algorithm that uses a learning rate equals to 0.1 and a momentum term equals to 0.3. The initial weights were randomly chosen from the range [-0.3, +0.3]. Besides, we trained the models using the examples in the order they appear in the training set. The validation set average errors were calculated after each elapsed epoch. Table 1 describes the ranking of the neural network models in our experiments. The error column refers to the average quadratic error observed in the validation set.

Ranking	Inputs	Hidden	Outputs	Error	Ranking	Inputs	Hidden	Outputs	Error
1	9 [0,1]	2 logistic	2 logistic	0.013	21	9 [-1,1]	1 logistic	2 tanh	0.084
2	9 [0,1]	3 logistic	2 logistic	0.014	22	9 [-1,1]	2 logistic	2 tanh	0.112
3	9 [-1,1]	2 logistic	2 logistic	0.014	23	9 [-1,1]	4 logistic	2 tanh	0.115
4	9 [0,1]	1 logistic	2 logistic	0.014	24	9 [-1,1]	5 logistic	2 tanh	0.118
5	9 [0,1]	5 logistic	2 logistic	0.015	25	9 [-1,1]	3 logistic	2 tanh	0.119
6	9 [-1,1]	2 tanh	2 logistic	0.015	26	9 [-1,1]	1 tanh	2 linear	0.145
7	9 [0,1]	4 logistic	2 logistic	0.015	27	9 [-1,1]	2 tanh	2 linear	0.749
8	9 [-1,1]	1 tanh	2 logistic	0.015	28	9 [-1,1]	2 logistic	2 linear	2.219
9	9 [-1,1]	1 logistic	2 logistic	0.015	29	9 [0,1]	2 logistic	2 linear	2.358
10	9 [-1,1]	4 logistic	2 logistic	0.017	30	9 [0,1]	4 logistic	2 linear	2.683
11	9 [-1,1]	3 logistic	2 logistic	0.018	31	9 [0,1]	3 logistic	2 linear	3.095
12	9 [-1,1]	3 tanh	2 logistic	0.018	32	9 [0,1]	5 logistic	2 linear	3.270
13	9 [-1,1]	5 logistic	2 logistic	0.019	33	9 [0,1]	1 logistic	2 linear	3.367
14	9 [-1,1]	4 tanh	2 logistic	0.020	34	9 [-1,1]	1 logistic	2 linear	4.466
15	9 [-1,1]	5 tanh	2 logistic	0.020	35	9 [-1,1]	3 tanh	2 linear	4.508
16	9 [0,1]	5 logistic	2 tanh	0.059	36	9 [-1,1]	4 tanh	2 linear	4.609
17	9 [0,1]	3 logistic	2 tanh	0.059	37	9 [-1,1]	5 logistic	2 linear	4.721
18	9 [0,1]	2 logistic	2 tanh	0.061	38	9 [-1,1]	3 logistic	2 linear	4.773
19	9 [0,1]	4 logistic	2 tanh	0.068	39	9 [-1,1]	4 logistic	2 linear	4.900
20	9 [0,1]	1 logistic	2 tanh	0.076	40	9 [-1,1]	5 tanh	2 linear	4.922

Table 1. Ranking of the neural network models – Wisconsin Breast Cancer Database.

As can be observed in Table 1, the best neural network model has 9 input units, using the linear [0,1] transformation function, 2 hidden units using the logistic function and 2 output units using the logistic function. This model provides an average quadratic classification error equals to 0.013. Considering this best neural network model, the hidden unit activation expressions are given by:

$$a_1 = 8.34 - 0.51A - 0.23B - 0.40C - 0.43D - 0.04E - 0.61F - 0.12G - 0.24H - 0.27I; \qquad (5)$$

$$a_2 = -5.52 + 0.36A - 0.23B + 0.15C + 0.12D + 0.10E + 0.19F + 0.23G + 0.12H + 0.32I; \qquad (6)$$

Using these expressions one can calculate the hidden units activation values for each example. Then, the examples that belong to the training set were separated according to their respective class and the CGA was applied separately in each different class. Thus one can find clusters that define the sub-classes. Considering the benign class, the CGA employed 52 representative objects (51 genotypes) and converged after 22 generations. The fitness function value was equal to 0.91. Silhouettes higher than 0.71 indicate that a strong structure has been found [36]. The CGA found two clusters. One cluster, called c5, has 254 objects and the other cluster, called c6, has 12 objects. Considering the malignant class, the CGA found 48 representative objects and converged after 27 generations. The fitness function value was equal to 0.77 and this solution represents two clusters: c7, which has 140 objects, and c8, which has 3 objects. The following rules describe these clusters:

If　　{ 0.43 ≤ a_1 ≤ 2.18　∧ -2.18 ≤ a_2 ≤ -0.43 }　　then c5;　　(benign)
If　　{-2.25 ≤ a_1 ≤ 0.05　∧ -0.05 ≤ a_2 ≤ 2.25 }　　then c6;　　(benign)
If　　{-2.38 ≤ a_1 ≤ -0.51　∧ 0.51 ≤ a_2 ≤ 2.38 }　　then c7;　　(malignant)
If　　{-0.20 ≤ a_1 ≤ 1.00　∧ -1.00 ≤ a_2 ≤ 0.20 }　　then c8;　　(malignant)

These rules classify correctly 79.56% of the validation set, where the neural network model provides an average absolute error equals to 0.12. Considering the test set, these rules classify 73% of the examples correctly and the neural network model provides an average absolute error equals to 0.13. The methodology of applying the CGA in the hidden units activation space provided good results. It happens because it is simpler to extract patterns in a space transformed by the hidden units, which are feature detectors. Basically, it happens because each hidden unit in the first layer (after the input layer) of a MP can create a hyperplane [3], because the input transformation function used is linear. Put it in another way, the hidden units are feature detectors because they try to map the classes, which in the original space can be not linearly separable, into the hidden unit activation space, where the classes are likely to be linearly separable [42]. From the geometrical point of view, these crisp logical rules correspond to a division of the feature space with hyperplanes perpendicular to the axes, into areas with symbolic names. If the classes in the input space are correctly separated with such hyperplanes, accurate logical description of the data is possible and worthwhile [1].

Considering data mining applications, the extracted rules must be evaluated not only on the basis of accuracy, but also on their comprehensibility. In this sense, one can think that the rules obtained by the CGA are not simple enough. We believe that it is possible to simplify these rules, mainly by using techniques to optimize the neural network architecture. Besides, these rules could also be used in the development of expert systems. More specifically, if one takes into consideration the neural network models listed in Table 1, it would be interesting to investigate whether the use of the fourth model, which has just one hidden unit, could provide better results or not. However, we did not worry about getting the best neural network model to this problem, which would demand extra computational efforts, but we do intend to show that our approach can be useful for data mining applications.

One can also observe that cluster c6 just represents 12 objects, which corresponds to 2.93% of the total number of examples, whereas cluster c8 just represents 3 objects, which corresponds to 0.73% of the total number of examples. The measure that relates the number of examples represented by the cluster to the total number of examples is usually called the *support of a rule* and it is used to evaluate the interestingness of association rules [43]. In this context, some methodologies consider that a rule is interesting if its support is greater than the user specified minimum support [44]. This fact leads us to study the possibility of not considering the objects belonging to these clusters. Thus, we eliminated these objects, which could represent outliers, and applied the CGA to the remaining datasets again. Then, the CGA found 32 clusters describing the benign class and 2 clusters describing the malignant one. This huge number of clusters found by the CGA, when applied to the benign examples, is comprehensible, because the algorithm was trying to find clusters that describe very similar objects. However, it is obviously not feasible to describe the dataset by means of 34 rules, although the average classification rates obtained in the validation set and in the test set were equal to 94.16% and 95.62% respectively. Thus, we decided to evaluate the rules obtained by just taking two clusters, one describing the benign cases and the other describing the malignant cases. To do so, we have just computed the range of the hidden unit activation values. This way, one can get the rules described by clusters c5 and c7, i.e we considered that the clusters c6 and c8 are not valid anymore. This approach leads to the following rules:

$$\text{If } \{0.43 \leq a_1 \leq 2.18 \wedge -2.18 \leq a_2 \leq -0.43\} \text{ then benign;}$$
$$\text{If } \{-2.38 \leq a_1 \leq -0.51 \wedge 0.51 \leq a_2 \leq 2.38\} \text{ then malignant;}$$

These rules classify correctly 97.81% of the examples both in the validation set and in the test set. This result is comparable to the best ones described in the literature. Therefore, it seems that the CGA can also be useful to detect outliers, but this property was not sufficiently investigated yet.

Duch et al.[1] reported some results obtained in the Wisconsin Breast Cancer Dataset. Their experiments were done considering a 10-fold cross-validation and their results provide average classification rates which vary from 92.7% to 99%. The authors argue that cross-validation is useful as a measure of generalization capability since classifiers may overfit the training data. However, such danger does not exist if a small number of simple rules is extracted. Besides, the accuracy in the validation set is similar to the one in the test set, what suggests that the rules provide good generalization capability. We also agree with Duch et al.[1] that the stratified 10-cross validation test is difficult to perform since rules have to be extracted many times. On top of that, when one uses this process, it is impossible not only to present a single set of rules and but also to compare rules extracted by different methods.

3.2 Australian Credit Approval Database

This database, which is available at *http://www.ncc.up.pt/liacc/ML/statlog/datasets.html*, concerns about credit card applications. The database contains 690 instances - 307 of people who were granted credit (class 0) and 383 of people who were not granted credit (class 1) - formed by 14 attributes and the class label. There are six continuous attributes and 8 categorical ones. However, all the attribute names and values have been changed to meaningless symbols in order to protect the confidentiality of the data. Thus, the involved attributes are:

A_1:0, 1	A_8:0,1
A_2:continuous: [13.75;80.25]	A_9:0,1
A_3:continuous: [0;28]	A_{10}:continuous: [0;67]
A_4:1,2,3	A_{11}:0,1
A_5:1,2,3,4,5,6,7,8,9,10,11,12,13,14	A_{12}:1,2,3
A_6:1,2,3,4,5,6,7,8,9	A_{13}:continuous: [0;2000]
A_7:continuous: [0;28.5]	A_{14}:continuous: [1;100001]

The methodology applied to evaluate the performance of the CGA in this database follows the approach previously employed in the Wisconsin Breast Cancer Database.

3.2.1 Applying the CGA to the original space of attributes

First, we applied the CGA to find the "right clusters" according to the original database, i.e. the clusters that correctly put the objects into their right group. In this example, we used a population formed by 50 genotypes. The CGA was not able to find the right clusters, but it just separated three objects from the other ones. The best fitness function value was obtained after 3 generations and it was equal to 0.98, which is a very high value. The same result was observed using both the Euclidean distance and the Manhattan distance. Considering that the best solution provided a very high objective function value, we decided to extract these three objects from the database. Then, we applied the CGA in this "new database", formed by 687 objects, but again the CGA just found a very small cluster, formed by 6 objects, and a big one, formed by 681 objects. We extracted these 6 objects from the database and applied the CGA again, but the CGA just separated 5 objects from the others. Thus, the CGA had a very poor performance in this database.

Second, we randomly extracted three datasets from the original database. The *training set*, which is used to get the knowledge structure, and the *validation* and *test* sets, which are used to evaluate the performance of the set of rules. The training set is formed by 229 objects of class *zero* and 183 objects of class *one*. The validation and test sets are formed by 77 objects from class *zero* and 62 objects form class *one*.

Considering the class *zero*, the CGA has found 107 seeds (106 genotypes), and the best solution was got after 21 generations. The best fitness function value was equal to 0.91 and this solution corresponds to two clusters: c1, formed by 225 objects, and c2, formed by 4 objects. Considering the class *one*, the CGA has found 91 representative objects. The best fitness function value was equal to 0.94 after 1,186 generations. This solution corresponds to two clusters, c3 and c4, which have 181 and 2 objects respectively. These clusters can be translated into the following rules:

If { $0 \leq A_1 \leq 1$ ∧ $15.17 \leq A_2 \leq 80.25$ ∧ $0 \leq A_3 \leq 26.33$ ∧ $1 \leq A_4 \leq 2$ ∧ $1 \leq A_5 \leq 14$ ∧ $1 \leq A_6 \leq 8$ ∧ $0 \leq A_7 \leq 13.88$ ∧ $0 \leq A_8 \leq 1$ ∧ $0 \leq A_9 \leq 1$ ∧ $0 \leq A_{10} \leq 11$ ∧ $0 \leq A_{11} \leq 1$ ∧ $1 \leq A_{12} \leq 3$ ∧ $0 \leq A_{13} \leq 2,000$ ∧ $1 \leq A_{14} \leq 2,198$ } then c1 ;

If { $A_1 = 1$ ∧ $22.5 \leq A_2 \leq 36.17$ ∧ $2.5 \leq A_3 \leq 18.13$ ∧ $1 \leq A_4 \leq 2$ ∧ $1 \leq A_5 \leq 9$ ∧ $1 \leq A_6 \leq 4$ ∧ $0.085 \leq A_7 \leq 1.5$ ∧ $A_8 = 0$ ∧ $0 \leq A_9 \leq 1$ ∧ $0 \leq A_{10} \leq 2$ ∧ $0 \leq A_{11} \leq 1$ ∧ $A_{12} = 2$ ∧ $0 \leq A_{13} \leq 320$ ∧ $3,553 \leq A_{14} \leq 5,553$ } then c2 ;

If { $0 \leq A_1 \leq 1$ ∧ $15.83 \leq A_2 \leq 76.75$ ∧ $0 \leq A_3 \leq 28$ ∧ $1 \leq A_4 \leq 2$ ∧ $1 \leq A_5 \leq 14$ ∧ $1 \leq A_6 \leq 9$ ∧ $0 \leq A_7 \leq 28.5$ ∧ $0 \leq A_8 \leq 1$ ∧ $0 \leq A_9 \leq 1$ ∧ $0 \leq A_{10} \leq 67$ ∧ $0 \leq A_{11} \leq 1$ ∧ $1 \leq A_{12} \leq 3$ ∧ $0 \leq A_{13} \leq 840$ ∧ $1 \leq A_{14} \leq 31,286$ } then c3 ;

If { $0 \leq A_1 \leq 1$ ∧ $34.17 \leq A_2 \leq 47.42$ ∧ $1.54 \leq A_3 \leq 8$ ∧ $A_4 = 2$ ∧ $10 \leq A_5 \leq 13$ ∧ $4 \leq A_6 \leq 5$ ∧ $1.54 \leq A_7 \leq 6.5$ ∧ $A_8 = 1$ ∧ $A_9 = 1$ ∧ $1 \leq A_{10} \leq 6$ ∧ $0 \leq A_{11} \leq 1$ ∧ $A_{12} = 2$ ∧ $375 \leq A_{13} \leq 520$ ∧ $5,001 \leq A_{14} \leq 51,101$ } then c4 ;

These rules just classify correctly 12.95% of the validation set examples and 10.07% of the test set examples. On the one hand, considering the class *one* as *default* (so considering the clusters c1 and c2 to describe the database), average classification rates equal to 66.91% and 63.31% are got in the validation and test sets respectively. On the other hand, considering the class *zero* as *default*, the results are worse, because the CGA just classify

45.32% and 42.45% of the examples in the validation and test sets respectively. Besides, we got the same results when the Manhattan distance is used as the dissimilarity measure. On top of that, one observes that the obtained rule set is very complex.

3.2.2 Applying the CGA to the hidden units activation space

Following the methodology applied in the Wisconsin Breast Cancer Database, we trained some MP and we apply the CGA in the hidden units activation values. The 40 neural network models used are listed in Table 2 where the error refers to the average quadratic error on the validation test (used to train the model). The best neural network model was employed to the rule extraction task. This model provides an average absolute error equals to 0.25 in the training set, 0.25 in the validation set and 0.29 in the test set.

Ranking	Inputs	Hidden	Outputs	Error	Ranking	Inputs	Hidden	Outputs	Error
1	14 [0,1]	1 logistic	2 logistic	0.1358	21	14 [-1,1]	2 logistic	2 tanh	1.3912
2	14 [-1,1]	1 logistic	2 logistic	0.1374	22	14 [0,1]	1 logistic	2 tanh	1.3914
3	14 [-1,1]	1 tanh	2 logistic	0.1447	23	14 [-1,1]	1 logistic	2 tanh	1.8709
4	14 [-1,1]	2 tanh	2 logistic	0.1571	24	14 [0,1]	2 logistic	2 tanh	2.1262
5	14 [-1,1]	3 tanh	2 logistic	0.1653	25	14 [-1,1]	3 logistic	2 tanh	2.3139
6	14 [-1,1]	4 tanh	2 logistic	0.1746	26	14 [-1,1]	4 tanh	2 linear	2.9735
7	14 [-1,1]	2 logistic	2 logistic	0.1773	27	14 [-1,1]	2 tanh	2 linear	3.1840
8	14 [0,1]	2 logistic	2 logistic	0.1934	28	14 [-1,1]	5 tanh	2 linear	3.1971
9	14 [-1,1]	3 logistic	2 logistic	0.2097	29	14 [-1,1]	3 logistic	2 linear	3.3108
10	14 [0,1]	3 logistic	2 logistic	0.2186	30	14 [-1,1]	1 logistic	2 linear	3.3146
11	14 [-1,1]	4 logistic	2 logistic	0.2227	31	14 [0,1]	1 logistic	2 linear	3.3411
12	14 [0,1]	5 logistic	2 logistic	0.2258	32	14 [-1,1]	5 logistic	2 linear	3.4981
13	14 [-1,1]	5 tanh	2 logistic	0.2304	33	14 [0,1]	3 logistic	2 linear	3.5000
14	14 [0,1]	4 logistic	2 logistic	0.2390	34	14 [-1,1]	4 logistic	2 linear	3.5740
15	14 [-1,1]	5 logistic	2 logistic	0.2446	35	14 [0,1]	2 logistic	2 linear	3.6471
16	14 [-1,1]	4 logistic	2 tanh	0.8710	36	14 [0,1]	4 logistic	2 linear	3.6600
17	14 [-1,1]	5 logistic	2 tanh	0.8832	37	14 [0,1]	5 logistic	2 linear	3.8028
18	14 [0,1]	3 logistic	2 tanh	0.9627	38	14 [-1,1]	3 tanh	2 linear	4.0081
19	14 [0,1]	5 logistic	2 tanh	1.0327	39	14 [-1,1]	1 tanh	2 linear	4.2428
20	14 [0,1]	4 logistic	2 tanh	1.2143	40	14 [-1,1]	2 logistic	2 linear	4.2486

Table 2. Ranking of the neural network models – Australian Credit Approval Database.

First, the hidden unit activation values were computed. Then, we applied the CGA to extract rules from the activation space. The hidden unit activation values are calculated by means of the following expression:

$$a_1 = -0.58A_1-0.02A_2+0.089A_3+1.23A_4+0.3A_5-0.086A_6+0.026A_7+7.02A_8$$
$$+3.27A_9+0.034A_{10}-0,02A_{11}-0.065A_{12}-0.003A_{13}+0.00007A_{14}-8.88; \quad (7)$$

Considering the class *zero*, the CGA employed 113 genotypes and found the best clustering in the first generation, with a fitness function value equals to 0.67, corresponding to clusters c9 (170 objects) and c10 (59 objects). Considering the class *one*, the CGA employed 86 genotypes and found the best clustering after 28 generations, with a fitness function value equals to 0.75, corresponding to clusters c11(169 objects) and c12 (14 objects). This solution provides the following rule set:

If $(-9.86 \leq a_1 \leq -2.83)$ then c9 – class zero;
If $(-2.65 \leq a_1 \leq 7.24)$ then c10 – class zero;
If $(-2.23 \leq a_1 \leq 9.27)$ then c11 – class one;
If $(-9.87 \leq a_1 \leq -4.01)$ then c12 – class one;

This rule set does not provide good classification rates. Considering the validation set, the rule set provides an average classification rate equals to 7.19% whereas in the test set just 3.6% of the examples are correctly classified. One can explain these low classification rates by looking at the rules. We can observe that, considering class *zero*, a_1 varies in the range [-9.86,7.24] whereas considering class *one*, it varies in the range [-9.87,9.24]. Thus, both classes are modelled by almost the same ranges of values. However, if one considers just the bigger clusters of each class (c9, which represents 74% of the class zero examples and c11, which represents 92% of the examples of class one) it is possible to get average classification rates equal to 86.33% and 76.26% in the validation and test sets respectively. On top of that, we believe that it is not worth to consider one of these clusters as *default*, since using cluster c9 as default we get average classification rates in the validation and test sets equal to 87.05% and 76.97% respectively. Considering cluster c11 as default, average classification rates equal to 86.33% and 76.62% are obtained in the validation and test sets respectively. Thus, the best result that the CGA got to this problem involves the following rules:

If $(-9.86 \leq a_1 \leq -2.83)$ then "Class Zero";
If $(-2.23 \leq a_1 \leq 9.27)$ then "Class One";

These rules describe 82% of the examples in the training set and classify correctly 86.33% of the examples in the validation set as well as 76.26% of the examples of the test set.

4. Conclusions and Future Work

This paper describes a way of using multilayer perceptrons in data mining applications, where it is important to have symbolic rules or other forms of knowledge structure. The proposed methodology makes use of a Clustering Genetic Algorithm, which is applied in the hidden units activation space in order to extract rules from multilayer perceptrons trained in classification problems. This algorithm allows both to find the *best* number of clusters and to get the best clustering according to the Average Silhouette Width Criterion. We have illustrated the method by means of two datasets: Wisconsin Breast Cancer and Australian Credit Approval. The CGA was employed both in the original space of attributes and in the hidden units activation space, and the experimental results show that the CGA provided better results when it is applied to the hidden units activation space.

Comprehensibility is often regarded as a very important issue in data mining applications. In this sense, one can think that the rules obtained by the CGA are not simple enough. We believe that it is possible to simplify these rules, mainly by using techniques to optimize the neural network architecture. Therefore, one important future development is to evaluate the complexity of the obtained rule set in other databases, because hundreds of logical rules provide an opaque description of the data and therefore are not more comprehensible than any black-box classification system [1]. Besides, there is a potential application that could be also investigated, which is the using of the rules in the development of expert systems.

References

[1] DUCH, W., ADAMCZAK, R., GRABCZEWSKI, K. *A New Methodology of Extraction, Optimization and Application of Crisp and Fuzzy Logical Rules*, IEEE Transactions on Neural Networks, vol.11, n.2, pp.1–31, March 2000.

[2] KEEDWELL, E., NARAYANAN, A., SAVIC, D. *Creating Rules from Trained Neural Networks Using Genetic Algorithms*, International Journal of Computers, Systems and Signals (IJCSS), vol.1,n.1, pp. 30-42, December 2000.

[3] FU, L. Neural Networks in Computer Intelligence, First edition, McGraw-Hill Inc., USA ,1994.

[4] ANDREWS, R., DIEDERICH, J., TICKLE, A. B. "A Survey and Critique of Techniques for Extracting Rules from Trained Artificial Neural Networks", Knowledge Based Systems, 8(6), pp. 373- 389, 1995.

[5] LU, H., SETIONO, R., LIU, H. "Effective Data Mining Using Neural Networks", IEEE Transactions on Knowledge and Data Engineering, v. 8, n. 6, pp. 957-961, 1996.

[6] CRAVEN, M. W., SHAVLIK, J. W., "Using Neural Networks for Data Mining", Future Generation Computer Systems, 1997.

[7] SHAVLIK, J. "An Overview of Research at Wisconsin on Knowledge-Based Neural Networks", *Proceedings of the International Conference on Neural Networks*, 1996.

[8] NARAZAKI, H., SHIGAKI, I., WATANABE, T. "A Method for Extracting Approximate Rules from Neural Network", IEEE Press, 1995.

[9] KANE, R., MILGRAM, M. "Financial Forecasting and Rules Extraction from Trained Networks", IEEE Press, pp. 3190-3195, 1994.

[10] NARAZAKI, H., WATANABE, T., YAMAMOTO, M., "Re-organizing Knowledge in Neural Networks: An Explanatory Mechanism for Neural Networks in Data Classification Problems", IEEE Transactions on Systems, Man. and Cybernetics, v.26, n. 1, pp. 107-117, Feb., 1996.

[11] SETHI, I, YOO, J., "Multi-valued Logic Mapping of Neurons in Feedforward Networks", Engineering Intelligent Systems, v.4, n.4, pp. 243-253, 1996.

[12] CRAVEN, M. W., SHAVLIK, J. W., "Extracting Tree-Structured Representations of Trained Networks", Advances in Neural Information Processing Systems, v.8, MIT Press, Cambridge, MA, 1996.

[13] TAHA, I., GHOSH, J. "Symbolic Interpretation of Artificial Neural Networks", IEEE Transactions on Knowledge and Data Engineering, Sep. 1996.

[14] TAHA, I., GHOSH, J. "Three Techniques for Extracting Rules from Feedforward Networks", Intelligent Engineering Systems Through Artificial Neural Networks, v.6, ASME Press, Nov. 1996.

[15] TAHA, I., GHOSH, J., "A Hybrid Intelligent Architecture and Its Application to Water Reservoir Control", Journal of Smart Engineering Systems, October 1996.

[16] BENÍTEZ, J.M., CASTRO, J.L., REUQENA, I. "Are Artificial N.N. Black Boxes?", IEEE Transactions on Neural Networks, v.8, n.5, Sep. 1997.

[17] SETIONO, R., LIU, H. "NeuroLinear: From neural networks to oblique decision rules", Neurocomputing, v.16, n. 3, Sep. 1997, pp. 225-235.

[18] SETIONO, R. "Extracting Rules from Neural Networks by Pruning and Hidden-unit Splitting" Neural Computation, v.9, n.1, pp. 205-225, Jan. 1997.

[19] HRUSCHKA, E.R., EBECKEN, N. F. F. *Rule Extraction from Neural Networks in Data Mining Applications*, In: Data Mining, pp. 289-301, Computational Mechanics Publications, 1998, Southampton.

[20] WEIJTERS, T., BOSH, A., HERIK, J., "Interpretable Neural Networks with BP-SOM", *10th European Conference on Machine Learning - ECML 98*, April 1998, Chemnitz, Germany.

[21] DAS, S., MOZER, M. , "Dynamic on-line clustering and state extraction: an approach to symbolic learning " , Neural Networks, n.11, pp. 53-64, 1998.

[22] HRUSCHKA, E.R. and EBECKEN, N.F.F. " Rule Extraction from Neural Networks: Modified RX Algorithm ", *Proceedings of the IEEE International Joint Conference on Neural Networks* (IJCNN'99), Washington DC, USA, July, 1999.

[23] HRUSCHKA, E.R. and EBECKEN, N.F.F. "*Applying a Clustering Genetic Algorithm for Extracting Rules from a Supervised Neural Network*", Proceedings of the *IEEE International Joint Conference on Neural Networks (IJCNN'2000),* Como, Italy, July, 2000.

[24] HRUSCHKA, E.R. and EBECKEN, N.F.F. "*Using a Clustering Genetic Algorithm for Rule Extraction from Artificial Neural Networks*" , Proceedings of *The First IEEE Symposium on Combinations of Evolutionary Computation and Neural Networks,* San Antonio, *TX, USA,* May, 2000.

[25] LUDERMIR, T.B., "*Extracting Rules from Feedforward Boolean Neural Networks*", Proceedings of V Brazilian Symposium on Neural Networks, IEEE Computer Society, December, 1998.

[26] ANDREWS,R., DIEDERICH, J. (eds), Rules and Networks, Proceedings of the Rule Extraction from Trained Artificial Neural Networks Workshop, April, 1996.

[27] ANDREWS,R., DIEDERICH, J. (eds), Proceedings of the NIPS96 Rule Extraction from Trained Artificial Neural Network Workshop, December, 1996.

[28] HRUSCHKA, E.R. and EBECKEN, N.F.F. "*A Clustering Genetic Algorithm for Extracting Rules from Supervised Neural Network Models in Data Mining Tasks*" , International Journal of Computers, Systems and Signals (IJCSS), Special Issue: Knowledge Discovery from Structured and Unstructured Data, vol.1, n.1, pp. 17-29, December 2000.

[29] MERZ, C.J., MURPHY, P.M., UCI Repository of Machine Learning Databases, [http://www.ics.uci.edu]. Irvine, CA, University of California, Department of Information and Computer Science.

[30] ARABIE, P., HUBERT, L. J., An Overview of Combinatorial Data Analysis (Chapter 1). *Clustering and Classification,* ed. P. Arabie, L.J. Hubert, G. DeSoete, World Scientific, 1999.

[31] COLE, R. M., *Clustering with Genetic Algorithms*, Master of Science Thesis, Department of Computer Science, University of Western Australia, 1998.

[32] FALKENAUER, E., Genetic Algorithms and Grouping Problems, John Wiley & Sons, 1998.

[33] PARK, Y., SONG, M., "*A Genetic Algorithm for Clustering Problems*", Proceedings of the Genetic Programming Conference, University of Wisconsin, July, 1998.

[34] HRUSCHKA, E.R. and EBECKEN, N.F.F. "*Credit approval by a clustering genetic algorithm*", in Data Mining II, N. Ebecken and C.A. Brebbia editors, Proceedings of the *Data Mining 2000 Conference,*p.p. 403-413, *Cambridge University,* Cambridge, UK, July, 2000.

[35] HRUSCHKA, E.R., EBECKEN, N.F.F., *A Genetic Algorithm for Cluster Analysis,* IEEE Transactions on Evolutionary Computation, submitted in January 2001, in review.

[36] KAUFMAN, L., ROUSSEEUW, P. J., *Finding Groups in Data* – An Introduction to Cluster Analysis, Wiley Series in Probability and Mathematical Statistics, 1990.

[37] DAVIS, L. *Handbook of Genetic Algorithms*, International Thompson Computer Press, 1996.

[38] GOLDBERG, D. E., *Genetic Algorithms in Search, Optimization, and Machine Learning*, Addison Wesley Longman, Inc., 1989.

[39] KOTHARI, R., PITTS, D., *"On finding the number of clusters "*, Pattern Recognition Letters 20, 405-416, 1999.

[40] SMITH, M., Neural Networks for Statistical Modeling, First edition, International Thomson Computer Press, USA, 1996.

[41] YILDIZ, N., "Correlation Structure of Training Data and the Fitting Ability of Back Propagation Networks: Some Experimental Results ", Neural Computing & Applications, v.5, pp. 14-19, 1997.

[42] HAYKIN, S. S, Neural Networks: A Comprehensive Foundation, 2 edition , Prentice Hall, 1998.

[43] LIU, B., HSU, W., CHEN, S., MA, Y. "Analyzing the Subjective Interestingness of Association Rules", *IEEE Intelligent Systems,* IEEE Press, pp.47-55, October 2000.

[44] LIU, B., HSU, W., CHEN, S., MA, "Pruning and Summarizing the Discovered Associations", Proceedings of the ACM International Conference on Knowledge Discovery & Data Mining (KDD-99), August 15-18, 1999, San Diego, CA, USA.

Logic, Artificial Intelligence and Robotics
J.M. Abe & J.I. da Silva Filho (Eds.)
IOS Press, 2001

Applications of Product Boolean Algebras in Cluster Analysis

González, Carlos G.[a]   Maia. E.C.[b]
[a]*Centro Universitário N. Sra. do Patrocínio, Itu–SP–Brazil*   [b]*Uniso, Sorocaba–SP–Brazil*

Abstract. A structural analysis of product Boolean algebras is used as an intuitive ground for the formulation of algorithms in pattern recognition. In this analysis, we show some mathematical results to elucidate the properties of product Boolean algebras relevant to understand how its structure can be used for classification in the feature space. In this sense, we discuss two algorithms for clustering selection in feature spaces where the clusters are not clearly delimited.

1 Introduction

In [2] was proposed that the product Boolean algebras are the most natural way to deal with bit sequences, as long as we want to use a structured algebraic framework. In that paper, a study of the relationship between the bit sequences and the basic structural properties of product Boolean algebras is carried out. The aim of this paper is to continue that work and of the [3] in showing the manner how a deeper structural analysis of these algebras can be used to yield new ways to generate algorithms.

The organization of this paper is as follows. In Section 2 we discuss the distance concept in a general, abstract way. In Section 3 we present the definition of product Boolean algebra and define a distance function into this algebra: the Boolean valued distance. This distance concept is fundamental in the approach of this paper, and the sequel of this paper is based on it. In Section 4 we show how the Boolean space can be partitioned into clusters by using the Boolean valued distances. Finally, in Section 5 we present two algorithms that show how to apply the theoretical concepts.

2 Similarity and Distance Concepts

Let us assume as known the basic concepts of pattern recognition as feature vector, feature space, and so on (see [6], 3). Following [6], let X be a data set. The (hard) clustering of X is a partition of X into n sets $X_0, X_1, \ldots, X_{n-1}$ such that the vectors in X_i are "more similar" to each other than the feature vectors of the other clusters. But "more similar" is a vague and imprecise expression and it should be replaced for a more precise one. In order to formalize the concept of similarity, several measure concepts were proposed (see, for instance, [4], 271–272 and [1], 130–136). The most used is a concept of *dissimilarity measure*, or *distance*, formalized by using the framework of the topology of metric spaces.

DEFINITION 2.1. *Let E be a feature space. A map $d \ : \ E \times E \ \longrightarrow \ R$ (the set of real numbers) is a dissimilarity measure or distance if:*

a) $d(x, x) = 0, \ \forall x \in E$;
b) $d(x, y) = d(y, x), \ \forall x, y \in E$;
c) $d(x, y) = 0 \ \Rightarrow \ x = y, \ \forall x, y \in E$;
d) $d(x, z) \leq d(x, y) + d(y, z), \ \forall x, y, z \in E$.

The question we want to discuss is: can we replace R by another set appropriate to give measures? This requires an abstract analysis of the conditions of Definition 2.1 and in this analysis we prefer to talk about an indetermined set A, instead of R, and about a distinguished element d_0, instead of 0[1]. Condition a) states that each element of the space is identical to itself, i.e., that the distance function takes a d_0 value, where d_0 denotes the non-dissimilarity. Furthermore, it can be necessary to establish certain properties of this d_0 w.r.t. the operation $+$ in d). Condition b) is only the symmetry of the distance function. The c) is the classic identity of the indiscernibles, i.e., the indistinct elements of the space must be the same. At first look, it seems that this is not a very important condition, since we can set the indistinct elements in the same class, and then we work with these classes in some kind of quotient space. But this condition prevents the collapse of d, in the sense that $\forall a, b \in A \ d(a, b) = d_0$.

Whereas conditions a), b) and c) do not implicate structural properties on the set A, condition d) requires a relation $\leq$ and an operation $+$. We will assume once more an abstract viewpoint and we will discuss about the relation ρ and the operation $\star$, instead of $\leq$ and $+$. First, we analyze the question: what about the relation ρ? We want to use ρ for clustering. The minimum for this purpose seems to be that ρ can be used to fix limits and then we can take the elements inside these limits to select a cluster. On the one hand, if ρ has cyclic subgraphs[2], we cannot take these limits in a simple way (this means that ρ is a too connected relation). On the other hand, if there exists many isolated points in A, then ρ also seems to be not appropriate for clustering. If ρ is an order or an order alike relation, then ρ seems to induce a structure of A appropriate to take these limits. With an "order alike relation" we indicate a relation that with a little work can be transformed into an order, for instance, making the transitive closure or taking out some points. This order represents the intuitive meaning of "more similar" in a natural way. In brief, the use of a relation ρ to select clusters can be extremely difficult when ρ is not an order.

Let us assume that ρ is an (partial) order. The operation $\star$ represents some kind of cumulative condition on the set A, since condition d) imposes that $\star$ must satisfy $\forall a \in A \ a \leq a \star a$ (at least, for the image of d). Other properties of $\star$ can be necessary in order to represent the cumulative properties of the distances: for instance, $\forall a, b \in A \ a \star b = b \star a$, $\forall a \neq d_0 \ \forall b \ b < a \star b$, and so on. If we pretend that ρ represents the degree of dissimilarity, it can be interesting that the d_0 will be the minimum of A with respect to ρ, and that $\forall a \in A \ a = a \star d_0$. In a general way, the condition d) implicates the lack of shortcuts by using $\star$. In other words, $d(x, y)$ is the shorter path between a and b, that is, the intuitive meaning of the triangular inequality. This premise can generate a further discussion, but it is outside the proposal of this paper.

Actually, the properties, relations and operations of the real numbers were used in clustering. For instance, a definition of the Euclidean distance is widely used in clustering, and requires not only the sum, but also the difference, the square and the square root. The purpose of this discussion is not how to eliminate the measures onto the reals, but in what cases an another set can be an appropriate candidate. In this sense, we propose a more general definition of a distance map:

---

[1] [6], 358, also uses d_0, but $d_0 \in R$.
[2] Non trivial cyclic subgraphs, i.e. with more than one element.

DEFINITION 2.2. *Let E be a feature space, A a set, $\leq$ a partial order on A, $\star$ a binary operation on A and d_0 a distinguished element of A. Then, $d : E \times E \longrightarrow A$ is a distance map if:*

a) $d(x,x) = d_0$, $\forall x \in E$;
b) $d(x,y) = d(y,x)$, $\forall x, y \in E$;
c) $d(x,y) = d_0 \Rightarrow x = y$, $\forall x, y \in E$;
d) $d(x,z) \leq d(x,y) \star d(y,z)$, $\forall x, y, z \in E$.

The use of this abstract definition is limited by the choice of appropriate A, $\leq$, $\star$, and d_0.

3 Bit sequences and product Boolean algebras

In order to minimize the processing time, bit sequences have been used a long time ago for encoding information. Thus, instead of dealing with a single information each time, a parallel processing of information is carried out. In spite of the widely divulged use of bit sequences, the structure of the product Boolean algebras was rarely applied as their theoretical framework. Actually, the algebras that can be used in this sense are products of several (but finite number) two-valued, $\{0,1\}$, algebras. The elements of these algebras are sequences of 0's and 1's, immediately related to bit sequences. In the sequel, we summarize the main concepts of the Boolean algebras needed in this paper[3]. First, we need to define a lattice.

DEFINITION 3.1. *A lattice $\langle L, \leq \rangle$ is a partially ordered set L with an order relation $\leq$, a join or supremum operation $(\vee)$ and a meet or infimum operation $(\wedge)$, such that for any $\{x, y\} \subseteq L$, has a supremum and an infimum.*

A lattice can be also denoted as $\langle L, \vee, \wedge \rangle$. Let us denote 0 the minimum of a lattice, and 1 the maximum of the lattice, if they exist. Now, we can define a Boolean algebra. For $x \in L$, a complement for x is a $y \in L$ such that $x \wedge y = 0$ and $x \vee y = 1$.

DEFINITION 3.2. *A Boolean algebra $\langle B, \wedge, \vee, \sim, 0, 1 \rangle$ is a distributive lattice with 0 and 1, such that each $x \in B$ has a (unique) complement.*

For a family of Boolean algebras, we can yield a new Boolean algebra that is the product of this family (see [5], 40–42):

DEFINITION 3.3. *Let $\langle A_0, \ldots, A_{n-1} \rangle$ be a family of Boolean algebras. The product Boolean algebra is $\langle A, \wedge, \vee, \sim, 0, 1 \rangle$, where $A = A_0 \times \ldots \times A_{n-1}$ and the operations defined point to point, i.e.,*

$$\langle a_0, \ldots, a_{n-1} \rangle \wedge \langle b_0, \ldots, b_{n-1} \rangle = \langle a_0 \wedge b_0, \ldots, a_{n-1} \wedge b_{n-1} \rangle$$

$$\langle a_0, \ldots, a_{n-1} \rangle \vee \langle b_0, \ldots, b_{n-1} \rangle = \langle a_0 \vee b_0, \ldots, a_{n-1} \vee b_{n-1} \rangle$$

and the order in the same way:

$$\langle a_0, a_1, \ldots, a_{n-1} \rangle \leq \langle b_0, b_1, \ldots, b_{n-1} \rangle \text{ iff } a_0 \leq b_0 \text{ and } a_1 \leq b_1 \text{ and } \ldots \text{ and } a_{n-1} \leq b_{n-1}.$$

---

[3]For a detailed treatment see [5].

The product Boolean algebras used in this paper are all (finite) products of $\{0,1\}$ algebras, and therefore their elements are always sequences of 0's and 1's.

In the context of product Boolean algebras, we can choose as $d : B \times B \longrightarrow R$ the function that $d(x,y) = n$ where n is the number of bits in x that are different from thoose in y. We will show that this d satisfies the conditions of Definition 2.1. The only condition that requires an argument is d): let x, y and z be bit sequences and let $d(x,z) = \varepsilon$. We can suppose without loss of generality that $d(x,y) = \sigma$, with $\sigma < \varepsilon$. Let us consider the bits with the same value in x and y. We see that y and z must differ at least in $\varepsilon - \sigma$ bits. Otherwise, let y and z differ in κ bits, with $\kappa \leq \varepsilon - \sigma$. Then, since these bits have the same value in x and in y, then only κ of the considered bits of x and of z were different, being $d(x,z) = \kappa + \sigma < \varepsilon$, a contradiction. Thus the function d is a distance in the sense of Definition 2.1.

The approach that is based on the number of bits that are different in both sequences is useful, but it does not matter *which* are the different bits. In [3] we have introduced another notion of distance that takes into account this fact. This can be necessary, for instance, if we are looking for vectors that differ from the representative in a determined way, i.e., in certain bits and not in n whichever bits. For these cases, the values of the function d can be taken in the same Boolean algebra which represents the feature space and we define $d(x,y) = x$ XOR y, the point to point XOR. We need also the ρ and the $\star$ of Definition 2.2. The order of the Boolean algebra is the natural candidate for ρ. For the $\star$ we choose the join $\vee$. Thus, we can define:

DEFINITION 3.4. *Let B be a product Boolean algebra. $d : B \times B \longrightarrow B$ is a Boolean valued distance if:*

    a) $d(x,x) = 0$, $\forall x \in B$;
    b) $d(x,y) = d(y,x)$, $\forall x,y \in B$;
    c) $d(x,y) = 0 \Rightarrow x = y$, $\forall x,y \in B$;
    d) $d(x,z) \leq d(x,y) \vee d(y,z)$, $\forall x,y,z \in B$.

PROPOSITION 3.1. *The operation XOR is a distance in the sense of Definition 3.4.*

Proof: As above, a), b) and c) are obvious. For d), notice that it suffices the argument for one bit, since all the operations and the order were defined point to point, i.e., bit to bit. Let x_n, y_n and z_n the n-th bit of x, y e z, respectively. We suppose, without loss of generality, that x_n XOR $z_n = 1$, the n-th bit is different in x from z. We have only two cases: or $x_n \neq y_n$ or $y_n \neq z_n$. Then, x_n XOR $y_n \vee y_n$ XOR $z_n = 1$.

4 Distances and clustering

The most natural way to select clusters into the feature space is to group in the same cluster the closest vectors, i.e. closest with respect to the distance map d in the sense that we fix a threshold of dissimilarity ε such that for x and y in the same cluster we have $d(x,y) \leq \varepsilon$. Another way consists in fixing once more a distance threshold ε, but defining that x and y belong to the same cluster if there exists a path $a = a_0, a_1, \ldots, a_n = b$ such that $d(a_i, a_{i+1}) \leq \varepsilon$. With the first approach, the simplest way is the existence of a set of representatives. Such a set of representatives can be conceived as having a central vector a in each cluster, so that it suffices to take every x such that $d(a,x) \leq \varepsilon$ to select each cluster. In the best case, this set of representatives partitions the feature space. Such a set is named *a complete set of representatives*, and defined:

DEFINITION 4.1. *Let A be a set, E a feature space, $\leq$ an order relationship on A, $\star$ an operation on $\star : E \times E \longrightarrow A$ and let ε be a distance threshold. Then, $D \subseteq E$ is a complete set of representatives if:*

a) $d(x,y) > \varepsilon \star \varepsilon$, $\forall x, y \in D$;
b) $\forall x \in E \; \exists y \in D \quad d(x,y) \leq \varepsilon$.

The existence of a complete representative set yields a partition of the feature space in clusters that group the closest vectors.

PROPOSITION 4.1. *Let D be a set satisfying the conditions of Definition 4.1. Then D yields a partition of the feature space.*

Proof: Let us assume the notation of Definition 4.1. We define $\bar{x} = \{y \in E \; : \; d(x,y) \leq \varepsilon\}$, for each $x \in D$. We will show that $P = \{\bar{x} \; : \; x \in D\}$ is a partition of E. First, note that $\bar{x} \neq \emptyset$, since $x \in \bar{x}$. For $\bigcup P = E$, only $E \subseteq \bigcup P$ needs an argument, since the other inclusion is immediate from the definition of the classes $\bar{x}$. Let $x \in E$. Hence, $\exists y \in D$, by b) of Definition 4.1, such that $d(x,y) \leq \varepsilon$, and then $x \in \bar{y} \in P$. We will show that $\bar{x} \cap \bar{y} = \emptyset$, for $\bar{x} \neq \bar{y}$. To get a contradiction, suppose $\exists z \; z \in \bar{x} \; e \; z \in \bar{y}$. Then we have $d(x,z) \leq \varepsilon \; e \; d(y,z) \leq \varepsilon$. We sum these inequalities and have $d(x,z) \star d(y,z) \leq \varepsilon \star \varepsilon$. By using b) and d) of Definition 2.1, we have $\varepsilon \star \varepsilon < d(x,z) \leq d(x,z) + d(z,y) = d(x,z) + d(y,z) \leq \varepsilon \star \varepsilon$.

If the feature space is isomorphic to R^n, $n \in N^+$, then we will not find frequently such a set of representatives. For a more general definition, we consider an ε_x for each representative x:

DEFINITION 4.2. *Let $E, A, \varepsilon, \leq$ and $\star$ be as above. Furthermore, let D be the set of the couples $\langle x, \varepsilon_x \rangle$ with $x \in E$ and ε_x a distance threshold depending of x. Then D is a relative set of representatives if:*

a) $d(x,y) > \varepsilon_x \star \varepsilon_y$, $\forall \langle x, \varepsilon_x \rangle, \langle y, \varepsilon_y \rangle \in D$;
b) $\forall x \in E \; \exists \langle y, \varepsilon_y \rangle \in D \quad d(x,y) \leq \varepsilon_y$.

The first problem that we find by using these definitions is that frequently the set of representatives does not exist. Another problem is that by using R to give distances, the feature spaces are partitioned into hyperspheres, so that it can be hard to put any vector in some hypersphere, since the space is not the (finite) union of hyperspheres. Alternatively, we can choose to partition the feature space into hypercubes, instead of hyperspheres. But if we use hypercubes of different sizes, we can have a vector in the vertices which is closer of the representative of another cluster than the cluster that it belongs. This is not the case with a Boolean algebra, since the distances that satisfy the Definition 3.4 yield clusters that are Boolean hypercubes.

Actually, once determined a distance threshold in a Boolean algebra, it is very easy to define a partition of the feature space. We note that:

$$H_{x,\varepsilon} = \{y \in B \; : \; d(x,y) \leq \varepsilon\}$$

is a Boolean hypercube and also a Boolean algebra[4] with the restriction of the Boolean operations to $H_{x,\varepsilon}$. We denote with $\bigwedge_{H_x} = x_0 \wedge x_1 \wedge \ldots \wedge x_n$, with $x_i \in H_{x,\varepsilon}$ the minimum of

---

[4]It is possible that H_x would not be a sub-algebra of B since the minimum and the maximum H_x can be different in B and in $H_{x,\varepsilon}$, but the Boolean operations of $H_{x,\varepsilon}$ are a restriction of the operations in B.

this algebra, and $\bigvee_{H_{x,\varepsilon}} = x_0 \vee x_1 \vee \ldots \vee x_n$ the maximum. Anytime that ε results obvious from the context, we use H_x. We will now define the useful concept of mask. Let x_i be i-th bit of x. Then

$$(\equiv_{H_x})_i = 1 \quad \text{iff} \quad \forall a, b \in H_x \; a_i = b_i$$

In other words, $\equiv_{H_x}$ has the i-th bit equal to 1 if and only if this bit is equal for all the vectors in H_x. We note that, for $a \in H_x$, it holds $a \wedge \equiv_{H_x} = \bigwedge_{H_x}$. For $A \subseteq B$, we state $\sim A = \{\sim a \; : \; a \in A\}$. Then we have $a \vee \sim\equiv_{H_x} = \bigvee_{H_x}$, and for $a, b \in H_x$, we have $a \wedge \equiv_{H_x} = b \wedge \equiv_{H_x}$. Each $x \in B$, whereas is considered a mask, partitions B into equivalence classes. As above, each such equivalence class is a Boolean algebra with the restricted operations and the minimum and maximum defined from the mask and any $x \in B$. Furthermore, for any class it can be found a distance threshold $\varepsilon \in B$ such that, $d(a, b) \leq \varepsilon$ if a and b belong to the same class. In brief, any vector of the Boolean algebra can partition the algebra into classes of similar vectors, whereas we consider it as a mask.

We have analyzed above the distance concept based on the number of different bits in two sequences. We can define now more formally:

DEFINITION 4.3. *Let $q(x)$ be the number of bits equal to 1 in $x \in B$.*

And $Q(x, y)$ as:

DEFINITION 4.4.

$$Q : B \times B \longrightarrow N \qquad Q(x, y) = q\left(\equiv_{\{x,y\}}\right)$$

5   Algorithms

The concepts that we have discussed above yield a new viewpoint useful for the development of algorithms, since the purely pragmatic manners to work with bit sequences can be replaced for general methods in a powerful framework. Consider, for instance, the binary morphology clustering algorithms (see [6], 516). The first step, the discretization of the feature space, has no sense, since the Boolean spaces are always discrete. It can be necessary to consider the "resolution" or granularity, because of the impossibility of the computational treatment of all the hypercubes of the Boolean space. In these cases we can proceed by using blocks of bits instead of all the elements in the Boolean space. Actually, we have implicitly defining a homomorphism of the initial algebra in the one formed by blocks of bits.

To present an algorithm (see [3]), we first need a definition:

DEFINITION 5.1. *Let B be a product Boolean algebra. For $n \in N$, let F_n be the family of distances of degree n, defined by:*

$$F_n = \{\varepsilon \in B \; : \; q(\varepsilon) = n\}$$

We can now present the following algorithm:

ALGORITHM 5.1. *Cluster Detection by Valley Creation Algorithm, CDVCA.*

```
Initially no element is consider as processed.
Fix b,n ∈ N.
Repeat
  Choose a nonprocessed x.
  Consider in the sequel only the nonprocessed elements.
  If there is some class H_{x,ε} with more than n elements
  for ε ∈ F_b:
    a) Choose the H_{x,ε}'s with the greatest number of
       elements.
    b) For H_{x,ε}, set H'_{x,ε} = H_{x,ε} − ⋂_{δ∈F_b} H_{x,δ}.
    c) Choose one H'_{y,ε}, with y ∈ H'_{x,ε}, with the greatest
       number of elements, create a new cluster H_{y,ε} and
       consider x and the elements of H_{y,ε} as processed.
  Else
    consider x as processed
  EndIf
Until all elements have been processed.
```

This algorithm is specially useful when, in the one hand, the feature space has regions with difference of density, but there is not "valleys", i.e., regions of the space without elements that make the cluster selection easier, but, in the other hand, we do no want too few clusters. This is the purpose when we take out the intersection of the classes selected in a first step (the $H_{y,\varepsilon}$'s): we want to create "artificial valleys" and then we will use these valleys to divide too big regions. In this sense, the CDVCA has a certain analogy with the k-closest neighbors algorithms (see [1], 88). Notice that the structure of the Boolean algebra allows that there exists a procedure that in other spaces (for instance R^n) would be trivial. The distances in F_b formalize the tentatives to define the classes of the element x in different directions of the Boolean space. If we conceive the feature space as encoding information by using bit sequences, then these tentatives are looking for sets with properties that are common to several objects. Furthermore, it can be interesting to set a weight for each element in the Boolean algebra, i.e. reformulate the CDVCA, so that in step b), we will sum the weight of elements, instead of the number of the elements in each class, and then we will select the classes with the greatest weight.

Another algorithm that results of the application of the Boolean ideas is:

ALGORITHM 5.2. *Minesweeper.*

```
Fix a k ∈ N and a distance threshold ε.
Divide the Boolean space in hypercubes (i.e. algebras with
the same number of elements).
For each hypercube, calculate the number of objects in it.
For each hypercube, calculate the sum of the elements of
itself and the other hypercubes in the ε−neighbor.
This result in a number n for each hypercube.
Consider the hypercubes with n < k as processed and the
others as nonprocessed.
```

```
Repeat
  Choose a hypercube c among the nonprocessed of the
  greatest n.
  If there exists nonprocessed elements in c then
    Choose a nonprocessed element x of c.
    Create the cluster A = {y ∈ B : d(x,y) ≤ ε},
    by using only nonprocessed y.
    Consider c and the elements in A as processed.
  Else
    Consider c as processed.
  EndIf.
Until all hypercubes have been processed.
```

To find the ε-neighbor hypercubes we can use, for instance, the minimum of each hypercube (we have seen that the hypercubes are Boolean algebras). Two hypercubes c and d are neighbors if the minimum x of c is different and contiguous to the minimum y of d:

$$x \leq y \quad \text{and} \quad \neg \exists z\, z \neq x, z \neq y\, x \leq z \leq y$$

or

$$y \leq x \quad \text{and} \quad \neg \exists z\, z \neq x, z \neq y\, y \leq z \leq x.$$

We can consider once more the weight for reformulate the algorithm.

6   Conclusions

In our opinion, the use of product Boolean Algebras to represent bit sequences seems to be a promissory field for research and development. The use of the Boolean valued distance that is exposed shows to be fruitful in creating new viewpoints and applications. Some well-known concepts can be moved into the framework of product Boolean algebras to yield new tools and techniques, and we have intended to show this in the algorithm section. The structure of the product Boolean algebras can seem complex in a first view, but then happens to be very motivating as a conceptual way of dealing with information encoded using bit sequences.

References

[1] Abdel Belaïd and Yolande Belaïd. *Reconnaissance des Formes.* Inter Editions, Paris, 1992.

[2] Carlos G. González. Aplicação de produtos de álgebras de Boole para codificar seqüências de bits. In Jair Minoro Abe, editor, *LAPTEC 2000. Anais do I Congresso de Lógica Aplicada à Tecnologia*, pages 679–689, São Paulo–SP, Brasil, setember 2000. Editora Plêiade.

[3] Carlos G. González and Edson Campos Maia. Aplicações de álgebras de Boole produto em reconhecimento de padrões. *Revista de Estudos Universitários*, 27(1):75–88, 2001.

[4] A. K. Jain, M. P. Murty, and P. J. Flynn. Data clustering: A review. *ACM Computing Surveys*, 31(3):264–322, 1999.

[5] Roman Sikorski. *Boolean Algebras.* Springer Verlag, Heildelberg, 1969.

[6] Sergios Theodoridis and Konstantinos Koutroumbas. *Pattern Recognition.* Academic Press, San Diego, 1998.

On Finite BCK with Condition (S)

Kiyoshi ISÉKI
Emeritus Professor of Kobe University
&
Emeritus Professor of Naruto University of Education

Abstract. In this paper, we we discuss a special class of BCK: those which satisfy the
condition (S).

In this paper, we discuss a special class of BCK. BCK is a relational system X with a
binary operation $*$ and a distinguished element 0 which is called a zero element. Then we
assume that the binary operation $*$ satisfies the following conditions:

BCK 1. $((x * y) * (x * z)) * (z * y) = 0,$
BCK 2. $(x * (x * y)) * y = 0,$
BCK 3. $0 * x = 0,$
BCK 4. $x * y = y * x = 0$ if and only if $x = y.$

We introduce $x \leq y$ by the following way: $x \leq y$ if and only if $x * y = 0$. To define a BCK
with condition (S), for any two elements $a, b \in X$, we consider the set

$$X_{a,b} = \{ x \in X / x * a \leq b \}.$$

The set $X_{a,b}$ is never empty, because it contains two elements, namely 0 and a. Then if for
any $a, b \in X$ the set $X_{a,b}$ has a greatest element (which is denoted aob), X is called a *BCK with
condition (S)*. Some basic properties of BCK are found in [1].

Remark 1. By a similar way, we can define a BCI with condition (S) (see [1]). By *BCI*, we
mean a system with a binary operation $*$ satisfying the following conditions:

BCI 1. $((x * y) * (x * z)) * (z * y) = 0,$
BCI 2. $S(x * (x * y)) * y = 0,$
BCI 3. $x * x = 0,$
BCI 4. $x * y = y * x = 0$ if and only if $x = y.$

In general, BCI may contain an element which is not comparable with 0. Hence there
exists at least one element x such that $0 * x \neq 0$ in proper BCI (which is not BCK). More
information about BCI will be found in Appendix of this paper.

An order on BCI X is defined as follows: $x \leq y$ is defined by $x * y = 0$. This definition is
quite similar with the case of BCK.

Let X be a BCK with condition (S). Then X has the following basic properties:

(1) $x, y \leq xoy = yox,$
(2) $xo(yoz) = (xoy)oz,$
(3) $x \leq y \Rightarrow xoz \leq yoz,$
(4) $x * (y * z) = x * (yoz).$

Hence a BCK with condition (S) is a partially ordered commutative semigroup. This semigroup is called the *associated semigroup* of a BCK (with condition (S)). Moreover, W. H. Cornish introduced a concept which is called a BCK with supremum ([2], [3]).

Let X be a BCK with condition (S). X is called a BCK with supremum if the following identity holds for any elements $C_y \in X$:

$$xo(y * x) = yo(x * y).$$

$yo(y * x)$ is denoted by $x \vee y$. It is just the least upper bound of x,y. Therefore a BCK with supremum is an upper semilattice.

Remark. This concept was introduced in my papers [4], [5]. H. Yutani proved that the class of BCK with condition (S) is equationally definable.

As easily seen, any non-trivial finite BCK with condition (S) has at least two idempotent *(xox = x)*. On the other hand, there exists an infinite BCK with condition (S) such that the associated semigroup has only one idempotent. Of course one of them is *0*. All BCK of order 3, 4 and 5 are already known (see [8]).

order	@2	@3	@4	@5
number	@1	@3	@14	@88

In this paper, refering [8], all BCK with condition (S) of order 3, 4 and 5 are mentioned.

(1) Case of order 3 (there exist three BCK, two BCK are linear and another is not linear). A collection of concrete examples are very useful and helpful to introduce new concepts and powerful to develop our theory.

1)

*	0	1	2
0	0	0	0
1	1	0	0
2	2	1	0

commutative

2)

*	0	1	2
0	0	0	0
1	1	0	0
2	2	2	0

positive implicative

These linear BCK has the condition (S) as follows:

o	0	1	2
0	0	1	2
1	1	2	2
2	2	2	2

ideal *(2), (1,2)*
proper subsemigroup *(0), (2)*
(0, 2), (1, 2)

o	0	1	2
0	0	1	2
1	1	1	2
2	2	2	2

ideal *(2), (1,2)*
proper subsemigroup *(0), (1), (2)*
(0, 1), (1, 2), (2, 0)

3)

*	0	1	2
0	0	0	0
1	1	0	1
2	2	2	0

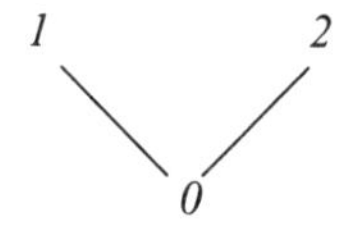

This is a non-linear BCK (tree) of order *3*. In this case, *1* and *2* are incomparable, so there is no associated semigroup (but if we replace the greatest element into local maximum, a multiple system is obtained. Following this new operation, from a pair *(1, 2)*, we have *1* and *2*.

(2) Case of order *4* (there exist only 6 linear BCK. Other BCK are not linear). Among them, each of linear BCK has the associated semigroup. The automorphism group istrivial. There exist two non-linear BCK with condition (S) and the automorphism groups have both two elements. The mentioned numbers correspond to the numbers of BCK given in [8].

We first mention the linear case.

1.

*	0	1	2	3
0	0	0	0	0
1	1	0	0	0
2	2	1	0	0
3	3	1	1	0

o	0	1	2	3
0	0	1	2	3
1	1	3	3	3
2	2	3	3	3
3	3	3	3	3

2.

*	0	1	2	3
0	0	0	0	0
1	1	0	0	0
2	2	1	0	0
3	3	2	1	0

o	0	1	2	3
0	0	1	2	3
1	1	2	3	3
2	2	3	3	3
3	3	3	3	3

4.

*	0	1	2	3
0	0	0	0	0
1	1	0	0	0
2	2	2	0	0
3	3	2	1	0

o	0	1	2	3
0	0	1	2	3
1	1	1	3	3
2	2	3	3	3
3	3	3	3	3

5.

*	0	1	2	3
0	0	0	0	0
1	1	0	0	0
2	2	1	0	0
3	3	3	2	0

o	0	1	2	3
0	0	1	2	3
1	1	1	3	3
2	2	2	3	3
3	3	3	3	3

7.

*	0	1	2	3
0	0	0	0	0
1	1	0	0	0
2	2	1	0	0
3	3	3	2	0

o	0	1	2	3
0	0	1	2	3
1	1	2	2	3
2	2	2	3	3
3	3	3	3	3

10.

*	0	1	2	3
0	0	0	0	0
1	1	0	0	0
2	2	2	0	0
3	3	3	3	0

o	0	1	2	3
0	0	1	2	3
1	1	1	2	3
2	2	2	2	3
3	3	3	3	3

Non-linear case.

6.

*	0	1	2	3
0	0	0	0	0
1	1	0	1	0
2	2	2	0	0
3	3	2	1	0

o	0	1	2	3
0	0	1	2	3
1	1	1	3	3
2	2	3	3	3
3	3	3	3	3

13.

*	0	1	2	3
0	0	0	0	0
1	1	0	1	0
2	2	2	0	0
3	3	3	3	0

o	0	1	2	3
0	0	1	2	3
1	1	1	2	3
2	2	2	2	3
3	3	3	3	3

2 and *6; 5* and *13* have same associated semigroups respectively.

(3) Case of order 5.

1.

*	0	1	2	3	4
0	0	0	0	0	0
1	1	0	0	0	0
2	2	1	0	0	0
3	3	1	1	0	0
4	4	1	1	1	0

o	0	1	2	3	4
0	0	1	2	3	4
1	1	4	4	4	4
2	2	4	4	4	4
3	3	4	4	4	4
4	4	4	4	4	4

2.

*	0	1	2	3	4
0	0	0	0	0	0
1	1	0	0	0	0
2	2	1	0	0	0
3	3	1	1	0	0
4	4	2	1	1	1

o	0	1	2	3	4
0	0	1	2	3	4
1	1	1	3	3	4
2	2	3	3	3	4
3	3	3	3	3	4
4	4	4	4	4	4

3.

*	0	1	2	3	4
0	0	0	0	0	0
1	1	0	0	0	0
2	2	1	0	0	0
3	3	1	1	0	0
4	4	3	2	1	0

o	0	1	2	3	4
0	0	1	2	3	4
1	1	3	3	4	4
2	2	3	4	4	4
3	3	4	4	4	4
4	4	4	4	4	4

4.

*	0	1	2	3	4
0	0	0	0	0	0
1	1	0	0	0	0
2	2	1	0	0	0
3	3	1	1	0	0
4	4	3	3	1	0

o	0	1	2	3	4
0	0	1	2	3	4
1	1	3	3	4	4
2	2	3	4	4	4
3	3	4	4	4	4
4	4	4	4	4	4

7.

*	0	1	2	3	4
0	0	0	0	0	0
1	1	0	0	0	0
2	2	1	0	1	0
3	3	1	1	0	0
4	4	1	1	1	0

o	0	1	2	3	4
0	0	1	2	3	4
1	1	4	4	4	4
2	2	4	4	4	4
3	3	4	4	4	4
4	4	4	4	4	4

9.

*	0	1	2	3	4
0	0	0	0	0	0
1	1	0	0	0	0
2	2	1	0	0	0
3	3	2	1	0	0
4	4	2	1	1	0

o	0	1	2	3	4
0	0	1	2	3	4
1	1	2	4	4	4
2	2	4	4	4	4
3	3	4	4	4	4
4	4	4	4	4	4

11.

*	0	1	2	3	4
0	0	0	0	0	0
1	1	0	0	0	0
2	2	1	0	0	0
3	3	2	1	0	0
4	4	3	2	1	0

o	0	1	2	3	4
0	0	1	2	3	4
1	1	2	3	4	4
2	2	3	4	4	4
3	3	4	4	4	4
4	4	4	4	4	4

12.

*	0	1	2	3	4
0	0	0	0	0	0
1	1	0	0	0	0
2	2	1	0	1	0
3	3	1	1	0	0
4	4	2	1	2	0

o	0	1	2	3	4
0	0	1	2	3	4
1	1	3	4	4	4
2	2	4	4	4	4
3	3	4	4	4	4
4	4	4	4	4	4

15.

*	0	1	2	3	4
0	0	0	0	0	0
1	1	0	0	0	0
2	2	2	0	0	0
3	3	2	1	0	0
4	4	2	1	1	0

o	0	1	2	3	4
0	0	1	2	3	4
1	1	1	4	4	4
2	2	4	4	4	4
3	3	4	4	4	4
4	4	4	4	4	4

17.

*	0	1	2	3	4		o	0	1	2	3	4
0	0	0	0	0	0		0	0	1	2	3	4
1	1	0	0	0	0		1	1	1	3	3	4
2	2	2	0	0	0		2	2	3	4	4	4
3	3	2	1	0	0		3	3	3	4	4	4
4	4	3	2	2	0		4	4	4	4	4	4

18.

*	0	1	2	3	4		o	0	1	2	3	4
0	0	0	0	0	0		0	0	1	2	3	4
1	1	0	0	0	0		1	1	1	2	4	4
2	2	2	0	0	0		2	2	2	3	4	4
3	3	3	2	0	0		3	3	4	4	4	4
4	4	3	2	1	0		4	4	4	4	4	4

20.

*	0	1	2	3	4		o	0	1	2	3	4
0	0	0	0	0	0		0	0	1	2	3	4
1	1	0	0	0	0		1	1	1	2	3	4
2	2	2	0	0	0		2	2	2	4	4	4
3	3	3	2	0	0		3	3	3	4	4	4
4	4	3	2	2	0		4	4	4	4	4	4

21.

*	0	1	2	3	4		o	0	1	2	3	4
0	0	0	0	0	0		0	0	1	2	3	4
1	1	0	0	0	0		1	1	1	2	3	4
2	2	2	0	0	0		2	2	2	3	4	4
3	3	3	2	0	0		3	3	3	4	4	4
4	4	4	3	2	0		4	4	4	4	4	4

23.

*	0	1	2	3	4		o	0	1	2	3	4
0	0	0	0	0	0		0	0	1	2	3	4
1	1	0	0	0	0		1	1	2	2	4	4
2	2	1	0	0	0		2	2	2	2	4	4
3	3	3	3	0	0		3	3	4	4	4	4
4	4	3	3	1	0		4	4	4	4	4	4

24.

*	0	1	2	3	4		o	0	1	2	3	4
0	0	0	0	0	0		0	0	1	2	3	4
1	1	0	0	0	0		1	1	2	2	3	4
2	2	1	0	0	0		2	2	2	2	3	4
3	3	3	3	0	0		3	3	3	3	4	4
4	4	4	4	3	0		4	4	4	4	4	4

25.

*	0	1	2	3	4		o	0	1	2	3	4
0	0	0	0	0	0		0	0	1	2	3	4
1	1	0	0	0	0		1	1	2	2	3	4
2	2	1	0	1	0		2	2	2	2	3	4
3	3	3	3	0	0		3	3	3	3	4	4
4	4	4	4	3	0		4	4	4	4	4	4

26.

*	0	1	2	3	4		o	0	1	2	3	4
0	0	0	0	0	0		0	0	1	2	3	4
1	1	0	0	0	0		1	1	2	2	4	4
2	2	1	0	1	0		2	2	2	2	4	4
3	3	3	3	0	0		3	3	4	4	4	4
4	4	3	2	1	0		4	4	4	4	4	4

31.

*	0	1	2	3	4
0	0	0	0	0	0
1	1	0	0	0	0
2	2	2	0	0	0
3	3	3	3	0	0
4	4	4	4	3	0

o	0	1	2	3	4
0	0	1	2	3	4
1	1	1	2	3	4
2	2	2	2	3	4
3	3	3	3	4	4
4	4	4	4	4	4

32.

*	0	1	2	3	4
0	0	0	0	0	0
1	1	0	0	0	0
2	2	2	0	0	0
3	3	3	3	0	0
4	4	3	3	1	0

o	0	1	2	3	4
0	0	1	2	3	4
1	1	1	2	4	4
2	2	2	2	4	4
3	3	4	4	4	4
4	4	4	4	4	4

33.

*	0	1	2	3	4
0	0	0	0	0	0
1	1	0	0	0	0
2	2	2	0	0	0
3	3	3	3	0	0
4	4	4	3	2	0

o	0	1	2	3	4
0	0	1	2	3	4
1	1	1	2	3	4
2	2	2	2	4	4
3	3	3	4	4	4
4	4	4	4	4	4

34.

*	0	1	2	3	4
0	0	0	0	0	0
1	1	0	0	0	0
2	2	2	0	0	0
3	3	3	0	0	0
4	4	4	3	2	0

o	0	1	2	3	4
0	0	1	2	3	4
1	1	1	2	3	4
2	2	2	2	4	4
3	3	3	4	4	4
4	4	4	4	4	4

35.

*	0	1	2	3	4
0	0	0	0	0	0
1	1	0	0	0	0
2	2	2	0	2	0
3	3	3	3	0	0
4	4	4	3	2	0

o	0	1	2	3	4
0	0	1	2	3	4
1	1	4	4	4	4
2	2	4	4	4	4
3	3	3	3	4	4
4	4	4	4	4	4

36.

*	0	1	2	3	4
0	1	0	0	0	0
1	1	0	1	0	0
2	2	2	0	0	0
3	3	3	3	0	0
4	4	4	4	3	0

o	0	1	2	3	4
0	0	1	2	3	4
1	1	1	2	3	4
2	2	2	2	3	4
3	3	3	3	4	4
4	4	4	4	4	4

37.

*	0	1	2	3	4
0	0	0	0	0	0
1	1	0	1	0	0
2	2	2	0	0	0
3	3	3	3	0	0
4	4	3	4	1	0

o	0	1	2	3	4
0	0	1	2	3	4
1	1	1	2	4	4
2	2	2	2	4	4
3	3	3	4	4	4
4	4	4	4	4	4

38.

*	0	1	2	3	4
0	0	0	0	0	0
1	1	0	0	0	0
2	2	1	0	0	0
3	3	3	3	0	0
4	4	4	4	4	0

o	0	1	2	3	4
0	0	1	2	3	4
1	1	2	2	4	4
2	2	2	2	3	4
3	3	3	3	3	4
4	4	4	4	4	4

42.

*	0	1	2	3	4
0	0	0	0	0	0
1	1	0	0	0	0
2	2	1	0	1	0
3	3	3	3	0	0
4	4	4	4	4	0

o	0	1	2	3	4
0	0	1	2	3	4
1	1	2	2	3	4
2	2	2	2	3	4
3	3	3	3	3	4
4	4	4	4	4	4

47.

*	0	1	2	3	4
0	0	0	0	0	0
1	1	0	0	1	0
2	2	1	0	2	0
3	3	3	3	0	0
4	4	4	4	4	0

o	0	1	2	3	4
0	0	1	2	3	4
1	1	2	2	3	4
2	2	2	2	3	4
3	3	3	3	3	4
4	4	4	4	4	4

50.

*	0	1	2	3	4
0	0	0	0	0	0
1	1	0	0	0	0
2	2	2	0	0	0
3	3	3	3	0	0
4	4	4	4	4	0

o	0	1	2	3	4
0	0	1	2	3	4
1	1	1	2	3	4
2	2	2	2	3	4
3	3	3	3	3	4
4	4	4	4	4	4

54.

*	0	1	2	3	4
0	0	0	0	0	0
1	1	0	0	0	0
2	2	2	0	2	0
3	3	3	3	0	0
4	4	4	4	4	0

o	0	1	2	3	4
0	0	1	2	3	4
1	1	1	2	3	4
2	2	2	2	3	4
3	3	3	3	3	4
4	4	4	4	4	4

57.

*	0	1	2	3	4
0	0	0	0	0	0
1	1	0	0	1	0
2	2	2	0	2	0
3	3	3	3	0	0
4	4	4	4	4	0

o	0	1	2	3	4
0	0	1	2	3	4
1	1	1	2	3	4
2	2	2	2	3	4
3	3	3	3	3	4
4	4	4	4	4	4

61.

*	0	1	2	3	4
0	0	0	0	0	0
1	1	0	1	0	0
2	2	2	0	0	0
3	3	3	3	0	0
4	4	4	4	4	0

o	0	1	2	3	4
0	0	1	2	3	4
1	1	1	2	3	4
2	2	2	2	3	4
3	3	3	3	3	4
4	4	4	4	4	4

67.

*	0	1	2	3	4
0	0	0	0	0	0
1	1	0	0	0	0
2	2	2	0	0	0
3	3	2	1	0	0
4	4	4	4	4	0

o	0	1	2	3	4
0	0	1	2	3	4
1	1	1	3	3	4
2	2	3	3	3	4
3	3	3	3	3	4
4	4	4	4	4	4

70.

*	0	1	2	3	4
0	0	0	0	0	0
1	1	0	0	0	0
2	2	2	0	0	0
3	3	3	2	0	0
4	4	4	4	4	0

o	0	1	2	3	4
0	0	1	2	3	4
1	1	1	2	3	4
2	2	2	3	3	4
3	3	3	3	3	4
4	4	4	4	4	4

75.

*	0	1	2	3	4		o	0	1	2	3	4
0	0	0	0	0	0		0	0	1	2	3	4
1	1	0	1	0	0		1	1	1	3	3	4
2	2	2	0	0	0		2	2	3	3	3	4
3	3	2	1	0	0		3	3	3	3	3	4
4	4	4	4	4	0		4	4	4	4	4	4

77.

*	0	1	2	3	4		o	0	1	2	3	4
0	0	0	0	0	0		0	0	1	2	3	4
1	1	0	0	0	0		1	1	3	3	3	4
2	2	1	0	0	0		2	2	3	3	3	4
3	3	1	1	0	0		3	3	3	3	3	4
4	4	4	4	4	0		4	4	4	4	4	4

83.

*	0	1	2	3	4		o	0	1	2	3	4
0	0	0	0	0	0		0	0	1	2	3	4
1	1	0	0	0	0		1	1	2	2	3	4
2	2	1	0	0	0		2	2	2	3	3	4
3	3	2	1	0	0		3	3	3	3	3	4
4	4	4	4	4	0		4	4	4	4	4	4

85.

*	0	1	2	3	4		o	0	1	2	3	4
0	0	0	0	0	0		0	0	1	2	3	4
1	1	0	0	0	0		1	1	4	4	4	4
2	2	1	0	1	0		2	2	4	4	4	4
3	3	1	1	0	0		3	3	4	4	4	4
4	4	4	4	4	0		4	4	4	4	4	4

BCK with condition (S) of order 3, 4 and 5 are completely described. Then N is a commutative BCK and it has the associated semigroup. In this semigroup, o is just the addition $+$ on N. From this result, we know that there exists a finite BCK with condition (S) of any finite order, and there exists a countable infinite BCK with condition (S), but all finite sub BCK have the condition (S).

Appendix

In this section, we give some results on a proper BCI. A simple example of BCI is with two elements *0, 1* in which the multiplication is given by the following:

$$x * x = 0, \ \ 0 * 1 = 1 * 0 = 1.$$

A generalization of both concepts of BCK, BCI was introduced by A. Ursini [9]. The system is called a subtractive algebra. It has a constant *0* and a binary operation *. We assume two axioms, namely $x * 0 = x$ and $x * x = 0$. Therefore the classes of substrative algebras makes a variety.

Let *X* be a BCK, and let $a \notin X$. For any element $x \in X$, consider $x * a$. Suppose $x * a$ is an element of *X*, namely $y = x * a$, and $y \in X$. If $X \cup \{a\}$ is a BCI, then $y = x * a$ implies $x * y \leq a$, since the permutation rule holds in any BCI. This means $a \in X$, which is impossible. So we have $x * a = a$ for every element $x \in X$.

We give a binary operation * on $Z = X \cup \{a\}$ by the following way: * operation on *X* is defined by the original BCK. For any $x \in X$ we define $x * a = a * x = a$, and $a * a = 0$. Under this definition, *Z* is a proper BCI. From this very simple proper BCI, we have some important results:

(1) There exists a proper BCI with any cardinality,

(2) If *X* is a BCK with condition (S), then $Z(=X \cup \{a\})$ has a BCI with condition (S).

(3) There exists a BCI which has the condition (S).

References

[1] M. Abe and K. Iseki, A Survey on BCK and BCI –Algebras, Anais do I Congresso de Lógica Aplicada à Tecnologia, LAPTEC'2000, 431-443, São Paulo.

[2] W. H. Cornish, BCK-Algebras with a Supremum, Math. Japonica, 27, 1981, 63-73.

[3] W. H. Cornish, BCK-Algebras with a Supremum II. Distributivity and Interpolations, Math. Japonica, 29, 1982, 439-447.

[4] K. Iseki, A Special Class of BCK-Algebras, Math. Seminar Notes, 5, 1977, 101-198

[5] K. Iseki, BCK-Algebras with Condition (S), Math. Japonica, 24, 1979.

[6] K. Iseki, On BCI-Algebras with Condition (S), Math. Seminar Notes, 8, 1980, 171-172

[7] K. Iseki, A way to BCK and related systems, Math. Japonica, 52, 2000, 163-170.

[8] J. Meng and Y. B. Jun, BCK-Algebras, Kyung Moon Ss. Co., 1994.

[9] A. Ursini, On Subtractive Algebras, I, Algebra Universalis, 31, 1994, 204-222.

Logic, Artificial Intelligence and Robotics
J.M. Abe & J.I. da Silva Filho (Eds.)
IOS Press, 2001

Some Fundamental Theorems on BCK

Dedicated to Professor Gabriel Thierrin on the occasion
of his 80th birthday.

Kiyoshi ISÉKI
Emeritus Professor of Kobe University
&
Emeritus Professor of Naruto University of Education

Abstract. In this paper, we present some new results on BCK, in particular, we are mainly concerned with finite BCK.

In this note, we present some new results on BCK, in particular, we are mainly concerned with finite BCK.

First, I give a new definition of BCK which is equivalent to an old definition (for example, see [1], [5], [7], [8]). Let P be a set with partial order $\leq$. We assume P has a least element 0, and a binary operation $*$ is defined on P. The partial order $x \leq y$ is reflexive, antisymetric and transitive.

We say that P is a BCK if a binary operation $*$ on P satisfies the following conditions:

BCK 1) $(x * y) * (x * z) \leq z * y$,
BCK 2) $x * (x * y) \leq y$,
BCK 3) $x * x = 0$,
BCK 4) $0 * x = 0$,
BCK 5) $x * y = 0$ is equivalent to $x \leq y$.

We know some special classes of BCK which are defined by the following way: Let X be a BCK.

(i) X is *positive implicative*, if $(x * y) * y = x * y$ for any x, y in X.
(ii) X is *commutative*, if $x * (x * y) = y * (y * x)$ for any x, y in X. Then $x * (x * y)$ gives the greatest lower bound of x,y. Let us denote it by $x \wedge y$. This operation defines a lower semi-lattice on x.
(iii) X is *implicative*, if X is positive implicative and commutative.
(iii) X is a BCK *with condition (S)*, if for any $a,b \in X$, non-empty set

$$\{ x : x * a \leq b \}$$

has the greatest element in X. Its element is denoted by aob.

Then X is a partially ordered commutative semigroup with respect to the just introduced operation o. This is refered as the *associated* (p.o.commutative) *semigroup* of X.

The operation o has the following basic properties:

6) $x,y \leq xoy = yox$,
7) $xo(yoz) = (xoy)oz$,

8) $x * (y * z) = x * (yoz)$,

9) $x \leq y \to xoz \leq yoz$ for any $z \in X$.

(v) Let X be a BCK with condition (S). X is a BCK *with supremum*, if $xo(y * x) = yo(x * y)$ for any x,y in X.

A supremum $x \vee y$ of x,y is defined by $xo(y * x)$. The operation $x \vee y$ gives an upper semilattice structure on X.

This concept was introduced by W.H. Cornish ([2], [3]).

We can find many basic and useful properties of these classes in [7] amd [8]. Among them, we only mention some results.

10) $x * 0 = x$,

11) $(x * y) * z = (x * z) * y$. (permutation rule).

12) $x * [x * (x * y)] = x * y$.

Remark 1. The class of all BCK structures is not a variety, so I do not use the terminology "BCK algebra". I only refer to it as a *BCK*.

A system I satisfying BCK 1)-3), BCK 5) and 'there is no element smaller than 0' (in symbol, $x \leq 0$ implies $x = 0$) is called *BCI*. Then 0 is not necessarily the least element in I.

Theorem 1. There exists at least one BCK structure on any partially ordered set with a least element 0. This BCK structure is given by the following way: $x * y = 0$, if $x \leq y$ and $x * y = x$, otherwise.

Proof: BCK 3) and BCK 4) are obvious from the definition of $x * y$. To prove BCK 2), let us consider the two cases: if $x * y = 0$, then $x \leq y$, and

$$x * (x * y) = x * 0 = x \leq y.$$

If $x * y = x$, then

$$x * (x * y) = x * x = 0 \leq y.$$

So, we have BCK 2). Finally, we prove BCK 1). First let us consider the case $x * y = 0$. Then

$$(x * y) * (x * z) = 0 * (x * z) = 0 \leq z * y.$$

Next let us consider the case: $x * y = x$. Then $x * z = x$, or $x * z = 0$. Hence, if $x * z = x$, then we have

$$(x * y) * (x * z) = x * x = 0 \leq z * y.$$

If $x * z = 0$, then $x \leq z$ and

$$(x * y) * (x * z) = x * 0 = x.$$

On the other hand, $z * y = 0$ or $z * y = z$. If $z * y = 0$, then $z \leq y$, which implies $x \leq y$. This is impossible. If $z * y = z$, then $x \leq z$ holds, so BCK 1) is true. This completes the proof.

Consequently this structure is positive implicative, as easily seen. Then we can not change $x(= x * y)$ into z greater than x. In this sense, the obtained BCK is *maximal*, namely if there is a new BCK structure on X, then $x * y$ can not be greater than $x(= x * y)$ in the first given BCK structure.

Theorem 2. There exists a maximal BCK-structure or any partially ordered set with a least element 0.

Proof: We show that the BCK-structure introduced in Theorem 1 is maximal.

If $x * y = 0$, this 0 can not be replaced by a non-zero element. If the replacement is possible, $x \leq y$ does not hold.

Suppose that $x * y = z$, $x < z$. Then

$$(x * y) * x = z * x \neq 0.$$

On the other hand, $(x * y) * x = (x * x) * y = 0 * y = 0$, which is impossible. Similarly, we know x, z are comparable. This complete the proof.

To define another important class of BCK, we introduce a simple concept on a partially ordered set.

Definition 1. A partially ordered set P with a least element 0 is called the *type Y* if there exists an element a $(\neq 0)$ such that $a \leq x$ for every non-zero element x in P. Let us denote by 1 such element a.

A very simple BCK is a fan shaped one which has two different kinds. We define two kinds of fans, namely Japanese style and Chinese style. One of them is of the type Y.

Definition 2. A partially ordered set P with a least element 0 is of a *Japanese (style) fan* if it is of type Y and for any a $(\neq 0, 1)$ in P, there are no elements x, y satisfying $1 < x < a$, $a < y$. P is called *Chinese (style) fan* if it is not a type Y and for any a $(\neq 0)$ there are no elements x,y such that $0 < x < a$, $a < y$.

In the theory of BCK, the distinctions are quite important. It is very easy to introduce at least one BCK (maximal) structure by applying Theorem 1. There exists no other BCK (maximal) structure on any Chinese style fan and all subsets including 0 are subalgebras. On the other, any Japanese style fan has some other BCK structures.

Problem 1. Find all (finite and infinite) BCK with only one BCK structure. Are there none in the finite BCK with only one structure except the Chinese style one? The answer may depend on the order.

Example 1. Let us consider the Japanese style fan of order 5 as an example. The *-table of the of the maximal BCK is given in the following left tables. In this case, there exist more three different BCK structures on it.

*	0	1	2	3	4		*	0	1	2	3	4
0	0	0	0	0	0		0	0	0	0	0	0
1	1	0	0	0	0		1	1	0	0	0	0
2	2	2	0	2	2		2	2	1	0	1	1
3	3	3	3	0	3		3	3	1	1	0	1
4	4	4	4	4	0		4	4	1	1	1	0

The above right structure is commutative. Other two structures are mentioned in the following tables, but it seems that no these structures have special kinds of properties.

*	0	1	2	3	4		*	0	1	2	3	4
0	0	0	0	0	0		0	0	0	0	0	0
1	1	0	0	0	0		1	1	0	0	0	0
2	2	1	0	1	1		2	2	1	0	1	0
3	3	3	3	0	3		3	3	1	1	0	1
4	4	4	4	4	0		4	4	4	4	4	0

All elements in a BCK X appear in the column containing 0 in the *-table X. All elements which appear in the diagonal of the *-table are 0. Let us consider the part of the triangle of the left side of the diagonal, and eliminate the first column (the 0-column). For example, from the above first two *-tables, we have the following parts:

$$
\begin{array}{ccc}
2 & & \\
3 & 3 & \\
4 & 4 & 4
\end{array}
\qquad\qquad
\begin{array}{ccc}
1 & & \\
1 & 1 & \\
1 & 1 & 1
\end{array}
$$

Such a triangle is called the *essential part* of X. Similarly, we can define the quasiessential part of X. Let us consider the triangle formed through the right side of the diagonal. We eliminate the first row (*0*-row) of the triangle. The triangle obtained is called the *quasiessential part* of X.

The quasiessential parts of the first two in Example 1 are as follows:

$$
\begin{array}{ccc}
0 & 0 & 0 \\
  & 2 & 2 \\
  &   & 3
\end{array}
\qquad\qquad
\begin{array}{ccc}
0 & 0 & 0 \\
  & 1 & 1 \\
  &   & 1
\end{array}
$$

For any BCK linearly ordered, all elements of its quasiessential part are *0*, as easily seen.

Definition 3. A BCK is called *minimal*, if all elements of the essential and quasiessential parts are *0* or *1*. The above Japanese style fan has a minimal structure, but for a Chinese style fan, it is not always true. Then we have the following important

Theorem 3. A BCK is the type Y if and only if it is minimal.

Proof. Let X be a partially ordered set with a least element *0*. We assume X is of type Y. $x * y$ $(x,y \in X)$ is defined as follows:

Of course, if $x \leq y$, then we define $x * y = 0$. For other cases, define $x * y$ as being x, if $y = 0$ and define $x * y$ as being *1*, if $y \neq 0$, and x is not less or equal than y. We check that $x * y$ gives a BCK-structure on X.

It is obvious that BCK 3)-5) hold in X. To prove BCK 2), we consider two cases: i) $x * y = 0$. Then $x \leq y$, so we have

$$x * (x * y) = x * 0 = x \leq y.$$

ii) $x * y = 1$. Then $y = 0$ or $y = 1$, or $1 \leq y$. If $y = 0$, then $x = 1$ and BCK 2) holds. For the other two cases, since $x * (x * y) \leq 1$, BCK 2) also holds. Finally for BCK 1), if $x * y = 0$ or $x * y = 1, x * z = 1$, then BCK 1) is true. Let $x * y = 1, x * z = 0$. Then $x \leq z$ and

$$(x * y) * (x * z) = 1 * 0 = 1.$$

If $z * y = 0$, then $z \leq y$. Hence $x \leq y$, which is impossible. So $z * y = 1$. Conversely, a BCK X with a minimal structure is of type Y. To prove this, we suppose that there exists another branch starting from *0*. Let $a\ (\neq 0\ )$ be an element of the branch. Then by the minimality, $a * 1 = 1 * a = 1$. Hence

$$(a * 1) * a = 1 * a = 1,$$

but by permutation rule, we have $(a * 1) * a = (a * a) * 1 = 0 * a = 0$, which is a contradiction. This completes the proof.

Hence any linear (finite or infinite) BCK has a minimal structure. The minimal linear BCK of order 3 is commutative, but such a BCK of order 4 is not commutative.

The essential parts of the later two structures are

$$
\begin{array}{ccc}
1 & & \\
3 & 3 & \\
4 & 4 & 4
\end{array}
\qquad\qquad
\begin{array}{ccc}
1 & & \\
1 & 1 & \\
4 & 4 & 4
\end{array}
$$

The first element of these essential parts are *1*. They are smaller than the first element *3* of the maximal structure, and greater than the first element *1* of the minimal structure. There are the same situations among the corresponding elements. The same fact holds for the quasiessential part. Therefore we can say that each structure is between the maximal and minimal structures (this is valid for a non-minimal case). For every finite BCK, we can theoretically find all BCK structures, but the calculation is not easy. This is very tedious work.

Problem 2. Find an algorithm to determine all finite BCK structures. Is there a Turing machine to describe the structures of all finite BCK? Next we consider a factor (branch) of a partially ordered set P with a least element 0.

Definition 4. Each member of family $\{P_n\}$ of subsets of P is called *factor (branch)* if it satisfies the following conditions:

(F-1) Each P_n is a partially ordered set with 0,
(F-2) P is the union of $\{P_n\}$,
(F-3) $P_m \cap P_n = \{0\}$ for $m\ (\neq n)$.

Then each P_n is a BCK, and for $x \in P_m$, $y \in P_n$ $(m \neq n)$ we have $x * y = x$. Conversely, let $\{P_n\}$ be a family of partially ordered sets. Let us suppose that the least element 0 is common in all P_n, and $\{P_n\}$ is a disjoint family except the element 0. Moreover, we suppose each of the $\{P_n\}$ has a BCK structure. Then the following result holds true.

Theorem 1. On the union P of P_n, a BCK structure is uniquely introduced, and each P_n is a subalgebra under the structure on P. For $x \in P_m$ and $y \in (m \neq n)$, $x * y = x$.
Proof. Under the above definition, BCK 3)-5) hold in P. If x,y are in the same branch, then BCK 2) holds. If x,y belong to distinct branches, $x * y = x$, then for any element y in P,

$$
x * (x * y) = x * x = 0 \leq y.
$$

Hence BCK 2) holds in P. To very BCK 1), three cases are considered, namely, for $m \neq n$,
1) $x,y \in P_m$, $z \in P_n$. 2) $x,z \in P_m$, $y \in P_n$. 3) $y,z \in P_m$, $x \in P_n$.
1) Let $x,y \in P_m$, $z \in P_n$. Then by $x * y \leq x$,

$$
(x * y) * (x * z) = (x * y) * x = 0 \leq z * y.
$$

2) Let $x,z \in P_m$, $y \in P_n$. Then

$$
(x * y) * (x * z) = x * (x * z)\, x * x = 0 \leq z * y.
$$

3) Let $y,z \in P_m$, $x \in P_n$. Then

$$
(x * y) * (x * z) = x * x = 0 \leq z * y.
$$

Therefore, BCK holds.
Conversely, we prove $x * y = x$ for $x \in P_m$, and $y \in P_n$ $(m \neq n)$. Since $x * y \leq x$, $x * y$ belongs to P_n. From $x * (x * y) \leq x$, $x * (x * y) \in P_m$. On the other hand, by BCK 2), $x * (x * y) \leq y$, so $x * (x * y) \leq y$. $y \in P_n$ implies $x * (x * y)$ belongs to P_n. Consequently, $x * (x * y) \in P_m \cap P_n = \{0\}$. Therefore we have $x * (x * y) = 0$. Hence $x \leq x * y$. Since $x * y \leq x$ holds, we obtain $x = x * y$. This completes the proof.

It follows from the proof that the BCK structure of each P_n is preserved in P. Theorem 4 is very much useful to construct various kinds of special BCK. For example, the union of several two elements BCK is a Chinese style fan and the BCK structure is given by the above equation. The uniqueness of the BCK structure on a Chinese style fan also follow from Theorem 4.

Moreover, the BCK structure of each branch is positive implicative (or commutative), if and only if P is positive implicative (or commutative). It follows from $(x * y) * y = x * y$ for $x \in P_m$, $y \in P_n (m \neq n)$.

On the other hand, we can easily construct many interesting commutative BCK using Theorem 4.

References

[1] M. Abe and K. Iseki, A Survey on BCK and BCI –Algebras, Anais do I Congresso de Lógica Aplicada à Tecnologia, LAPTEC'2000, 431-443.

[2] W. H. Cornish, BCK-Algebras with a Supremum, Math. Japonica, 27, 1981, 63-73.

[3] W. H. Cornish, BCK-Algebras with a Supremum II. Distributivity and Interpolations, Math. Japonica, 29, 1982, 439-447.

[4] J. M. Harper and J. E. Rubin, Variations of Zorn's Lemma, Principles of cofinality, and Hausdorff's maximal principle I, II, Notre Dame Journal of Formal Logic, 17 , 1977, 161-163.

[5] K. Iseki, A Way to BCK and Related Systems, Math. Japonica, 52, 2000, 163-170.

[6] K. Iseki, On Finite BCK with condition (S), Proc. Algebra, Languages and Computation, 2000, 16-25.

[7] K. Iseki and S. Tanaka, An Introduction to the Theory of BCK-Algebras, Math. Japonica, 23, 1978, 1-21.

[8] J. Meng and Y. B. Jun, BCK-Algebras, Kyung Moon Ss. Co., 1994.

Logic, Artificial Intelligence and Robotics
J.M. Abe & J.I. da Silva Filho (Eds.)
IOS Press, 2001

Note on the structure of weak interlaced bilattice $\mathcal{K}(\mathrm{L})$

Michiro KONDO

Department of Mathematics and Computer Science
Shimane University, Matsue, 690-8504, Japan
e-mail:kondo@cis.shimane-u.ac.jp

Abstract

We study fundamental properties of weak interlaced bilattices $\mathcal{K}(L)$
and show that for any weak interlaced bilattice $\mathcal{W}$ there exists a lattice L such that $\mathcal{W}$ can be embedded into a weak interlaced bilattice
$\mathcal{K}(L)$. Hence, any interlaced bilattice can be embedded into the weak
interlaced bilattice $\mathcal{K}(L)$ for some lattice L.

1 Introduction

It is well-known the Kleene's 3-valued logic in the field of multiple-valued logics. The
logic has three values *false, true*, and $\perp$ (*unknown*) as truth values. These values have
two informal orderings concerning "amount of knowledge" and "degree of truth". For
example, if we think of a certain proposition such as *Goldbach's conjecture* assigned $\perp$
as truth value, then it is possible that we can conclude the truth value of the proposition
as *true* or *false* with increasing knowledge. Thus in the ordering of knowledge, $\perp$ is
smaller than *true* and *false*. A sentence with $\perp$ is between *false* and *true* in the ordering
of degree of truth. In this way it can be considered that the three valued logic has two
orderings. Belnap ([2]), Ginsberg([5]), and others proposed concept of a *bilattice* which
has two orderings and proved some fundamental results ([1, 3, 4]). It is shown by
Fitting ([3]) that bilattices can give a uniform semantics for many lanuages of logic
programming. Since then the theory of bilattices is a hot reserach field.

On the other hand, as in *Fuzzy logics*, a truth value can be taken as a closed interval
$[a, b]$. Let L be a lattice and $\mathcal{K}(L)$ be the set of all closed intervals of L. In this case we
also define two orderings. For $[a, b], [c, d] \in \mathcal{K}(L)$, if $[a, b] \subseteq [c, d]$ then the knowledge in
$[a, b]$ is greater than that in $[c, d]$. Thus we set $[a, b] \sqsubseteq_k [c, d]$ if $[a, b] \subseteq [c, d]$. Likewise
we also define $[a, b] \sqsubseteq_t [c, d]$ if $a \leq c$ and $b \leq d$, because $[c, d]$ is greater than $[a, b]$
in the ordering degree of truth. The structure $\mathcal{K}(L) = <\mathcal{K}(L), \sqsubseteq_t, \sqsubseteq_k>$ which precise
definition is given below has the property of *weak interlaced bilattice*.

In [3, 4], Fitting, Font and Moussavi have investigated the strucutre of $\mathcal{K}(L)$ and
proved that if L is a bounded lattice, then $\mathcal{K}(L)$ is a weak interlaced bilattice ([4]).

Now does the converse hold?, that is, is there a lattice L such that $\mathcal{W} \cong \mathcal{K}(L)$ for every weak interlaced bilattice $\mathcal{W}$?

Clearly we answer "No". Because we have a simple counterexample. Let $\mathcal{B}$ be a weak interlaced bilattice with 5 elements, for example, a set $\{0, p, \bot, q, 1\}$ with $0 \leq_t p \leq_t \bot \leq_t q \leq_t 1$, $\bot \leq_k p \leq_k 0$ and $\bot \leq_k q \leq_k 1$. It is obvious that $\mathcal{B}$ is a weak interlaced bilattice. Suppose that there is a lattice L such that $\mathcal{B} \cong \mathcal{K}(L)$. If $|L| \geq 3$, then there exists an element $a \in L$ such that $0 < a < 1$. For that element we have $[0,0], [0,a], [0,1], [a,1], [a,a], [1,1] \in \mathcal{K}(L)$ and $|\mathcal{K}(L)| \geq 6$. Since $|\mathcal{B}| = 5$, it must be $|L| \leq 2$. But, in this case, we have $|\mathcal{K}(L)| \leq 3$. This means that there is no lattice L such that $\mathcal{B} \cong \mathcal{K}(L)$.

Now we settle a more general question.

> **Question** : Is there a lattice L such that any weak interlaced bilattice $\mathcal{W}$ can be embedded to $\mathcal{K}(L)$?

In this note we study properties of $\mathcal{K}(L)$ and answer the question.

2 Definition of $\mathcal{K}(L)$

We define a structure $\mathcal{K}(L)$ for any lattice L. Let $L = (L, \leq)$ be a lattice and $K(L)$ be the set of all closed intervals of L, that is,

$$K(L) = \{[a,b] \mid a \leq b, a, b \in L\}$$
$$[a,b] = \{x \mid a \leq x \leq b\}.$$

For any $[a,b], [c,d] \in K(L)$, we define two orderings $\sqsubseteq_t, \sqsubseteq_k$ on $K(L)$ as follows :

$$[a,b] \sqsubseteq_t [c,d] \iff a \leq c, b \leq d$$
$$[a,b] \sqsubseteq_k [c,d] \iff a \leq c, b \geq d$$

We set $\mathcal{K}(L) = <K(L), \sqsubseteq_t, \sqsubseteq_k>$. It is obvious from definition that $[0,0]$ ($[1,1]$) is the minimum (maximum) element with respect to $\sqsubseteq_t$. On the other hand, while $[0,1]$ is the minimum element, there is no maximum element with respect to the ordering $\sqsubseteq_k$. This means that $\mathcal{K}(L)$ is a lattice with respect to $\sqsubseteq_t$ and is a semi-lattice concering $\sqsubseteq_k$. Four operators $\sqcap_t, \sqcup_t, \sqcap_k, \sqcup_k$ are defined by

$$\inf\nolimits_{\sqsubseteq_t}\{a,b\} = a \sqcap_t b$$
$$\sup\nolimits_{\sqsubseteq_t}\{a,b\} = a \sqcup_t b$$
$$\inf\nolimits_{\sqsubseteq_k}\{a,b\} = a \sqcap_k b$$
$$\sup\nolimits_{\sqsubseteq_k}\{a,b\} = a \sqcap_k b \quad \text{(if it is defined)}$$

Next we give definitions of an interlaced bilattice and of a weak interlaced bilattice. A relational system $<B, \leq_t, \leq_k>$ is called an *interlaced bilattice* if it satisfies

1. B is a non-empty set

2. $<B, \leq_t>, <B, \leq_k>$ are bounded lattices and satisfy

(a) $x \leq_t y \implies x \otimes z \leq_t y \otimes z,\ x \oplus z \leq_t y \oplus z$

(b) $x \leq_k y \implies x \wedge z \leq_k y \wedge z,\ x \vee z \leq_k y \vee z$

where four operators are defined by

$$\inf_{\leq_t}\{x, y\} = x \wedge y$$
$$\sup_{\leq_t}\{x, y\} = x \vee y$$
$$\inf_{\leq_k}\{x, y\} = x \otimes y$$
$$\sup_{\leq_k}\{x, y\} = x \oplus y$$

By $0(1)$, we mean the minimum (maximum) element with respect to the ordering $\leq_t$. We also denote by $\perp(\top)$ the minimum (maximum) element concering to $\leq_k$.

A map $\neg$ from B into itself is called a *negation* if

$x \leq_t y \implies \neg y \leq_t \neg x$

$x \leq_k y \implies \neg x \leq_k \neg y$

$\neg\neg x = x.$

For lattices $L_1 =< L_1, \wedge_1, \vee_1 >$ and $L_2 =< L_2, \wedge_2, \vee_2 >$, we define operations $\wedge, \vee, \otimes, \oplus$ on the product $L_1 \times L_2$: For $(a, b), (c, d) \in L_1 \times L_2$,

$$(a, b) \wedge (c, d) = (a \wedge_1 c, b \vee_2 d)$$
$$(a, b) \vee (c, d) = (a \vee_1 c, b \wedge_2 d)$$
$$(a, b) \otimes (c, d) = (a \wedge_1 c, b \wedge_2 d)$$
$$(a, b) \oplus (c, d) = (a \vee_1 c, b \vee_2 d).$$

The structure $L_1 \odot L_2 =< L_1 \times L_2, \wedge, \vee, \otimes, \oplus >$ is called a *Ginsberg product*. There are some fundamental results about the structure :

Proposition 1 (Fitting). *If L_1, L_2 are bounded lattices then the Ginsberg product $L_1 \odot L_2 =< L_1 \times L_2, \wedge, \vee, \otimes, \oplus >$ is an interlaced bilattice. Espectially, $L \odot L$ is an interlaced bilattice with negation $\neg$, where $\neg$ is defined by $\neg(a, b) = (b, a)$.*

It is proved that the converse holds by Avron ([1]).

Proposition 2. *[Avron] For any interlaced bilattice $\mathcal{B}$, there are bounded lattices L_1, L_2 such that $\mathcal{B} \cong L_1 \odot L_2$. In particular, for any interlaced bilattice $\mathcal{B}$ with negation, there is a bounded lattice L such that $\mathcal{B} \cong L \odot L$.*

It is clear from definition that orderings $\sqsubseteq_t, \sqsubseteq_k$ on $\mathcal{K}(L)$ are the same as $\leq_t, \leq_k$ on Ginsberg product $L \odot L$, respectively :

$\sqsubseteq_t$ in $\mathcal{K}(L) \iff \leq_t$ in $L \odot L$

$\sqsubseteq_k$ in $\mathcal{K}(L) \iff \leq_k$ in $L \odot L$

Hence in the following we use the same symbols $\wedge, \vee, \otimes, \oplus$ in $\mathcal{K}(L)$ and in $L \odot L$.

Next we give a definition of a *weak interlaced bilattice* according to Font ([4]). A structure $\mathcal{W} =< W, \leq_t, \leq_k >$ is called a *weak interlaced bilattice* if

1. $< W, \leq_t >\cdot$ lattice

2. $< W, \leq_k >\cdot$ meet semilattice

3. $a \leq_k b, c \leq_k d \implies a \wedge c \leq_k b \wedge d, a \vee c \leq_k b \vee d$

4. $a \leq_t b, c \leq_t d \implies a \otimes c \leq_t b \otimes d,$

5. $a \leq_t b, c \leq_t d \implies a \oplus c \leq_t b \oplus d$ if $a \oplus c$ and $b \oplus d$ exist.

3   Properties of weak interlaced bilattices

For any weak interlaced bilattice $\mathcal{W}$, if we define

$$L_1 = \{x \in \mathcal{W} \mid x \leq_k 0\} = [\bot, 0]_k$$
$$L_2 = \{x \in \mathcal{W} \mid x \leq_k 1\} = [\bot, 1]_k,$$

then we have

Proposition 3.

$$L_1 = [\bot, 0]_k = [0, \bot]_t$$
$$L_2 = [\bot, 1]_k = [\bot, 1]_t$$

Proof. Let $x \in [\bot, 0]_k$. Since $\bot \leq_k x \leq_k 0$, we have $\bot \vee \bot \leq_k x \vee \bot \leq_k 0 \vee \bot$ by definition of weak interlaced bilattice. From $\bot \vee \bot = 0 \vee \bot = \bot$, it follows that $x \vee \bot = \bot$ and hence that $x \leq_t \bot$. This means $[\bot, 0]_k \subseteq [0, \bot]_t$.

Conversely, suppose $x \in [0, \bot]_t$. If we put $u = 0 \otimes x$, then it is clear that $u \leq_k 0$ and $u \leq_k x$. Since $0 \leq_t x$, we have $0 \otimes x \leq_t x \otimes x = x$ and hence $u \leq_t x$. It follows from $\bot \leq_k u$ that $x \wedge \bot \leq_k x \wedge u$. Since $x \leq_t \bot$, we also have $x \wedge \bot = x$. On the other hand, since $u \leq_t x$, we get $u \wedge x = u$. Theses imply that $x \leq_k u$ and hence that $x = u$. Thus we have $x \leq_k 0$. Namely, we have $[0, \bot]_t \subseteq [\bot, 0]_k$.

The second equation can be proved similarly.

$\square$

The result implies that L_1 and L_2 are lattices with ordering $\leq_1$ and $\leq_2$ in $\mathcal{B}$, respectively, where $\leq_1$ and $\leq_2$ are defined by

$$\leq_1 = \leq_t = \geq_k$$
$$\leq_2 = \leq_t = \leq_k$$

Thus we can consider the Ginsberg product $L_1 \odot L_2$, which becomes an *interlaced bilattice*. Moreover we can prove

Proposition 4. *Let $\mathcal{W}$ be any weak interlaced bilattice. For any $x \in \mathcal{W}$, we have*
$$x = (x \otimes 0) \oplus (x \otimes 1) = (x \wedge \bot) \vee (x \vee \bot)$$

Proof. See Avron [1] Cor.3.8

$\square$

Now we investigate a realtion between a weak interlaced bilattice $\mathcal{W}$ and an interlaced bilattice $L_1 \odot L_2$ constructed by $\mathcal{W}$.

Lemma 1. *A map $\xi : W \to L_1 \times L_2$ defined by $\xi(x) = (x \otimes 1, x \otimes 0) = (x \vee \perp, x \wedge \perp)$ is an embedding.*

This means that

Theorem 1. *Any weak interlaced bilattice can be embedded into an interlaced bilattice.*

4 Answer to the question

In this section we give a positive answer to the question above. Since any weak interlaced bilattice W can be embedded to an interlaced bilattice, it sufficies to show that any interlaced bilattice of a form $L_1 \odot L_2$ is embeddable into a weak interlaced bilattice $K(L)$ for some lattice L. Because, from proposition 2, every interlaced bilattice has a form of $L_1 \odot L_2$ for some lattices L_1, L_2. Let $L_1 \odot L_2$ be any interlaced bilattice and L be a set $(L_1 \times \{0\}) \cup (L_2 \times \{1\})$. We define an order $\sqsubseteq$ on L. For any element $(a, i), (b, j) \in L$, we define

$$(a, i) \sqsubseteq (b, j) \iff i < j \text{ or } i = j \text{ and } a \leq b$$

It is easy to show that the relation $\sqsubseteq$ is a partially order on L and that

$$(a, i) \wedge (b, j) = \inf\{(a, i), (b, j)\} = \begin{cases} (a \wedge b, i) & \text{if } i = j \\ (a, i) & \text{if } i < j \\ (b, j) & \text{if } i > j \end{cases}$$

$$(a, i) \vee (b, j) = \sup\{(a, i), (b, j)\} = \begin{cases} (a \vee b, i) & \text{if } i = j \\ (b, j) & \text{if } i < j \\ (a, i) & \text{if } i > j \end{cases}$$

Hence L is a lattice with this order. Let $K(L)$ be the set of all elements $[(a, i), (b, j)]$ such that $(a, i) \sqsubseteq (b, j)$ for $(a, i), (b, j) \in L$. In this case, four operators $\wedge, \vee, \otimes, \oplus$ on $K(L)$ are defined as follows:

$$[(a, i), (b, j)] \wedge [(a', i'), (b', j')] = [(a, i) \wedge (a', i'), (b, j) \wedge (b', j')]$$
$$[(a, i), (b, j)] \vee [(a', i'), (b', j')] = [(a, i) \vee (a', i'), (b, j) \vee (b', j')]$$
$$[(a, i), (b, j)] \otimes [(a', i'), (b', j')] = [(a, i) \wedge (a', i'), (b, j) \vee (b', j')]$$
$$[(a, i), (b, j)] \oplus [(a', i'), (b', j')] = [(a, i) \vee (a', i'), (b, j) \wedge (b', j')]$$

Of course, the last equation is defined when $(a, i) \vee (a', i') \leq (b, j) \wedge (b', j')$.
Now we define a map $\xi : L_1 \odot L_2 \to K(L)$ by

$$\xi(a, b) = [(a, 0), (b, 1)]$$

It is obvious that ξ is well-defined and injective. We only show that ξ is a homomorphism. We only think of two cases. For the case of $(a, b) \wedge (a', b')$, we have

$$\begin{aligned}
\xi((a,b) \wedge (a',b')) &= \xi(a \wedge a', b \vee b') \\
&= [(a \wedge a', 0), (b \vee b', 1)] \\
&= [(a,0) \wedge (a',0), (b,1) \vee (b',1)] \\
&= [(a,0), (b,1)] \otimes [(a',0), (b',1)] \\
&= \xi(a,b) \otimes \xi(a',b')
\end{aligned}$$

For another case of $(a,b) \oplus (a',b')$, we also have

$$\begin{aligned}
\xi((a,b) \oplus (a',b')) &= \xi(a \vee a', b \vee b') \\
&= [(a \vee a', 0), (b \vee b', 1)] \\
&= [(a,0) \vee (a',0), (b,1) \vee (b',1)] \\
&= [(a,0), (b,1)] \vee [(a',0), (b',1)] \\
&= \xi(a,b) \vee \xi(a',b')
\end{aligned}$$

Hence the map $\xi : L_1 \odot L_2 \to \mathcal{K}(L)$ is an embedding, that is,

Theorem 2. *For every interlaced bilattice $L_1 \odot L_2$, there exists a lattice L such that it is embedded into a weak interlaced bilattice $\mathcal{K}(L)$.*

From these results, we have have a main theorem.

Theorem 3. *Every interlaced bilattice $\mathcal{W}$ can be embedded into a weak interlaced bilattice $\mathcal{K}(L)$ for some lattice L.*

References

[1] A.Avron, The structure of interlaced bilattices, *Math. Struct. in Comp. Science*, **6** (1996) 287-299.

[2] N.D.Belnap, Jr., A useful four-valued logic, In : J.M.Dunn and G.Epstein (eds), Modern Uses of Multiple-Valued Logic. D.Reidel, 1977.

[3] M. Fitting, Kleene's Logic, generalized, *Jour.Logic Comput.*, **1** (1991) 77-810.

[4] J.M.Font and M.Moussavi, Notes on a six-valued extension of three-valued logic, *Jour. of Applied Non-Classical Logic*, **3** (1993) 173-187.

[5] M.Ginsberg, Multivalued logic : a uniform approach to reasoning in artificial intelligence, *Computational Intelligence*, **4** (1988) 265-316.

[6] M.Kondo, Filter theory of bilattices in the semantics of logic programming, *Far East Jour. of Mathematical Sciences*, **3** (2001) 177-189.

Logic, Artificial Intelligence and Robotics
J.M. Abe & J.I. da Silva Filho (Eds.)
IOS Press, 2001

135

Decision-Making System based on Fuzzy and Paraconsistent Logics

Germano Lambert-Torres   Cláudio Inácio de Almeida Costa   Helga Gonzaga
Escola Federal de Engenharia de Itajubá
Av. BPS 1303 – Itajuba – 37500-000 – MG – Brazil - germano@iee.efei.br

Abstract – The decision-making processes are getting each time more complex and the classic logic has not been offering the necessary support to contemplate all existing cases. A way for the present controllers is to set up simplifications or not consider some facts that may be important in the whole context. This paper presents a process controller based upon two-non classic logic's: the fuzzy and the paraconsistent. The proposed ideas were introduced into a computer program in order to make their test possible.

1. Introduction

Time after time happen in the real world inconsistent situations, undefined, uncertain knowledge or uncompleted knowledge.

As an example for the first two cases it is possible to have superfluous information supplied by two sensors, which make the supervision upon a same system. This frequently happen in very important circuits as the ones for heavy water of atomic reactors. Some of them have three sensors for the measuring of a same value, based on three different physic principles. Let us suppose that two sensors indicate high temperature, by a numeric value, for example, 220°C, indicating that the temperature is really high and equal to the indicated value. However, if one sensor indicates high temperature (for example 220 °C) and the other indicate low temperature (for example, -50°C) it makes a case of inconsistency. Furthermore, if two sensors are having no communication with the supervisory system, the conveyed information is none, generating an undefined case. The computer programs based on classic logic's show problems in working with this type of information. More sophisticated systems, but having yet the classic logic's as a basis, try through state estimation programs, to conclude which is the right information. This is not an ordinary procedure, once depends upon system's topology and other available measures. It is not always possible to discover the right values, so this information is frequently disregarded by decision take system.

As an example for the last two cases, uncertain knowledge, and uncompleted knowledge, it is possible to present the linguistic values that have a large quantity of information, but that are not kept by the classic computer programs. For example, the weather information is subject to this kind of problem. Suppose that a specialist system has the following information: "if the environment temperature is low, then an overloaded transformer operation time may be large". This information is good for any electric system in the world. However, once the classic logic does not accept the linguistic values, it is required that numeric values be introduced into the rule, that will change to: "if the environment temperature be less than 15 °C, then an overloaded transformer operating time may be up to 1 hour", where 15°C as substituted the linguistic value "low", and "up to one hour" substituted the value "large". Let us suppose now that this system is helping the Canadian electric grid operation. The low temperature in that country starts in 0°C. Some of its transformers may operate for a large time of the year with the environment temperature rate at 15°C (or even though lower than that), once they are designed accordingly. That is, the rule that has the numeric values would not express well the transformer's operation knowledge; however, the rule with linguistic values would continue to be good, once for low temperature values, the overload time would be large.

Looking for a decision take structure that may deal with the four types of situations above, an inference and knowledge representation hybrid technology was developed, based upon fuzzy [1] and paraconsistent [2] logic's. This hybrid methodology was tested on a computer package developed before, which was used for teaching fuzzy logic [3]. This paper presents this hybrid methodology and how it was introduced into the computer package for the solution of one problem for which no adequate solution was known. Examples of some simulations made are shown.

2. Fuzzy Logic

The Fuzzy Sets Theory came from a Professor Lofti Zadeh's practical necessity for solving a US Army's problem. By the end of the 50's decade the US Army was looking for the development of an anti-missile system able for doing the interception of any missile aiming the US territory. For that purpose another missile would be used, comprising a sophisticated control system.

By that time Professor Zadeh was already known as a notorious investigator of this subject, and was assigned for finding the adequate solution for the anti-missile control. Beginning to work after to find a model for this type of system, the renowned professor realized the existence of an uncommon fact regarding the control systems design: the lack of numeric data.

In the meetings held with ballistics and weapons specialists he was always having vague information's such as: "if a missile be very loaded it will fly slowly, but if it carries a small explosives load and flying very high then it will fly faster". Doubts were in evaluating what was understood as slow or fast, heavy or light, and also high or low. These values change very much from person to person and their margins were not well defined.

So, the pertinent questions were: "how to develop a control system based only on such linguistic information?" and "would these information really be important for modeling the control?"

The second question may be answered in a more or less simple way. If an expert operator would be monitoring full time the possible invading missiles, he will succeed to detect and to hit them, and for that he would only use the rules that were told to Professor Zadeh. In other words, the "human" control system uses these rules for hitting the missile. Thus, this information is important and should be used in the construction of the control modules.

For answering the first question, Professor Zadeh fostered the Fuzzy Sets Theory, which allowed to represent mathematically linguistic values, and to handle them with traditional numeric values.

2.1 Fuzzy Sets Definition

The exact fuzzy set definition is as follow:
"Be E an universe set, denumerable or not , and be x one element of E. Thus a fuzzy set A from E is a set of ordered pairs in the form: $\{(x \mid \mu_A(x))\}, \forall\, x \in E$
where $\mu_A(x)$ is the membership function of x in A, being taken this value in a lattice set M".

The difference between the fuzzy set theory and the classic set theory (a particular case of fuzzy sets theory) is that the values of M are taken in the set $\{0,1\}$, that is to say, or one element x belongs to set A, $x \in A$, and hence $\mu_A(x)$ is equal to 1, or it do not belongs to A, $x \notin A$, thus $\mu_A(x)$ is equal to 0. In the fuzzy set theory, the set M can take several values.

Be, for example, one set E defined by { x_1 , x_2 , x_3 , x_4 , x_5 } and M taken in the interval [0,1], so it is possible to relate x_i with A as follow:

$$A = \{ (x_1 \mid 0) , (x_2 \mid 0,2), (x_3 \mid 0,5), (x_4 \mid 0,8), (x_5 \mid 1) \}$$

where:

x_1 belongs to set with degree 0, that is to say, it is not a member of A.,

x_2 belongs to the set with degree 0,2, that is to say, it is less member of A,

x_3 belongs to the set with degree 0,5, that is to say, it is "more or less" a member of A,

x_4 belongs to set with degree 0,8, is that to say, it is a quite member of A,

x_5 belongs to the set with degree 1, is that to say, it is totally a member of A.

Numberless examples of this type of set may be shown. One of those can be obtained if the set E be given by natural numbers in the interval between [0,10] and the set M defined between [0,1], so it is possible to define the following fuzzy sets:

a) small: { (0|1), (1|0,8), (2|0,6), (3|0,4), (4|0,2), (5|0), (6|0), (7|0),(8|0), (9|0), (10|0) }

b) medium: { (0|0), (1|0), (2|0,25), (3|0,5), (4|0,75), (5|1), (6|0,75), (7|0,5), (8|0,25), (9|0), (10|0) }

c) large: { (0|0), (1|0), (2|0), (3|0), (4|0), (5|0), (6|0,2), (7|0,4), (8|0,6), (9|0,8), (10|1) }

If it happen the association of the interval [0,10] with height, temperature or grades obtained by the students, it is possible to have the linguistic values: small, medium and large; as shown on Figure 1.

In above example, the charting set M was used in the interval [0,1]. The majority of the works under development uses this interval, and now and then is defined as the only one for the M values, as in the reference [4]. However, this is not true. Any lattice type set, that is to say, anyone that has a lattice structure, may represent the set M.

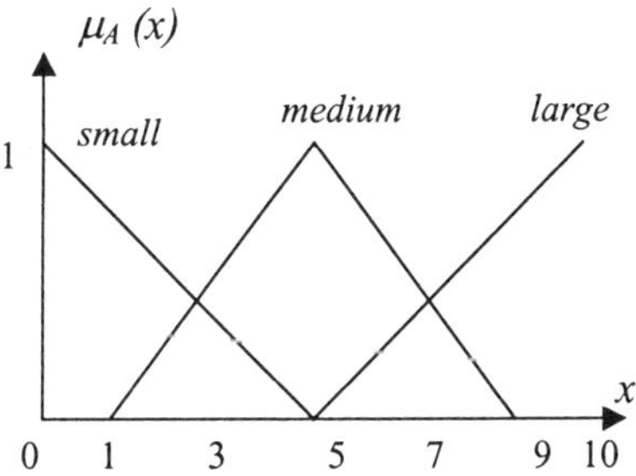
Figure 1 – Values of x: small, medium and large.

For one structure to be of the lattice type it has to met two main requirements:

$$\exists! \; x_k = x_i \, \Delta \, x_j \; e \; x_k \in M$$

$$\exists! \; x_m = x_i \, \nabla \, x_j \; e \; x_m \in M$$

is that to say, that for each pair of values always exist in the set M, an upper value , x_m, and a lower value, x_k, and that it be unique.

2.2 Heightening linguistics

The several adjectives may be yet modified by adverbs that can heighten or reduce its intensities. "Very small", "more or less small" and "not small" are examples of adverbs that may change the adjective small. To express knowledge people use them frequently and they have to be added to the Fuzzy Sets Theory.

There are several ways to represent these heightens, and one is to operate upon the original membership values, for example as per the rules bellow (Figure 2):

$$\mu \text{very small}(x) = [\mu \text{small}(x)]^2$$

$$\mu \text{more or less small}(x) = [\mu \text{small}(x)]^{1/2}$$

$$\mu \text{not small}(x) = 1 - \mu \text{small}(x)$$

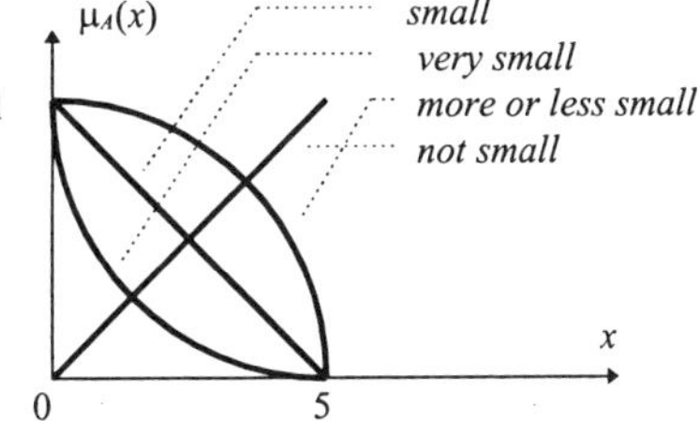
Figure 2 – Heightening linguistics in small values.

2.3 Operations and conjunctions

Same as in the Classic Sets Theory, the Fuzzy Sets Theory also has many operations between sets. These operations, besides having a mathematical meaning, also have a linguistic meaning. They represent the conjunctions that link the variable's values. The main operations (conjunctions) are:

2.3.1 Intersection
Definition: Be E the universe set and the fuzzy sets A and B linked to them. The intersection between $A \cap B$, is defined as being the minimum rate (MIN) between the many membership values, is that to say:
$$\forall\, x \in E : \mu_{A \cap B}(x) = \text{MIN}(\,\mu_A(x)\,,\,\mu_B(x)\,)$$
For linguistics purpose, this operation represents the conjunction e. For example, for the Figure 1 values bellow, the value *small E medium* (is that to say, the variable must answer to these two values at the same time), so it is possible to write in numbers:
Small and medium: { (0 |0) , (1 |0), (2 |0,25) , (3 |0,4), (4 |0,2) , (5 |0) , (6 |0) , (7 |0) , (8 |0) ,
(9 |0) , (10 |0) }

2.3.2 Union
Definition: Be E the universe set and the fuzzy sets A e B linked to them. The union between $A \cup B$ is defined as the maximum rate (MAX) between the many membership values, is that to say:
$$\forall\, x \in E : \mu_{A \cup B}(x) = \text{MAX}(\,\mu_A(x)\,,\,\mu_B(x)\,)$$
Linguistically, this operation represents the conjunction *or*. For example, for the values of Figure 1 bellow, the value *small OR medium* (is that to say, the variable should answer to at least one of these two values at the same time), thus numerically it is possible to write:
Small or medium: { (0 |1), (1 |0,8), (2 |0,6), (3 |0,5), (4 |0,75), (5 |1), (6 |0,75), (7 |0,5),
(8 |0,25), (9 |0), (10 |0) }

2.3.3 Disjunctive addition
Definition: Be E the universe set and the fuzzy sets linked to them A and B. The disjunctive addition between $A \oplus B$ is defined as being the following equation:
$$A \oplus B = (\,A \cap \neg B\,) \cup (\,\neg A \cap B\,)$$
Where $\neg A$ represents not A.

Linguistically, this operation represents the conjunction *or exclusive*. For example, for the values bellow from Figure 1, the value *small OR (exclusive) medium* (that is to say, the variable must answer to but one of the two values at the same time), so numerically it is possible to write:
Small or exclusive medium: { (0 |1), (1 |0,8), (2 |0,6), (3 |0,5), (4 |0,75), (5 |1), (6 |0,75),
(7 |0,5), (8 |0,25), (9 |0), (10 |0) }

Notes:
1. There are many other operations, such as: algebraic difference, product and addition and distances (how of Hamming and Euclidean) that will not be presented, once they will not be used in this paper.
2. The presented operations are also wholly valid in the Classic Sets Theory.

2.4 Rules and Mapping

Having above elements, it is possible to write complex affirmatives that may represent the real world situations. Through fuzzy logic, it is possible to write the control actions, by using the terms of these imprecise ideas, which comprise the variable's state. As an example, in the ABS brake system it is possible to write:

IF *vehicle velocity is **low** **AND** its location is **near the wall*** **THEN** *brake pressure* be
medium.

In one conventional PI this rule should be far more specific.
IF *vehicle velocity **lower than 20 km/h** **AND** its location **lower than 5 (m)***
THEN *brake pressure be **20 bar.***

Because the fuzzy rules act over a wide range, usually being more comprehensive than the conventional, the number of fuzzy rules that represents a determined control is smaller than with conventional rules.

In mathematical terms, a production rule may be considered as the charting between different universe sets. Being E_1, E_2 and E_3 the universe sets describing the possible values (or states) for the variables x, y and z, respectively. If in the rule's premises are founded fuzzy sets from E_1 and E_2 and in the consequence there are fuzzy sets from E_3, it is possible to denote the relation by:

$$E_1 \times E_2 \sim_\Gamma \to E_3$$

The charting Γ allows those values from x and y be converted for z values, that is to say:

$$z \in \Gamma \{ x \times y \}$$

2.5 Fuzzy Logic

The name "Fuzzy Logic" in the beginning seems to refer to something confuse (cloudy), even providing skepticism with respect to equipment's control increasing their capacity to perform the work.

Fuzzy Logic is a way to use inherent analogical process data that shift trough a continuous digital computer band, which works with well-defined numeric data, i.e., discrete values. For example, considering a brake system directed by a micro-control: the micro-control takes the decisions according the brake temperature, speed, and other system variables.

The temperature variable in this system may be divided in a "states" band: "cold", "fresh", "normal", "tepid", "hot". Though, it is not easy to set up the transition from one state to the next one; an arbitrary limit should be defined as to divide, for example, the "tepid" from the "hot", but this would drive to a discontinue change when the entry value pass trough the limit. The micro-control should be able to detect it.

The way would be the creation of the " fuzzy states" or more common "fuzzy memberships", which allows the graduated change from a state to the next. You could define the entry temperature by the use of intermediate functions.

In this way, the entry variable state will no longer to jump suddenly from one state to the next one, it looses gradually the value in one state while it gains value in the next state, instead. Up to in a moment the brake temperature "true value" will almost always be in some point between two consecutive functions: 0.6 normal and 0.4 tepid, or 0.7 normal and 0.3 fresh, and so on. Figure 3 shows a representation of such consecutive functions.

The entry variables of a fuzzy control system are generally map out inside consecutive function sets – the process for conversion of an intermediate entry value into a fuzzy value is called "fuzzification". We should note that a control system may have entry keys (on/off) together with analogical entries, and such entries (on/off) of the path always will have a true value equal to 1 or 0 – but such entries are in fact a simplified case of a

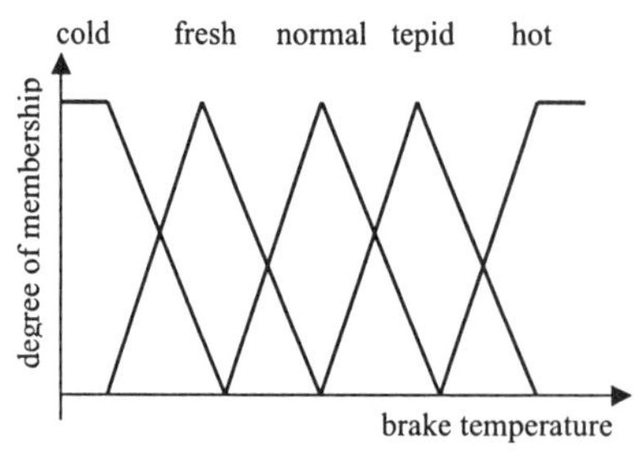

Figure 3 – Fuzzy Memberships

fuzzy variable, so the system may work with tem with no difficulties.

Setting up the entry variables map out inside the consecutive functions and true values, the micro-control then takes the decisions for making the actions to follow the rules:

IF the brake temperature is tepid **AND** the speed is not too fast **THEN** the brake pressure is slightly reduced

Where, in this case, the two entry variables are ''brake temperature'' and "speed". The outing variable, "brake pressure", is by the same token, generated from a fuzzy set as a starting point, which may have values as "static", "slightly reduced", "slightly increased", and so on.

The decision is based on a set of rules: all the applied rules are invoked, using the consecutive functions and true values which come from the entries, for the determination of the rule result – that in return will be map out inside the consecutive function and the true value, controlling the outing variable – and after these results are combined as to generate a specific answer, the present brake pressure, a procedure known as "defuzzification". The combination of fuzzy operations with rules based on "conclusion" describes an "expert fuzzy system".

2.6 Fuzzy Control and Inference

In concept fuzzy controls are very simple; they consist of one entry stage, a processing stage, and an outing stage. The entry stage map out sensors or another type of entries (numerical data upon which the system will base its decisions) in a proper way the consecutive functions and the true values; The processing stage selects each adequate rule and provides one result for each one of them, and then they combine the results of these rules; and finally the outing stage converts the combined result from the previous stage into the control.

The most usual shape of consecutive functions is the triangular (as in Figure 2), even though trapeze shaped curves, besides other shapes, are also used, but the shape is generally less important than the number of curves and where thy are located. From 3 up to 7 curves are in general appropriate to cover the required entry values band.

The processing stage is, as already discussed, based upon a collection of logic rules shaped on statements as IF – THEN, where IF is called the "antecedent" and THEN is called "consequence." Typical fuzzy control systems have dozens of rules. For practical purposes, the set of rules usually have several antecedents which are combined using fuzzy operators, such as AND, OR, and NO, (besides again the definitions are varying): And (in an informal definition) uses in a simple way the minimum weight for all the antecedents, whereas OR uses the maximum values [5]. (Also there is an operator NO that subtracts one consecutive function from 1 resulting the complementary function.

There are several ways for the definition of a rule's result, but one more usually used and simple is called "max-min", in which the outing of the consecutive function is given by the true value generated by the premise. Rules may be solved in parallel hardware or in sequential software. All rule's results are "defuzzificated" for a concise value by one of the several methods; there are several theoretical ones, all of them with advantages and disadvantages.

The centered method is very popular, in which the result's "mass center" supply the concise value; in the process of defuzzification by pondered average, the centered of each area is calculated separately and the outing value is calculated through the average of these centered and pondered with the maximum value of the pertinence function. In the process by maximum average of pertinence function, the pertinence function maximum value average is calculated, being this outing value [5].

The fuzzy control systems project is based in empirical methods – basically a methodical approach for try-and-error. There are few pre-defined rules in the present, once it is a new technology yet; generally the process follows the steps below:

- The system operational specifications, the entries and outings are documented.
- The fuzzy sets for the entries are documented.
- The set of rules is documented.
- The defuzzification method is determined.
- It is executed through a test, for checking the system, making the detail adjustments as required.

3. Paraconsistent Logic

Be an axiomatic system **F** with an underlying logic **L**. In the classic logic there is always a concern for checking whether they are consistent and complete. Defining the symbol ¬ in the sentential calculus as a denial and being one of the symbols of **F** and **A** language's a **F** formula, this system is consistent provided it will be possible to demonstrate **A** or ¬**A**, but never both, otherwise is called inconsistent. This system is complete if for the whole **A** it will be always possible to prove **A** or ¬**A**, otherwise is called uncompleted. For example, the propositional of 1st order is consistent and complete.

Up to the beginning of 20th century many mathematic schools were proposing the axiomatic treatment, it was having the so-called formalist school, supported by many mathematicians, including Hilbert. Also it was having the intuition school supported by Brouwer and the logistic school sponsored by Russell and Whitehead.

Hilbert was proposing that after laid down an axiomatic system, it should be possible to show its consistency. However, this hope was excluded by Gödel's discovery, in 1931, that the consistency of an axiomatic system cannot be demonstrated, unless using more powerful methods than the ones for demonstration from the system itself. Gödel also have shown that the elementary arithmetic used by Peano's axioms if are consistent then are incomplete. Gödel's conclusions have set up a panic climate into formalist school.

When we have an axiomatic theory with inconsistencies, it is possible to show that any formula **A** is true and this type of theory is called trivial, hence, without practical interest. However, the fact that a theory has one inconsistency and be trivial, it is linked directly to Aristotle's non-contradiction principle, which is almost always present into subjacent logic's. Thus, there is a hope that the theory that contains inconsistencies may become useful, provided that in the subjacent logic the non-contradiction principle be discarded. This was a proposal from Newton da Costa [2].

The possibility of working with inconsistent systems started about 1910 by the logic's Lukasiewicz and Vasil'ev, that have published independent works proposing the construction of a logic in which the non-contradiction principle would be discarded.

In 1948, Jaskowski worked out a new logic system named discursive logic, dealing only with the propositional calculus. Independently, Newton da Costa also worked out a logic system in which the non-contradiction principle was eliminated. Finally, in 1964, da Costa published the seminal work named "Inconsistent Formal Systems", generating a subjacent logic for inconsistent systems, without making them trivial. This logic was named paraconsistent in the 1976 Latin-American Congress of Logic.

3.1 Propositional Paraconsistent Calculus P_τ

To represent the many propositional calculus states from paraconsistent logic's, it is necessary to set up a previous structure for obtaining the true tables and the inference rules. The proper structure is a grid with the true values. For the classic logic such a structure is not necessary, because there is the simplicity of having but two terms, *false* and *true*, to which we associate arbitrarily the values of 0 and 1, respectively. This set is trivially ordered. Thus, a set $|\tau|$ is required, and also an order relation ($\geq$) that makes it ordered, at least partially, and that have one superior and one lower. This structure is called a grid and is represented by $\tau = <|\tau|, \geq>$.

At this situation, with the states false (F), true (V), inconsistent (T) and undefined ($\perp$), the elements of set $|\tau|$ are $|\tau| = \{F, V, T, \perp\}$, where is taken $\perp$ as lower and T as superior.

By taking yet one operator for the denying, for example ~, results:

$$\sim(F) = V \; ; \; \sim(V) = F \; ; \; \sim(T) = \perp \; ; \; \sim(\perp) = T$$

The proposition of a language of P_τ may be found in [6]. Hence, a logic L is called paraconsistent if it may be used as a basis for inconsistent theories, but not trivial. Furthermore, a logic L is called paracomplete, it could be the subjacent logic to the theories in which the law of the excluded third is infringed in the following way: from two contradictory propositions, one of them is true. In a precise way, one logic is called paracomplete if it comprises maximal non-trivial theories, into which do not belong one given formula and its denying. At least, a logic L is named non-athletic whenever is paraconsistent and paracomplete.

3.2 Paraconsistent Logic of Notation with Two Values - LPA2v

The paraconsistent logic of notation with two values was presented in [7] and extended in [8]. This extension may be represented by a Cartesian plane Unitarian square, which presents x and y values varying into a closed actual interval [0,1], in such a way that these values show the belief and unbelief grades. Figure 4 presents this cartesian plane.

Table 1 – Truth Table

P	V	V	V	V	V	F	F	F	F	T1	T1	T1	$\perp$	$\perp$	T0
Q	V	F	T1	$\perp$	T0	F	T1	$\perp$	T0	T1	$\perp$	T0	$\perp$	T0	T0
P and Q	V	F	T1	$\perp$	T0	F	F	F	F	T1	qF	F	$\perp$	q¬V	T0
P or Q	V	V	V	V	V	F	T1	$\perp$	T0	T1	qV	V	$\perp$	q¬F	T0

P	V	V	V	V	F	F	F	F	T1	T1	T1	T1	$\perp$	$\perp$	$\perp$
Q	qV	q¬V	qF	q¬F	qV	q¬V	qF	q¬F	qV	q¬V	qF	q¬F	qV	q¬V	qF
P and Q	qV	q¬V	qF	q¬F	F	F	F	F	T1	F	qF	QF	$\perp$	q¬V	qF
P or Q	V	V	V	V	qV	q¬V	qF	q¬F	qV	qV	T1	V	qV	$\perp$	$\perp$

P	$\perp$	T0	T0	T0	T0	qV	qV	qV	qV	q¬V	q¬V	q¬V	qF	qF	q¬F
Q	q¬F	qV	q¬V	qF	q¬F	qV	q¬V	qF	q¬F	q¬V	qF	q¬F	qF	q¬F	q¬F
P and Q	$\perp$	q¬V	q¬V	F	T0	qV	q¬V	qF	$\perp$	q¬V	F	q¬V	qF	qF	q¬F
P or Q	q¬V	V	T0	q¬F	q¬F	qV	qV	qV	V	q¬V	$\perp$	q¬F	qF	q¬F	q¬F

Some propositions may be true and false at the same time, generating the inconsistency points T_1 (belief 1 and unbelief 1) and T_0 (belief 0 and unbelief 0). Also, other notable points can be checked, as: qV – almost true, V - true, q¬V – almost non-true, qF – almost false, and q¬F – almost non-false. With this, it is possible to have the True Table for propositions in this extension of paraconsistent logic (Table 1).

4. Description of Computer Package

The computer package into which the paraconsistent logic was introduced has the aim of promoting the learning of fuzzy logic for the students, mainly the development of a fuzzy control (decision making). At present it is in its 4.0 version. The description of the computer package so as a description of the problems that have generated each version, will be made as follow.

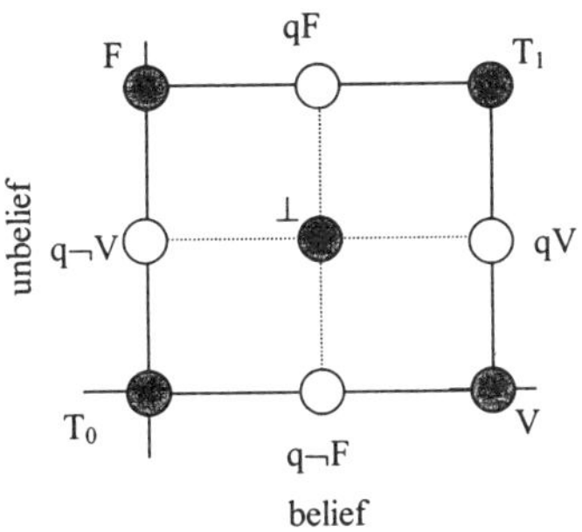

Figure 4 – Important Points

4.1 The Computer Package

After some years of fuzzy logic teaching, and its introduction into the actual controls, it was realized that the students, even observing through technical papers that the application of this logic was making easier the development of controls and decision take systems, were not totally acquainted with, once it was lacking an application tool in order to demonstrate its whole power.

For that purpose it was thought to develop a computer program in which fuzzy controls could be introduced. The problem to be solved had to be complex enough for turning its solution not obvious; otherwise the student would have a false feeling regarding the true power of the mathematics tool under study. Otherwise, the problem should be of easy understanding, and if possible, to be familiar for the students.

Several alternatives were considered to become the solution's main object of this computer program. Industrial systems, motor controls, temperature, pressure or brake control systems were considered. Because the students could belong to any course and be in any graduation year for the application of the Fuzzy Systems course, all these examples show some type of problem regarding the premises of the problem be chosen. As a first approach, it was chosen to develop the displacement of a robot inside a platform. This idea changed fast for the parking of a vehicle in a parking lot.

The parking of a vehicle is a complex problem to be solved by the conventional systems. Otherwise, the students are familiar with driving, or at least have already been playing with car toys by their early years. So, the problem of parking a vehicle would attend very well to the proposed objectives.

The main purpose of the package is to park a vehicle in a garage, beginning from any start position. In order to make it, the user should first to develop a set of fuzzy control rules and membership functions, which will define the vehicle path. Several windows and numeric routines can be found in the program for helping the users in making the rules. The variables fuzzing and defuzzing process are performed by the program with no action required from the user. To represent the vehicle parking problem the program has the basic picture shown in Figure 5. It shows the garage layout, the existing obstacles (the walls in this case) and the coordinates limits values. Also we can see the problem entry variables, i.e. *(x , y)* measured from the vehicle back central point, and the car angle (φ). The user may define a starting position for the vehicle (Coordinate X, Y and Angle). The simulation begins with a simulation option. Figure 6 shows the vehicle course using the set of 148 rules for the displacement.

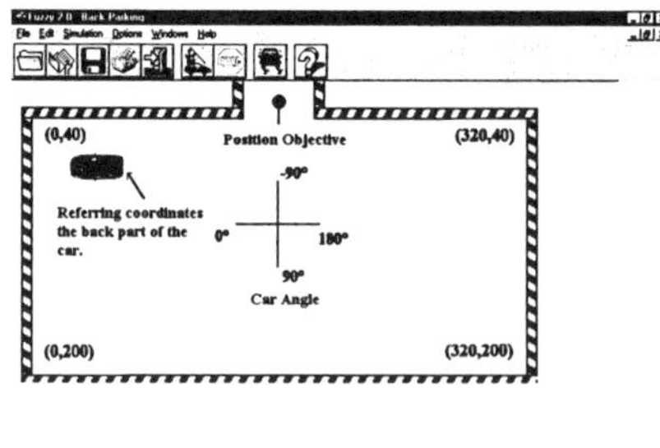

Figure 5 - Program basic picture

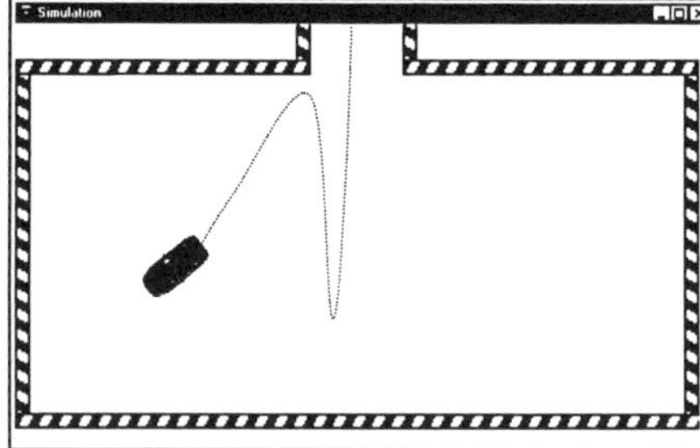

Figure 6 - Simulation Example

In the simulation example of Figure 6, it is shown the vehicle track left when routing. Each point means an iteration (i.e. a whole passage through the rules set) being the counting registered in the Variables window. In this example the fuzzy control has made 256 iterations.

4.2 The Previous Versions of the Package

The Version 1.0 of this computer package was developed in DOS environment using the elements of the first Visual-Basic program versions. Routines and graphic elements had to be developed for generating the supporting elements for this program [3].

Following, with the arrival of the Windows operational system and the evolution of the Visual-Basic language, a new version was made and new components were attached to it, aiming to increase the students comprehension with regard to the decision's take fuzzy systems. In this Version 2.0, different defuzzification methods and vehicle sizing were attached, beside others [9].

However, the computer package had a fundamental weak point, the pertinence functions and the control rules set up by students looking for parking the vehicle, still were based on a trial and error process, what some times could drive the students into wrong suppositions concerning the control action. As to minimize this mistake, in the Version 3.0 was introduced a training routine that has the objective of adjusting the pertinence functions. This training is based in evolution techniques and genetic algorithms, allowing the proposed control's [10] performance improvement.

4.3 The Problem to Be Solved by Version 4.0

In true life, the parking of a vehicle in a parking area always happens with present obstacles, which can be fixed or mobile ones. The fixed may be columns or garage physical limitations, while the mobile obstacles are temporary access restrictions or other vehicles already parked.

The past versions from computer package did not consider these problems, but if we want to set up a true case control they have to be taken into consideration. The paraconsistent logic together with the fuzzy logic will be used for solving this problem.

4.4 The Problem's Solution

The problem's solution will be achieved by using both the paraconsistent and the fuzzy logic's in the course of the inference process. For this purpose, two different sensors, generating in these way belief and unbelief degrees will measure each entry variable.

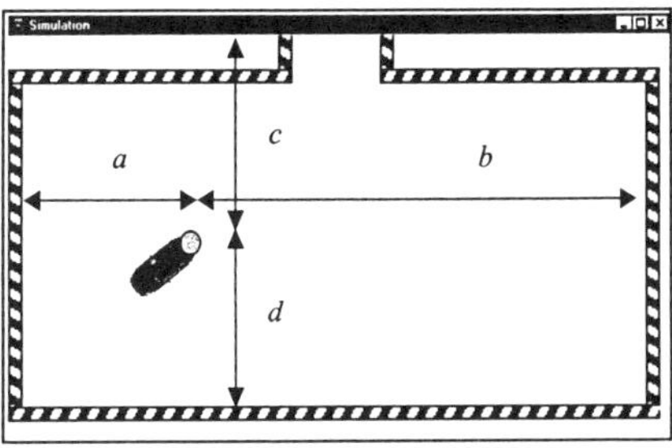

Figure 7 – Input Values

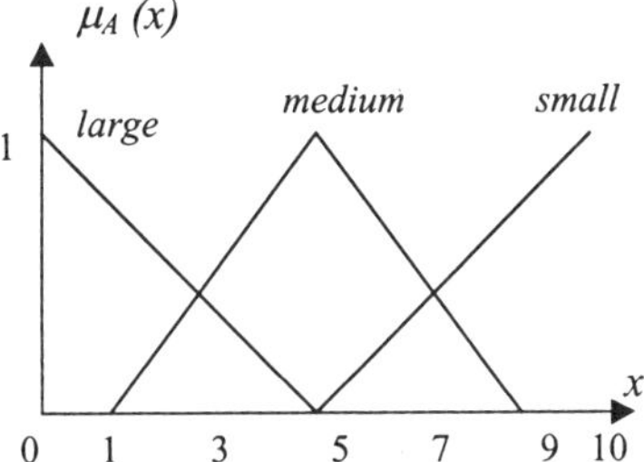

Figure 8 – Mirror Values of x according to Figure 1.

Figure 7 shows a presentation of such idea for the variables x and y. Where a represents the distance of variable x to the origin, and b the complementary distance. Where there are no obstacles, the sum of these two values will be the garage size. When there are obstacles this value will be reduced.

To get the exit variable's end value two different inference processes are provided, one using the belief degrees (traditional values from the entry variables, for example: a and c), and the other using the unbelief degrees (complementary values from entry variables, for example: b and d).

Thus, the inference process for the traditional values uses the pertinence functions and the rules edited by the user, as shown before. This generates the first value for the exit variable. Using the complementary entry values gets the second value. But, the entry variable's pertinence functions are mirrors of the traditional values, as shown, for example, in Figure 8, according Figure 1. By having these two exit values, which will be the same without obstacles, its combination is made by pondering the true exit values with its pertinence values, getting in this way the problem's exit end value. Figure 9 presents some illustrative examples of the proposed control.

5. Conclusions

The fuzzy systems are a convenient and efficient alternative for solution of problems where the fuzzy state are well defined. Nevertheless, the project of a fuzzy system may became difficult for large and complex systems, when the control quality depends of "try-and-error" methods for defining the best membership functions to solve the problem.

The main purpose of the Computing Package for The Fuzzy Logic Teaching, as used in this work, is to provide the students with the learning of this logic. The choice of a vehicle parking lot is justified because they don't need to have a previous knowledge (at least in mathematics terms) regarding the subject, for making the control.

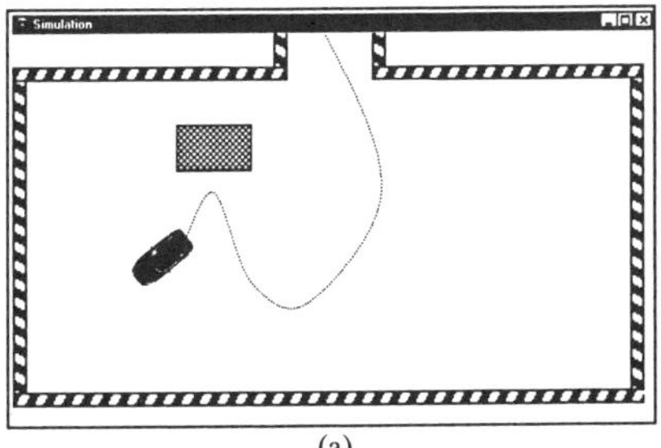

(a)

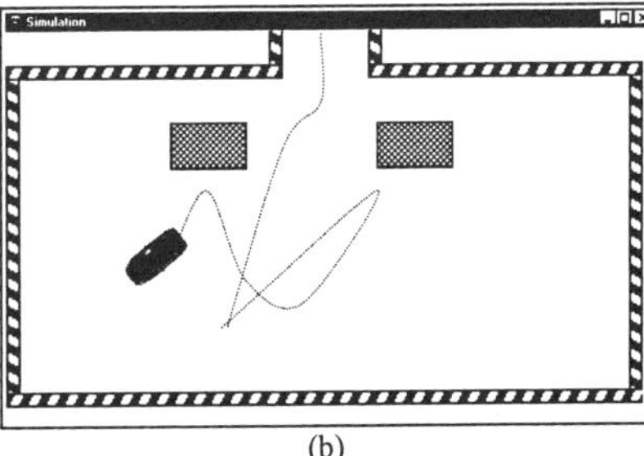

(b)

Figure 9 – Illustrative Examples

The genetic training modulus developed in this work added this program with an automatic technique for the adjustment of the membership functions parameters. This technique shows that the performance of a fuzzy control may be improved through the genetic algorithms, substituting for the "try-and-error" method, as used before by students for this purpose, with no good results.

The genetic algorithms provided distinctive advantages for the optimization of membership functions, resulting in a global survey, reducing the chances of ending into a local minimum, once it uses several sets of simultaneous solutions. The fuzzy logic supplied the evaluation function, a stage of the genetic algorithm where the adjustment is settled.

References

[1] Zadeh, Lotfi, *"Fuzzy sets"*, Information and Control, vol. 8, 338 - 353, 1965.
[2] Da Costa, Newton C. A., *"Sistemas Formais Inconsistentes"*, presented as a Thesis in 1963 - Faculdade de Filosofia, Ciências e Letras da Universidade do Paraná, Curitiba, Paraná, 1993 (in Portuguese).
[3] Lambert-Torres, G.; Quintana, V.H. & Borges da Silva, L.E. - "A Fuzzy Control Lab for Educational Purposes", Canadian Conference on Engineering Education, pp. 117-124, Kingston, Canada, Jun. 16-18, 1996.
[4] Zadeh, L. A., *"A Theory of Approximate Reasoning (AR)"*, Memo UCB/ ERL   M77/ 58. Univ. of California, Berkeley.
[5] Butler, K. & Lambert-Torres, G. - "Fuzzy Set Fundamentals", Tutorial on Fuzzy Logic Applications in Power Systems, by K. Tomsovic & M.Y. Chow, TP 140-0, IEEE Press, 1999.
[6] Da Costa, N. C. A. & Subrahmanian, V. S. & Vago, C. *"The Paraconsistent Logic Pτ"*, Zeitschrift fur Mathematische Logik und Grundlagen der Mathematik, Vol. 37, 139 - 148, 1991.
[7] Da Costa, N. C. A .& Abe, J. M. & Subrahmanian, V. S. *"Remarks on Annotated Logic"*, Zeitschrift fur Mathematische Logik und Grundlagen der Mathematik, Vol. 37, pp.561-570, 1991.
[8] Da Silva Filho, J. I., *"Métodos de Aplicações da Lógica Paraconsistente Anotada com Dois Valores - LPA2v com Construção de Algoritmo e Implementação de Circuitos Eletrônicos"*, Ph.D. Thesis, EPUSP, São Paulo, 1999 (in Portuguese).
[9] Tomsovic, K. & Lambert-Torres, G. – On the Use of Fuzzy Logic Techniques for Addressing Uncertainty in Power System Problems, PMAPS-RIMAPS Tutorial, 83 pages, 2000.
[10] Lambert-Torres, G.; Carvalho, M.A. & Borges da Silva, L.E. – "A Fuzzy-Genetic Hybrid System for Educational Purposes", International Conference on Intelligent System Applications to Power Systems, ISAP'2001, Budapest, Hungary, Jun. 17-21, 2001.

Logic, Artificial Intelligence and Robotics
J.M. Abe & J.I. da Silva Filho (Eds.)
IOS Press, 2001

A Polysynthetic Theory of Sets

E. G. K. LÓPEZ-ESCOBAR
Mathematics Department, University of Maryland

Abstract

It is common practice in computer languages to consider an expression as belonging to more than one type. For example, in an *IF* statement where one would expect "*IF* (t > 0)", where t is of type `int`, one often finds "*IF(t)*". That is t can also be considered as being of type `boolean`. Such premeditated confusion is frowned upon in the formal languages of mathematical logic; an *expression* is either a *term* or a *formula*. In addition in the case of mathematical theories, for example set-theory, one is careful to distinguish the purely logical part (usually the first order, classical predicate calculus) from the set theoretical assumptions (expressed in the language of first order logic using the binary relation symbol ε).

And yet, already in the early 1940's, A. P. Morde had developed a foundational system for set theory in which the formal expressions could be simultaneously be interpreted as propositions or sets and which had such polysynthetic expressions as "$(a \leftrightarrow (0 \in a))$". A watered down version of this system, i. e., a traditional system formulated in the first order predicate calculus appears in the Appendix to J. Kelley's "*General Topology*", it usually goes under the name of **KM** and is probably the best known formalization of *impredicative set theory*. Morse's original system appeared in 1965 as "*A Theory of Sets*" and although it is an oft quoted book, very few people actually use the formalization therein.

One can easily discern that "*A Theory of Sets*" is a product of the early 40's , for example it starts:

> A *mark* is a more or less connected inscription . . . we agree that c is a *symbol* if and
> only if c is a mark which is not a quotation mark . . . An *expression* is a linear array of
> symbols . . . c is a constant if c is '$\equiv$' or c is fixed by some definition . . . α is a
> *variable* if and only if α is a symbol which is not a constant . . .

This paper is a part of the project to develop **LMP**, a polysynthetic theory of sets, which, unlike **KM**, is faithful to the concets developed in "*A Theory of Sets*" and yet it is not unfamiliar to the modern reader. The syntax of **LMP** is reminiscent of a computer language but **LMP** will also have Gentzenian rules of inferences as well as definitional and conceptual aximos. In the paper we shall show how the method introduced by Scott and Solovay can be used to give a natural interpretation of **LMP**. At a later time we shall show how to apply Gödel's concepts of inner models and constructibility within **LMP**.

Logic, Artificial Intelligence and Robotics
J.M. Abe & J.I. da Silva Filho (Eds.)
IOS Press, 2001

Reasons and Ways to Cope with a Spectrum of Logics

Alfio Martini   Uwe Wolter
Edward Hermann Haeusler
Dept. Theoretical Computer Science - UFRGS - Porto Alegre - Brazil
University of Bergen - Dept. Informatics - Bergen, Norway
Dept. Informatics, PUC - Rio de Janeiro - Brasil

Abstract.[1] The integration in a sound and structured way of several specifications (or views) of a system is a key research area in (modern) software specification and development. In this paper we concentrate in developing the abstract mathematical structures to define both the syntactical and semantical domains of heterogenous structured specifications. We show that they happen to be grothedieck constructions over suitable indexed categories. Syntax and semantics are then related in a functorial way. A small case study, showing the naturality of the approach is also presented.

1 Introduction

Nowadays, it is widely recognizable that (modern) software specification and development takes place in a platform covering a wide range of rigorous semantic-based descriptions of diverse aspects or views of the system [15]. This includes, for example, both executable and non-executable formal specifications, models of concurrency and interaction, abstractions suitable for systems analysis and model checking, specification of security, dependability, and real-time requirements. As a consequence, a key unresolved problem is the interoperability problem, namely, how these several pieces could be connected and how their multiple and possibly heterogeneous semantic descriptions impose constraints on each other. In this paper we concentrate in developing the abstract mathematical structures to define both the syntactical and semantical domains of heterogenous structured specifications. We show that they happen to be grothedieck constructions over suitable indexed categories. A small case study, showing the naturality of the approach is also presented. These ideas capitalize on basic concepts found in the elegant and well-known theory of algebraic specifications [6, 3]. Specially, our attention will be focused on those kind of formalisms having an underlying well defined logic. In this way, heterogenous specifications can be achieved once we are able to offer a construction linking theories from different logics and the same for their respective (categories of) models.

2 Basic Concepts

Institutions were introduced in [8, 9] in order to describe in an uniform way several notions of a logical system used for specification purposes. An even stronger motivation was, how-

<hr>

[1]Work partially supported by Fapergs and CNPq.

ever, to give a mathematical semantics for Clear [4] independent of any particular underlying logic. Institutions concentrate on the model-theoretic aspects of a logic. Here a logic consists of a *syntax* and of a *semantics* that *fit together nicely*. These ideas can be traced back at least to [2]. Imposing a categorical structure on the syntax, one sees that the categorical concept of a functor lies at the heart of a formalization of syntax and semantics. All this is collected in the following concept

An *institution* $\mathcal{I} = (\mathsf{Sign}, Sen, Mod, \models)$ consists of: a category Sign whose objects are called *signatures*; a functor $Sen : \mathsf{Sign} \to \mathsf{Set}$; giving for each signature a set whose elements are called *sentences* over that signature; a functor $Mod : \mathsf{Sign}^{op} \to \mathsf{Cat}$, giving for each signature Σ a category whose objects are called Σ-models, and whose arrows are called Σ-morphisms; and a function $\models$ associating to each signature Σ a relation $\models_\Sigma \subseteq \mid Mod(\Sigma) \mid \times Sen(\Sigma)$, called Σ-*satisfaction relation*, such that for each arrow $\phi : \Sigma_1 \to \Sigma_2$ in Sign the *satisfaction condition*

$$M_2 \models_{\Sigma_2} Sen(\phi)(\varphi_1) \iff Mod(\phi)(M_2) \models_{\Sigma_1} \varphi_1 \quad \text{(Satisfaction Condition)}$$

holds for any $M_2 \in \mid Mod(\Sigma_2) \mid$ and any $\varphi_1 \in Sen(\Sigma_1)$.

Almost every presentation of a logic is an institution. First order-logic, and its well-known fragments (both in unsorted and many-sorted cases), lambda-calculus, higher-order logics see [9], variant of temporal logics [1] and meta-logical frameworks, like rewriting logic, for instance [12], are some examples, to name a few.

Given a arbitrary institution $\mathcal{I}$, we can derive a number of concepts. A theory is a pair $T = \langle \Sigma, \Gamma \rangle$, where $\Sigma \in \mid \mathsf{Sign} \mid, \Gamma \subseteq Sen(\Sigma)$. Given theories $\langle \Sigma, \Gamma \rangle$ and $\langle \Sigma', \Gamma' \rangle$, $\phi : \langle \Sigma, \Gamma \rangle \to \langle \Sigma', \Gamma' \rangle$ is a *theory morphism* if $\Gamma' \models Sen(\phi)(\Gamma)$. This defines a category Th in the expected way. Now let, for each $\langle \Sigma, \Gamma \rangle$, $Mod_\models(\langle \Sigma, \Gamma \rangle)$ be the full subcategory induced by all models $M \in Mod(\Sigma)$, such that $M \models_\Sigma \Gamma$. Then, he satisfaction condition implies that for each $\phi : \Sigma \to \Sigma'$ the model functor $Mod(\phi) : Mod(\Sigma') \to Mod(\Sigma)$ can be restricted to a functor $Mod(\phi) : Mod_\models(\langle \Sigma', \Gamma' \rangle) \to Mod_\models(\langle \Sigma, \Gamma \rangle)$, for each theory morphism $\phi : \langle \Sigma, \Gamma \rangle \to \langle \Sigma', \Gamma' \rangle$ This means, globally, that we have the generalized model functor $Mod_\models : \mathsf{Th}^{op} \to \mathsf{Cat}$.

3 An Introduction to Bridges

The idea of this section is to introduce the reader into the way bridges work. It can be read from anyone with a reading knowledge of formal specifications. The nice point, as we will see, is that they work as a straightforward generalization of the usual idea of inclusion of specifications.

To begin with, let us consider the classical example of the specification of lists of naturals, written in the sequel in the well-known Z-notation [14].

Since natural numbers are used everywhere, it is perfectly reasonable to assume that we have a single module specifying it, so that we can include it whenever we need it. The specification bellow is standard. Note that we use the *specification, theory*, and *module* to the refer to the same syntactical object.

ThNat, MSEqtl

Sorts
 nat
Opns
 $0 :\rightarrow nat$
 $succ : nat \rightarrow nat$
 $plus : nat, nat \rightarrow nat$

 $[x : nat]\ plus(x, 0) = x : nat$
 $[x : nat, y : nat]\ plus(x, succ(y)) = succ(plus(x, y)) : nat$

Since we are interested in lists of naturals, we can extend the module above by a sort *list* and the usual operations of *header*, *tail*, and of *construction* of lists. The clause **Include**, as it is hinted by its name, includes the module *ThNat* in the module *ThList*.

ThList, MSEqtl

Include [*Nat, MSEqtl*]
Sorts
 List
Opns
 emptylist $:\rightarrow List$
 head $: List \rightarrow Nat$
 tail $: List \rightarrow List$
 cons $: Nat, List \rightarrow List$

 $[x : nat, y : list]head(cons(x, l)) = x : nat$
 $[x : nat, y : list]tail(cons(x, l)) = l : list$
 $[\ \ ]tail(emptylist) = emptylist : list$

Note that, in the specification above, the operation *head* is underspecified, since we don't have a definition for *head(emptylist)*. This is mainly because, many-sorted algebras are unsuitable to model partial operations like this one. There are many efforts in the literature to cope with this burden, like error algebras, logic of partial algebras [13], and order-sorted logic [7], which will be considered in the next specification.

As an example of the use of this logic, consider the same specification of lists, but now using order-sorted logic (see below).

Note that the the problem of undefinedness of *head* is now solved by introducing a new sort, *nelist* to denote non-empty lists.

However, now we either specify natural numbers again from scratch (well, we agree that it would not be that difficult) or we might consider reusing it, just the way it is written in many-sorted logic. In fact, we can proceed exactly like that (since, as we will see, we have a sound way to connect the two logics). However, since the imported module is actually written in another logic, we have the clause **Bridge** which denotes a codification or representation of the (old) specification into the new logic. Note that in this case, and as in most practical applications as well, the signature morphism component of the bridge is a simple inclusion of (coded) natural numbers. In fact, the "bridged" specification of natural numbers would only have the harmless declaration *nat* < *nat*, saying that *nat* is a subsort of itself.

```
┌─ ThList, OSEqtl ──────────────────────────────────────────
│ Bridge [ThNat, MSEqtl]
│ Sorts
│   list, nelist
│ Subsorts
│   nat < nelist < list
│ Opns
│   emptylist :→ list
│   head : nelist → nat
│   tail : nelist → list
│   cons : nelist, list → list
├──────────────────────────────────────────────────────────
│ [x : nat, y : list]head(cons(x, l)) = x : nat
│ [x : nat, y : list]tail(cons(x, l)) = l : list
└──────────────────────────────────────────────────────────
```

It is also important to stress that the user, from the point of of the modeling of an application, can keep in mind the natural understanding of the specification in the old logic, since the bridge clause is completely defined once a map is established between the underlying institutions involved. But what is more important, he can keep in mind its many-sorted model understanding of the specification, which is much simpler that its order-sorted counterpart. Moreover, such step could be even achieved by a mechanical procedure, coded in some powerful meta-logical frameworks, like Rewriting Logic [12], for instance.

A more challenging situation is provided if we specify now lists using unsorted horn logic with equality, while still keeping in mind the desire to reuse the module of natural numbers in many-sorted logic. The specification of lists with a bridge for natural numbers is shown bellow:

In general, the well-definedness of the this situation is ensured by the fact that we have a map between the logics involved in the specification (see next section).

```
┌─ ThList, UHEqtl ──────────────────────────────────────────
│ Bridge [Nat, MSEqtl]
│ Sorts u
│ Pred nat, list : u
│ Opns emptylist :→ u, head : u → u, tail : u → u, cons : u, u → u
├──────────────────────────────────────────────────────────
│ [ ]list(emptylist)
│ [x, y : u]nat(x), list(y) ⇒ list(cons(x, y)), nat(x), list(y) ⇒ nat(head(cons(x, y))
│ [x, y : u]nat(x), list(y) ⇒ list(tail(cons(x, y))
│ [ ]tail(emptylist) = emptylist
│ [x, y : u]nat(x), list(y) ⇒ head(cons(x, y)) = x, nat(x), list(y) ⇒ tail(cons(x, y)) = y
└──────────────────────────────────────────────────────────
```

As remark, it is easy to see that *Bridge[Nat, MSEqtl]* it is just the specification of natural numbers where the sort *nat* is coded as a unary predicate.

> _____ *ThList, UHEqtl* ____
> **Sorts** u
> **Pred** $nat : u$
> **Opns** $0 :\to u, succ : u \to u, plus : u, u \to u$
> ____
> $[\ ]nat(0), [x : u]nat(x) \Rightarrow nat(succ(x)))$
> $[x, y : u]nat(x), nat(y) \Rightarrow nat(plus(x, y)), [x : u]nat(x) \Rightarrow plus(x, 0) = x$
> $[x, y : u]nat(x), nat(y) \Rightarrow plus(x, succ(y)) = succ(plus(x, y))$

4 Maps between Logics

A very flexible way to relate logics can be given by the following concept, due to [11]. It says that the target logic $\mathcal{I}'$ is rich enough to model the semantic requirements of the source logic $\mathcal{I}$. In fact, there are many situations where we can translate signatures and sentences of an institution $\mathcal{I}$ into signatures and sentences of another institution $\mathcal{I}'$, but where only subclasses of the corresponding model classes in $\mathcal{I}'$ can be translated back into models of $\mathcal{I}$. Fortunately, we are able in most cases to axiomatize these subclasses within $\mathcal{I}'$. This is the central idea of the following notion.

Let $\mathcal{I}$ and $\mathcal{I}'$ be institutions. A *simple map of institutions* $(\Phi, \Psi, \alpha, \beta) : \mathcal{I} \to \mathcal{I}'$ is given by a functor $\Phi : \mathsf{Sign} \to \mathsf{Sign}'$; a functor $\Psi : \mathsf{Sign} \to \mathsf{Th}'$ such that $\Psi(\Sigma) = \langle \Phi(\Sigma), \varnothing'_\Sigma \rangle$; a natural transformation $\alpha : Sen \Rightarrow \Phi$; $Sen' : \mathsf{Sign} \to \mathsf{Set}$, and a natural transformation $\beta : \Psi^{op}; Mod'_{\models} \Rightarrow Mod : \mathsf{Sign}^{op} \to \mathsf{Cat}$, such that the *simple map of institutions condition*

$$\beta(\Sigma)(M') \models_\Sigma \varphi \Longleftrightarrow M' \models'_{\Phi(\Sigma)} \alpha(\Sigma)(\varphi) \quad \text{(Simple Condition)}$$

holds for each $\Sigma \in | \mathsf{Sign} |$, $M' \in | Mod'_{\models}(\Psi(\Sigma)) |$, and $\varphi \in Sen(\Sigma)$.

The examples of bridges introduced in the previous section are all based on maps between the underlying logics in which those specifications were written. They are classical in the theory of logical systems and can be found, for instance, in [10].

It is not difficult to see that simple maps compose. Then institutions as objects and simple maps as morphisms define the category Logics. This category can be as an index for writing actual specifications, according to the following

Proposition[Indexed Category of Theories] Let $(\Phi, \Psi, \alpha, \beta) : \mathcal{I} \to \mathcal{I}'$ be a simple map. Then the assignments $\mathcal{I} \mapsto \mathsf{Th}(\mathcal{I})$ $(\Phi, \Psi, \alpha, \beta) : \mathcal{I} \to \mathcal{I} \mapsto \Psi_{\models} : \mathsf{Th}(\mathcal{I}) \to \mathsf{Th}(\mathcal{I}')$ defines an indexed category $\mathsf{Th} : \mathsf{Logics} \to \mathsf{Cat}.\square$

5 Bridges

Building on the previous proposition, we can now link in a natural way theories from these different logics. This is provided by the concept of a *bridge* right below.

Definition[Bridges] Let $(\Phi, \Psi, \alpha, \beta) : \mathcal{I} \to \mathcal{I}'$ be a simple map, and $\langle \Sigma, \Gamma \rangle, \langle \Sigma', \Gamma' \rangle$ theories from Th and Th'. Then a *bridge* between $\langle \Sigma, \Gamma \rangle$ and $\langle \Sigma', \Gamma' \rangle$ is a an arrow $\psi : \langle \Sigma, \Gamma \rangle \to$

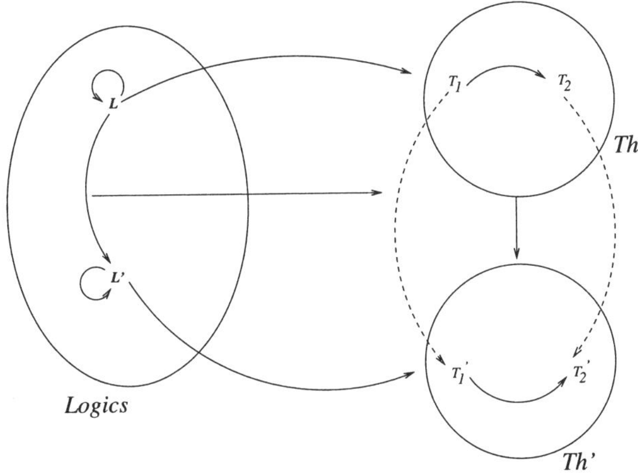

Figure 1: Bridges between theories

$\langle \Sigma', \Gamma' \rangle$ such that $\psi : \Phi(\Sigma) \to \Sigma'$ is an arrow in **Sign'** and such that $\Gamma' \models_{\Sigma'} Sen(\Phi(\phi))(\alpha(\Sigma))$
$(\Gamma) \cup \varnothing'_\Sigma).\square$

Bridges compose and build a category which is on top of the category **Logics**. It is essentially an indexed construction. This situation is best depicted in Figure 1. Each logic defines a category of theories in which diagrams model structured specifications. Now, each map induces a corresponding translation between the corresponding categories of theories. The solid arrows that link two given theories (inside the circles) represent the usual notion of theory morphisms, while the dotted ones denote bridges. Note that, in the figure, the loops denote identity maps, i.e., and by translating them, we recover the special case of specification in one fixed, arbitrary logic. This discussion is formalized below:

Proposition[Category of Bridges] Theories as objects (from arbitrary institutions) and bridges as arrows define the category **Bridges**.$\square$

Definition[Split Opfibration Grothendieck Construction] Given a functor $F : \mathsf{C} \to \mathsf{Cat}$, the Grothendieck construction constructs the opfibration induced by F, a category $\mathsf{G}(\mathsf{C}, F)$, defined as follows:

- An object of $\mathsf{G}(\mathsf{C}, F)$ is a pair $\langle A, x \rangle$, where A is an object of C and x an object of $F(A)$.

- An arrow $\langle f, u \rangle : \langle A, x \rangle \to \langle A', x' \rangle$ has $f : A \to A'$ an arrow of C and $u : F f(x) \to x'$ an arrow of $F(A')$ (note that, by definition, F; $f(x)$ is an object of $F(A')$).

- If $\langle f, u \rangle : \langle A, x \rangle \to \langle A', x' \rangle$ and $\langle g, v \rangle : \langle A', x' \rangle \to \langle A'', x'' \rangle$, then $\langle f, u \rangle$; $\langle g, v \rangle : \langle A, x \rangle \to \langle A'', x'' \rangle$ is defined as:

$$\langle f, u \rangle; \langle g, v \rangle = \langle f; g, F g(u); v \rangle$$

Theorem[Characterization of Bridges] The category Bridges is essentially the Grothendieck (flattened) category $G(\mathsf{Logics}, Th)$, where $Th : \mathsf{Logics} \to \mathsf{Cat}$.$\square$

Since ordinary theory morphisms have as denotation reindexing model functors (see section 2), we expect the same for bridges which are a kind of generalized theory morphisms. In fact, this is put precisely in the next

Proposition[Heterogenous model functor] Given a bridge $\phi : \langle \Sigma, \Gamma \rangle \to \langle \Sigma', \Gamma' \rangle$ one can show that its denotation is the reindexing functor $Mod(\phi) : Mod'_{\models}(\langle \Sigma', \Gamma' \rangle) \to Mod_{\models}(\langle \Sigma, \Gamma \rangle)$, which maps models from the target to the source logic.

This information can be collected and organized in a more general way.

Proposition[Indexed Category of Heterogenous Models] Let $G(\mathsf{Logics}, Th)$ be the grothedieck category of the previous theorem. Let also $\langle \Sigma, \Gamma \rangle$ and $\langle \Sigma', \Gamma' \rangle$ be, respectively, theories of $Th(\mathcal{I})$ and $Th(\mathcal{I}')$. Then the assignments $\langle \Sigma, \Gamma \rangle \mapsto Mod_{\models, I}(\langle \Sigma, \Gamma \rangle)$, $\phi : \langle \Sigma, \Gamma \rangle \to \langle \Sigma', \Gamma' \rangle \mapsto Mod(\phi) : Mod_{\models, I'}(\langle \Sigma', \Gamma' \rangle) \to Mod_{\models, I}(\langle \Sigma, \Gamma \rangle)$ define an indexed functor

$HMod : G(\mathsf{Logics}, Th)^{op} \to \mathsf{Cat}.$

The following is an analogous Grothendieck construction(actually this is the original one) as the one defined above.

Definition[Split Fibration Grothendieck Construction] Given a functor $G : \mathsf{C}^{op} \to \mathsf{Cat}$, the Grothendieck construction constructs the fibration induced by G, a category $\mathsf{F}(\mathsf{C}, \mathsf{G})$, defined as follows:

- An object of $\mathsf{F}(\mathsf{C}, \mathsf{G})$ is a pair $\langle A, x \rangle$, where A is an object of C and x an object of $G(A)$.

- An arrow $\langle f, u \rangle : \langle A, x \rangle \to \langle A', x' \rangle$ has $f : A \to A'$ an arrow of C and $u : x \to Gf(x')$ an arrow of $F(A)$.

- If $\langle f, u \rangle : \langle A, x \rangle \to \langle A', x' \rangle$ and $\langle g, v \rangle : \langle A', x' \rangle \to \langle A'', x'' \rangle$, then $\langle f, u \rangle; \langle g, v \rangle : \langle A, x \rangle \to \langle A'', x'' \rangle$ is defined as:

$$\langle f, u \rangle; \langle g, v \rangle = \langle f; g, u; Gf(u) \rangle$$

The above construction provides us also with a way to gave a domain where objects are very general logical structures and whose arrows are compatible relations between them.

Definition[Category of Heterogenous Homomorphisms] We define the category of heterogenous models and heterogenous generalized homomorphisms as the split fibration induced by $HMod : G(\mathsf{Logics}, Th)^{op} \to \mathsf{Cat}$, i.e. the category $\mathsf{F}(G(\mathsf{Logics}, Th), HMod)$.

In general, the indexed category of heterogenous models and heterogenous model functors $HMod : G(\mathsf{Logics}, Th)^{op} \to \mathsf{Cat}$ as the semantical functor, which maps structured (heterogenous) specifications to their model theoretical counterparts.

6 Concluding Remarks & Further Work

A related work can be found in the *extra-theory* morphism concept of Diaconescu [5]. Some initial examinations [10] make us expect that our proposal is simpler and somehow more flexible. However a detailed comparison between these two approaches should be given elsewhere. As a further and immediate work we intend to lift the classical theory of *Goguen & Burstall* on specification theory to heterogenous structured specifications. This mainly will involve special properties on the semantical functor presented above.

References

[1] M . Arrais and J. L. Fiadeiro. Unifying theories in different institutions. In *Recent Trends in Type Specification*, pages 81–101. Springer, LNCS 1130, 1996.

[2] K. J.. Bairwise. Axioms for Abstract Model Theory. *Annals of Mathematical Logic*, 7:221–265, 1974.

[3] R. M. Burstall and J. A. Goguen. Putting theories together to make specifications. In *Proc. Int. Conf. Artificial Intelligence*, 1977.

[4] R. M. Burstall and J. A. Goguen. The semantics of Clear, a specification language. In D. Bjøner, editor, *Abstract Software Specification, Proc. 1979 Copenhagen Winter School, LNCS 86*, pages 292–332. Springer, 1980.

[5] R. Diaconescu. Extra-theory morphisms in institutions: logical semantics for multi-paradigm languages. *Journal Applied Categorical Structures*, 1998. To appear.

[6] H. Ehrig and B. Mahr. *Fundamentals of Algebraic Specification 1: Equations and Initial Semantics*, volume 6 of *EATCS Monographs on Theoretical Computer Science*. Springer, Berlin, 1985.

[7] J. Goguen and J. Meseguer. Order-sorted algebra I: equational deduction for multiple inheritance, overloading, exceptions and partial operations. *Theoretical Computer Science*, 105:217–273, 1992.

[8] J. A. Goguen and R. M. Burstall. Introducing institutions. In *Proc. Logics of Programming Workshop*, pages 221 – 256. Springer LNCS 164, 1984.

[9] J. A. Goguen and R. M. Burstall. Institutions: Abstract Model Theory for Specification and Programming. *Journal of the ACM*, 39(1):95–146, January 1992.

[10] A. Martini. *Relating Arrows between Institutions in a Categorical Framework*. PhD thesis, Tecnhical University of Berlin, 1999. Dept. of Computer Science.

[11] J. Meseguer. General logics. In H.-D. Ebbinghaus et. al., editor, *Logic colloquium '87*, pages 275–329. Elsevier Science Publishers B. V.,North Holland, 1989.

[12] J. Meseguer. Conditional rewriting logic as a unified model of concurrency. *TCS*, 96:73–155, 1992.

[13] H. Reichel. *Initial Computability, Algebraic Specifications, and Partial Algebras*. Oxford University Press, Oxford, 1987.

[14] J . M. Spivey. *Understanding Z: A Specification language and its formal semantics*, volume 3 of *Cambridge Tracts in Theoretical Computer Science*. Cambridge University Press, Cambridge, 1988.

[15] A. Tarlecki. Moving between logical systems. In *Recent Trends in Data Type Specification*, pages 478–502. Springer, LNCS 1130, 1996.

Logic, Artificial Intelligence and Robotics
J.M. Abe & J.I. da Silva Filho (Eds.)
IOS Press, 2001

The Logic of Medical Diagnosis

Eduardo Massad and Neli Regina S. Ortega
School of Medicine, University of São Paulo
Av. Dr. Arnaldo 455, São Paulo, 01246-903, SP, Brazil

Abstract. In the history of medicine many theoretical aspects of the medical cognition have been studied, and mathematical structures have been used to modelled this kind of human thinking. The increase development of expert systems, decision making models, clinical guidelines, epidemiological studies and diagnostic systems are just some examples of how mathematics and artificial intelligence techniques could help the professionals of heath in them daily work, and in the acquisition of deep knowledge of the medical procedures and epidemiological dynamics. In this work is presented a overview of the logic of medical diagnosis and some mathematical techniques is discussed.

1. Introduction

Doctors have always been fascinated by diagnosis and the means by which it can be reached, but the purpose of studying diagnostic logic has simply been to improve thought processes [1]. More recently, however, a second purpose is becoming more important: the design of expert systems and computer modelling able to perform medical diagnosis. In addition, the paradigm shift represented by the emergence of Evidence Based Medicine, the conscientious, explicit and judicious use of current best evidence in making decision about the care of individual patients [2] is unearthing new problems related to the logic behind diagnosis.

Medical diagnosis has been defined as *the crucial process that labels patients and classifies their illnesses, that identifies their likely prognosis, and that defines the best treatment available* [3]. It is, actually, a complex process characterized by uncertainty in many stages [4].

We may think of diagnosis proceedings from symptoms and signs (and laboratory tests) to focus on the documentation of maladaptive alterations in structure, function, and/or response to stimuli [3]. Alternatively, diagnosis can proceed from symptoms and signs (and laboratory tests) to focus on prognosis. Finally, diagnosis may focus on a therapeutic trial of identifying the target disorder on the basis of its response to specific therapy [3].

Several attempts have been made to identify the possible cognitive pathways that lead to diagnosis: *pattern recognition, multiple-branching* or *arborization* strategy, *strategy of exhaustion*, and the *hypothetical-deductive* strategy. The hypothetical-deductive approach has been considered the most appropriate diagnostic process in the sense that it is time saving and it has the greatest accuracy.

In summary, diagnostic approaches can usefully be described as one or a combination of four types: the pattern recognition approach of the seasoned clinical, the multiple-branching method of the delegate, the exhaustion method of the novice, and the most widely used strategy, the hypothetical-deductive approach [3].

2. Computer Models and Expert Systems for Medical Diagnosis

The development of computer models of medical diagnosis as applied to the construction of expert systems is part of the field of investigation called *Artificial Intelligence*, which may be defined as *the branch of computer science that is concerned with the automation of intelligent behaviour* [5].

The basic assumption of artificial intelligence in medicine is that the logical analysis of the diagnostic process makes the medical diagnosis amenable to be mimicked (and if possible, surpassed) by computer systems. This sort of approach is based on the assumption that diagnosis is a highly desirable end and that doctors want a system that will do better than the best clinician.

2.1 The "Turing Doctor"

Suppose we have an experts system with diagnostic capacity and we want to check whether it is capable of passing in the Turing Test [6]. We may think of an adaptation of the classical Turing test for a diagnostic system as follows. A patient is interviewed by the diagnostic system and by a real, bone and flesh doctor. Both are out of view of the patient and an interface between the patient and both the computer system and the real doctor serves as an input-output communicator. A second doctor assists the interview and she also does not know which interviewer is which (see figure 1 below).

Figure 1: Turing Doctor Test

If after some period of time the second doctor is not able to differentiate the computer system from the real doctor we say that the system has passed the Turing Doctor test. Several diagnostic computer systems have already demonstrated such a capacity. Just to mention one of the most famous we highlight de Dombal's system [7].

2.2 Mathematical Models in Medical Diagnosis

The vast capacity of modern computers to process and store data and to carry out, almost instantaneously, complex logical manipulations has encouraged the description of the diagnostic process in mathematical terms. This formalization of the diagnostic logic, in turn, has permitted the development of computer programs that can aid physicians in their attempts to solve their patients' problems accurately and safely through the application of ever-expanding medical knowledge [8].

     E. Massad and N.R.S. Ortega / Logic of Medical Diagnosis

The mathematical and statistical techniques already applied in classical expert systems include decision tress and other logical schematics, likelihood ratio, Baye's theorem, discriminant analysis, and cluster analysis. In addition, subsidiary mathematical techniques has been developed for helping machine-learning systems, like the backpropagation algorithm of connectionist networks. Some of these techniques are aimed at the classificatory aspect of diagnosis, like discriminant analysis and cluster analysis.

Bayesian Reasoning Using probability, we can determine, often from a priori argument, the chances of events occurring. In knowledge-based problem solving, like in medical diagnosis, we often find ourselves reasoning with limited knowledge and incomplete information. Several techniques have been designed to deal with such a limited knowledge and information and, before we start the discussion of fuzzy reasoning we describe some of the techniques of probabilistic reasoning.

Bayesian reasoning is based in formal probability theory and is used extensively in several current areas of research, including pattern recognition and classification, both of paramount importance in diagnostic applications.

The Bayesian approach is based on *prior probabilities*, the unconditioned probability assigned to an event in the absence of knowledge supporting its occurrence or absence, and *posterior probability*, the condition probability of an event given some evidence. The usual notation for prior probability is **p(event)** and for posterior probability is **p(event|evidence)**. So, for instance, the prior probability of a person having a disease is the number of people with the disease divided by the number of people in the domain of concern. The posterior probability of a person having a disease **d** with symptom **s** is given by:

$$p(d\,|\,s) = \frac{|d \cap s|}{|s|}$$

$$(1)$$

where the bars brackets meaning the number of elements in that set. Therefore, the posterior probability given by equation 1 is the number of people having both (intersection) the disease **d** and symptom **s** divided by the total number of people having the symptom **s**. Equation 1 can also be written as

$$p(d\,|\,s) = \frac{p(d) \times p(s\,|\,d)}{p(s)}$$

$$(2)$$

also known as *Bayes equation* or *Bayes theorem*.

In clinical terms, the priori probability **p(d)** is called the *prevalence* of the disease, and the posteriori **p(s|d)** the *sensitivity* of the diagnostic test that revealed s (by the way, **s** may be a symptom, a sign, a laboratory test or any other diagnostic information). In real situation, however, rarely a final diagnostic is reached by a single symptom (sign, and/or laboratory test). We, therefore, need a generalized version of equation 2, a form of Bayes with multiple symptoms (signs and/or laboratory tests):

$$p(d\,|\,s_1\,\&\,s_2\,\&\,...\,\&\,s_n) = \frac{p(d) \times p(s_1\,\&\,s_2\,\&\,...\,\&\,s_n\,|\,d)}{p(s_1\,\&\,s_2\,\&\,...\,\&\,s_n)}$$

$$(3)$$

Now, for **m** diseases and **n** symptoms (signs and/or laboratory tests) there will be about ($m{\times}n^2$ conditional probabilities) + (n^2 symptom probabilities) + (m disease probabilities) or about ($m{\times}n^2{+}n^2 + m$) pieces of information to collect. In a realistic medical system with 100 diseases and 1000 symptoms (signs and/or laboratory tests), this value is $100{\times}1000^2{+}1000^2{+}100 = 10^8$, that is, over 100 millions! This simple calculation illustrates the difficulties involving such an approach.

The Dempster-Schafer Theory of Evidence Uncertainty often results from a combination of missing evidence, the inherent limitations of heuristic rules and the limitation of our own knowledge. The Dempster-Schafer theory of evidence considers sets of propositions and assigns to each of them an interval *[belief, plausibility]* within which the degree of belief for each proposition must lie. This belief measure, denoted **bl** ranges from zero (no evidence) to one (certainty) [9, 10]. Its complement is called *plausibility*, and is denoted **pl**. So, for a proposition **a**:

$$pl(a) = 1 - bl(\neg a)$$

(4)

where ¬a means *not a*. Plausibility also ranges between zero and one and reflects how evidence of **not(p)** relates the possibility for belief in **p**. Dempster-Schafer address the problem of measuring certainty by asking for a fundamental distinction between lack of certainty and ignorance. Belief functions allow us to use our knowledge to bound the assignment of probabilities to events in the absence of exact probabilities. The Dempster-Schafer theory is based on the idea of obtaining degrees of belief for one question from subjective probabilities for related questions and the use of a rule for combining the degrees of belief when they are based on independent items of evidence.

Let us suppose that we have a diagnosis domain **H**, containing some diagnostic hypotheses that a patient has tuberculosis (T), pneumonia (P), or common cold (C). We have to associate measures of beliefs with the hypotheses sets within the domain H. Evidence need not support individuals hypotheses exclusively. So, for instance, the presence of fever would support our three hypotheses simultaneously. On the other hand, evidence in favor of some hypotheses may affect belief in others.

The next step is to define a *probability density function, d*, for all subsets of the set H, where **d(h$_i$)** represents the belief that is currently assigned to each **h$_i$** of **H** (where in this case $\sum d(h_i) = 1$). If **H** has **n** elements then there are 2^n subsets of H. Since many of the subsets will never occur, it is possible to deal with the remaining subsets. The plausibility of H is:

$$pl(H) = 1 - \sum d(h_i)$$

(5)

where the **h$_i$** are the sets of hypotheses that have some supporting belief. Whenever we start a diagnosis, it is often the case that we have no information about any hypotheses, then **pl(H)=1.0**.

Suppose our first evidence is that the patient has cough, and that this supports **{T,P}** at 0.8. If this is our only hypotheses, then d$_1${T,P}=0.8 and d$_1$(C)=0.2 to account for the remaining distribution of belief, that is, all other possible beliefs across H (and not our belief in the complement of {T,P,C}. We next proceed by amplifying our investigation space and have now that the patient also has headache, which has the support level of {P,C} to 0.6, and so we have d$_2${P,C}=0.6 and d$_2$(H)=0.4. These two beliefs may now be combined by the Dempster's rule [5]:

$$d_3(Z) = \frac{\sum_{X \cap Y = Z} d_1(X)d_2(Y)}{1 - \sum_{X \cap Y = \Phi} d_1(X)d_2(Y)}$$

(6)

so that $[d_1\{T,P\}=0.8] \cap [d_2\{P,C\}=0.6]=[d_3\{P\}=0.48]$. As there are no sets $X \cap Y$ that are empty, the denominator of equation 6 is 1. We may, therefore, assign a belief of 0.48 that the patient in our example has pneumonia. The Dempster-Schafer approach is a very useful tool when the stronger Bayesian conclusions may not be justified [5].

3. Fuzzy Diagnostic Systems

The process of classifying different sets of symptoms, signs and laboratory tests under a single name is getting increasingly difficult. Several factors are contributing to this fact like the enormous amount of medical information available for clinicians and, most important, the great variety of uncertainties, vagueness and ambiguities involved in the diagnostic process. Therefore, alternative methods are desperately needed for the design of diagnostic systems. Fuzzy logic is one of the best, current candidate for this role [11].

Let us assume that the aim of the diagnostic process is to assign a label to patients who, in addition to their complaints (often presented in a fuzzy way), present a cluster of clinical signs and laboratory (complementary) tests alterations. The clinical purpose then, is to differentiate a "normal" from a "sick" individual. A classical view is to consider a crisp divide between a healthy and a non-health individuals. So, we need to start by defining what we understand by "health". The World Health Organization defines health as *the complete absence of physical, mental or social wellbeing*. Have you ever heard a fuzziest definition?

In a classical, standard approach, the diagnostic process should be, ideally a classificatory process able to determine the crisp divide between healthy and non-healthy individual. Let us imagine, like Bellamy [4], that a reliable measure of health is available and that a normal threshold, below which an individual is classified as non-healthy exists. A good clinician should be then, the one able to classify individuals below (diseased) or above (healthy) such a threshold. Figure 2 illustrates this measure:

An individual with a certain score 2.8 would be then considered as not healthy, according with this approach. Individuals in the borderline, however, would present additional classification problems.

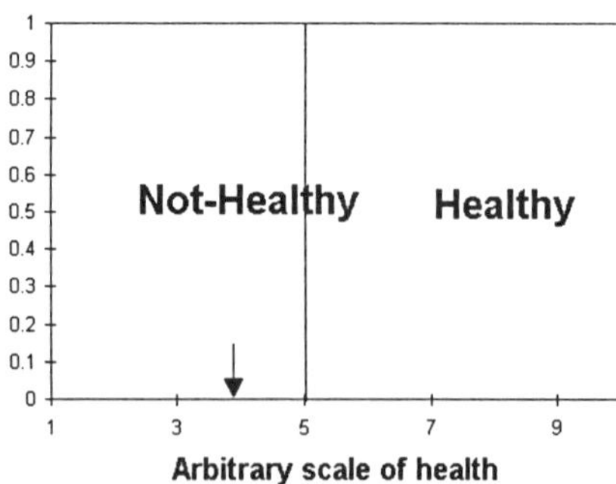

Figure 2: Hypothetical threshold to Healthy and Not Healthy.

The most fundamental aspect of fuzzy set theory is the idea of graded membership. Classifying individuals as healthy and not healthy by use fuzzy sets offers several advantages over use of crisp sets. The 2 sets "healthy" and "not healthy" can be represented by overlapping triangles with the y-axis indicating the grade of membership in the fuzzy set, like in figure 3 (after Bellamy [4]):

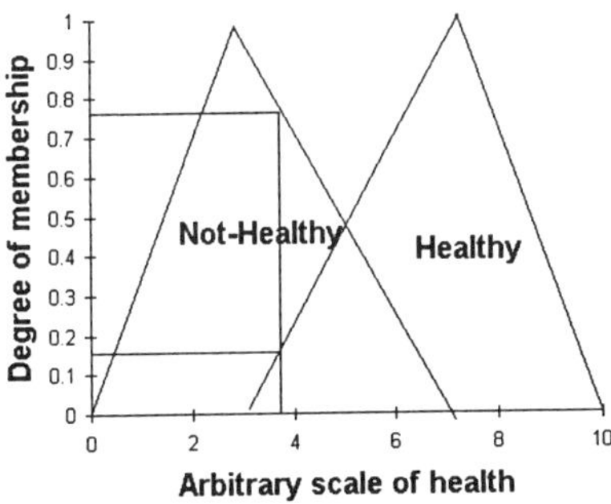

Figure 3: Membership functions for Healthy and Not Healthy

The same individual who scored 2.8 in the health scale now has 0.4 degree of membership to the set of healthy individuals and 0.55 degree of membership to the set of non-healthy individuals. It is indeed much more functional!

The fuzzy logic approach of modelling the diagnostic process is useful in all the diagnostic approaches: the pattern recognition approach; the multiple-branching method; the exhaustion method; and the hypothetical-deductive approach. The pattern recognition approach, due to its intrinsic subjectiveness is probably the most prone to be aided by fuzzy logic modelling. However, all the other approaches have several uncertainties and vagueness associated with the classificatory process of diagnosis.

3.1 Fuzzy Relations Diagnostic Models

The fuzzy logic framework has been utilised in several different approaches to modelling the diagnostic process [12]. In one of the first attempt to apply fuzzy logic to medical diagnosis, Sanchez [13] proposes a model in which the medical knowledge is represented as a fuzzy relation between symptoms and diseases. So that, given the fuzzy set **A** of the symptoms related by the patient and the fuzzy relation **R** representing the medical knowledge that relates the symptoms in the set **S** to the diseases in the set **D**, then the fuzzy set **B** of the possible diseases on the patient can be inferred by means of the compositional rule of inference [12]

$$B = A \circ R$$

$$(7)$$

or

$$B(d) = \max_{s \in S}[\min(A(s), R(s,d))]$$

$$(8)$$

for each **d** in **D**. The membership grades of related symptoms in the fuzzy set **A** represent the degree of possibility of the presence of the symptom or its severity. The membership grades in fuzzy set **B** denote the degree of possibility with which we can attach each relevant diagnostic label to the patient. The fuzzy relation **R** of medical knowledge should constitute the greatest relation such that given the fuzzy relation **Q** on the set **P** of patients and **S** of symptoms and the fuzzy relation **T** on the sets **P** of patients and **D** of diseases, then

$$T = Q \circ R$$

(9)

Relations **Q** and **T** may then represent, respectively, the symptoms that were present and diagnosis made for a certain number of known cases.

Let us see a practical example of such an approach. We know, from hundreds of years of medical practice, that three clinical findings, *headache* and *cough*, two symptoms, and *fever*, a sign, are associated, at different levels, with several possible diagnostic. Let us take, for instance, *Endocarditis* (End.), *Pneumonia* (Pn.), *Pertussis* (Pt.), *Tuberculosis* (Tb) and *Common Cold* (C.C.). The fuzzy relation **R** of medical knowledge that relates those symptoms and signs to the set of possible diseases may be like the matrix:

$$R = \begin{array}{c} \\ headache \\ fever \\ cough \end{array} \begin{array}{ccccc} End. & Pn. & Pt. & Tb. & CC. \\ \begin{bmatrix} 0.0 & 0.0 & 0.3 & 0.0 & 0.8 \\ 0.9 & 1.0 & 0.3 & 1.0 & 0.2 \\ 0.2 & 0.4 & 0.7 & 1.0 & 0.1 \end{bmatrix} \end{array}$$

(10)

Then a specific patient presents itself with *persistent* fever, *intense* and *constant* cough and *no complain* of headache (the italicised words are meant to emphasise the vagueness of clinical findings and complains). We may assign to this specific patient the following fuzzy set **A**:

$$A = \begin{array}{ccc} Head. & Fev. & Cou. \\ \begin{bmatrix} 0.0 & 0.7 & 1.0 \end{bmatrix} \end{array}$$

(11)

Relation 9 allows us to estimate the possible diseases of this patient, so that

$$\mu_{A \circ R} = \max_{s \in S}[\min(\mu_A(s), \mu_R(s,d))]$$

(12)

which can be calculated by operating the matrices

$$\begin{bmatrix} 0.0 & 0.7 & 1.0 \end{bmatrix} \circ \begin{bmatrix} 0.0 & 0.0 & 0.3 & 0.0 & 0.8 \\ 0.9 & 1.0 & 0.3 & 1.0 & 0.2 \\ 0.2 & 0.4 & 0.7 & 1.0 & 0.1 \end{bmatrix}$$

(13)

which results in the following possibilities of diagnosis for this patient

$$\begin{array}{ccccc} End. & Pn. & Pt. & Tb. & CC. \\ [0.7 & 0.7 & 0.7 & 1.0 & 0.2] \end{array}$$

$$(14)$$

that is, the highest diagnostic possibility for this hypothetical patient is tuberculosis, although a final decision for discriminating this diagnosis from endocarditis, pneumonia and pertussis should require further investigations.

One of the first fuzzy diagnostic systems, CADIAG-2 [11, 14-16] incorporates relations between symptoms and diseases and also between diseases themselves, between symptoms themselves, and between combinations of symptoms and diseases. It demonstrated an accuracy of 94.5% in achieving correct diagnosis in rheumatological diseases.

3.2 Fuzzy Cluster Analysis Models

Another example provided by Klir and Yuan [12] is a set of models which utilise a technique proposed by Fordon and Bezdek [17] and Esogbue and Elder [18-20]. Models of diagnosis using fuzzy cluster analysis examine the similarity of the presence and severity of symptoms patterns, which can be designated with degrees of memberships in fuzzy sets representing each symptom category. The patient is clustered to varying degrees with the prototypical patients whose symptoms are most similar. The most likely diagnostic candidates are those disease clusters in which the patient's degree of membership is greatest. The specific patient **x** presents itself displaying a set of symptoms **s$_i$** (again by symptoms we mean in addition to symptoms, signs and laboratory tests) at levels of severity given by a fuzzy set, **A$_x$** denoting the grade of membership the fuzzy set characterising the patient and defined on the set of symptoms **S** which indicates the severity level of each symptom presented by the patient. Each of the likely diseases is described by a matrix **B$_i$** giving the upper and lower bounds of the normal range of severity of each of the symptoms that should be expected in a patient with the disease. We further define a fuzzy relation **R** on the set of symptoms and diseases that specifies the pertinence or importance of each symptom **s$_i$** in the diagnosis of the matrix of each likely disease **d$_j$**. The clustering is performed by computing a similarity measure between the patient's symptoms and those typical of each disease **d$_j$**. The example provided by Klir and Yuan [12] uses a distance measure based on the Minkowski distance given by:

$$D_p(d_j,x) = \left[\sum_{i \in I_l} | R(s_i,d_j)(B_{il}(s_i) - A_x(s_i)) |^p + \sum_{i \in Iu} | R(s_i,d_j)(B_{iu}(s_i) - A_x(s_i)) |^p \right]^{1/p}$$

$$(15)$$

where

$$I_l = \{ i \in N_m \mid A_x(s_i) < B_{il}(s_i) \},$$
$$I_u = \{ i \in N_m \mid A_x(s_i) < B_{iu}(s_i) \}$$

and **m** denotes the total number of symptoms. The most likely disease candidate is the one for which the similarity measure attains the minimum value.

3.3 The State-Space Approach

In a recent, and one of the most interesting article on fuzzy medical diagnostic systems, Bellamy [4] proposes modelling diagnosis as a mapping between subsets of property measurements and subsets of diagnostic categories, described within multidimensional diagnostic spaces. This multidimensional space, combined with a fuzzy sets representation of the variables and states of the patient, leads to a fuzzy systems model of the diagnostic process.

Bellamy's model also assumes a multidimensional symptom space and plots the specific patient as a point in such a space. In addition, he adds a dynamic, representing how the system changes in time. As time passes, the point in the multidimensional space moves around and the trajectory of the point describes the behaviour of the system.

Certain combinations of fuzzy subsets correspond to particular diagnostic categories. For example, if the blood neutrophils counting is high, fever is moderate, and low-back pain is intense, then the regions overlap in the pyelonephritis region, and so on. In addition, each bounded region representing a specific diagnostic category is equivalent to one of the fuzzy linguistic rules used to relate inputs to outputs, that is, the diagnostic region for pyelonephritis is equivalent to the fuzzy rule described in that region of the figure 4:

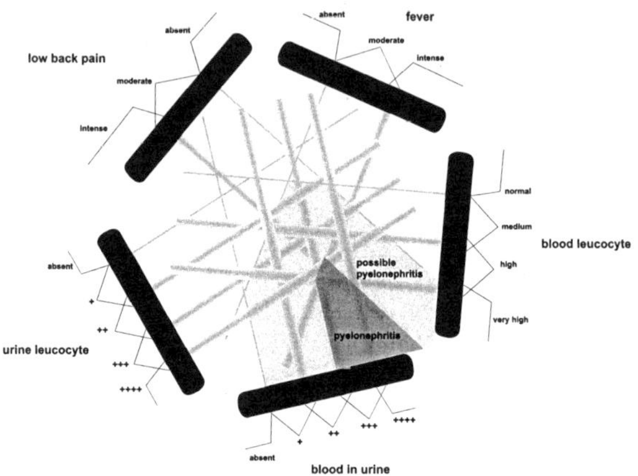

Figure 4: Example of the state space approach.

Finally, the interested reader may explore the technical details of the recent applications of fuzzy logic in medical diagnosis by reading the recent books by Adlassnig [21], Szczepaniak, Lisboa and Kacprzyk [22] and by Morderson, Malik and Cheng [23].

References

[1] F. Macartney, Logic in Medicine, *British Medical Journal* **295** (1987), 1325-1332.

[2] D.L. Sackett, W.S. Richardson, W. Rosenberg, R.B. Haynes, Evidence-based medicine. How to practice and teach EBM. New York: Churchill Livingstone, 1997.

[3] D.L. Sackett, R.B. Haynes, G.H. Guyatt, P. Tugwell, Clinical Epidemiology a Basic Science for Clinical Medicine, Boston: Little, Brown and Company, 1991.

[4] J.E. Bellamy, Medical Diagnosis, Diagnostic Spaces, and Fuzzy Systems, *J Am Vet Med Assoc.* **210**(3) (1997), 390-396.

[5] G.F. Luger and W.A. Stubblefield, Artificial Intelligence, Harlow, England. Addison Wesley Longman, Inc. 1998.

[6] A. Turing, Computing machinery and intelligence, *Mind* **59** (1950), 453-460.

[7] F. de Dombal, F.D. Leaper, J. Staniland et al, Computer-aided diagnosis of acute abdominal pain, *British Medical Journal* **1** (1972), 376-380.

[8] M.C. Miller, M.C. Westphal-Jr., J.R. Reigart, Mathematical Models in Medical Diagnosis, New York: Praeger Publishers, 1981.

[9] A.P. Dempster, Upper and lower probabilities induced by a multivalued mapping, *Ann. Math. Stat.* **38** (1967), 325-329.

[10] G.A. Shafer, Mathematical Theory of Evidence, Princeton University Press, Princeton, 1976.

[11] K-P. Adlassnig, A survey on medical diagnosis and fuzzy subsets, In: Gupta, M.M. and E. Sanchez, eds. Approximate Reasoning in Decision Analysis. North-Holland, New York, 1986, 203-217.

[12] G.J. Klir, B. Yuan, Fuzzy Sets and Fuzzy Logic, Upper Saddle River: Prentice Hall PTR, 1995.

[13] E. Sanchez, Inverses of Fuzzy Relations. Application to Possibility Distributions and Medical Diagnosis, *Fuzzy Sets and Systems* **2** (1979), 75-86.

[14] K-P. Adlassnig and G. Kolarz, CADIAG-2: Computer-assisted medical diagnosis using fuzzy subsets, In: Grupta, M.M. and Sanchez, E. eds. Approximate Reasoning in Decision Analysis. North-Holland, New York. 1982, 219-247.

[15] K-P. Adlassnig, G. Kolarz, W. Scheithauer, H. Effenberger, G. Grabner, G. CADIAG: Approaches to Computer-Assisted Medical Diagnosis, *Comput. Biol. Med.* **15**(5) (1985), 315-335.

[16] G. Kolarz and K-P. Adlassning, Problems in Establishing the Medical Expert Systems CADIAG-2 in Rheumatology, *J. Med. Syst.* **10**(4) (1986), 395-405.

[17] W.A. Fordon and J.C. Bezdek, The application of fuzzy set theory to medical diagnosis. In: Grupta, M.M., Ragade, R.K. and Yager, R.R. eds. Advances in Fuzzy Set Theory and Applications. North-Holland, New York. 1979, 445-461.

[18] A.O. Esogbue and R.C. Elder, Fuzzy sets and the modelling of physician decision processes: Part I: The initial interview-information gathering process, *Fuzzy Sets and Systems* **2** (1979), 279-291.

[19] A.O. Esogbue and R.C. Elder, Fuzzy sets and the modelling of physician decision processes: Part II: Fuzzy diagnosis decision models, *Fuzzy Sets and Systems* **3**(1) (1980), 1-9.

[20] A.O Esogbue and R.C. Elder, Measurement and valuation of a fuzzy mathematical model for medical diagnosis, *Fuzzy Sets and Systems* **10** (1983), 223-242.

[21] K-P. Adlassnig, Fuzzy Diagnostic and Therapeutic Decision Support, Österreichische Computer Gesellschaft, Wien, 2000.

[22] P.S. Szczepaniak, P.J.G. Lisboa and J. Kacprzyk, Fuzzy Systems in Medicine. Springer-Verlag, Berlin. 2000.

[23] J.N. Mordeson, D.S. Malik, S.C. Cheng, Fuzzy Mathematics in Medicine, Heildelberg: Physica-Verlag, 2000.

Logic, Artificial Intelligence and Robotics
J.M. Abe & J.I. da Silva Filho (Eds.)
IOS Press, 2001

Representation of Incompleteness and Inconsistency in Possible-Worlds Semantics

Tetsuya MURAI[1], Germano RESCONI[2] and Yoshiahru SATO[1]

1 Division of Systems and Information Engineering
   Graduate School of Engineering, Hokkaido University
   Kita 13, Nishi 8, Kita-ku, Sapporo 060-8628, JAPAN
2 Department of Mathematical and Phisics, Catholic University
   Via Trieste 17, 25121 Brescia, ITALY

Abstract. Examined are conditions of minimal models for modal logic, which are appropriate for dealing with two kinds of uncertainty, i.e., incompleteness and inconsistency. Such models with the conditions are called *general uncertain models*. In the models, incompleteness is represented by plurality of accessibile worlds, while inconsistency is represented by plurality of sets of accessible worlds.

1 Introduction

Two kinds of uncertainty can be often found in information which is dealt with in various areas: *incompleteness* and *inconsistency*. As many authors such as Dubois-Lang-Prade[3] suggested, incompleteness is derived from *lack* of information and is manipulated by, e.g., possibility theory[4], while inconsistency is generated by *excess* of information and is solved by, e.g., a method selecting maximal consistent sets. However, there seems no formal logical model which can explicitly deal with both kinds of uncertainty at the same time.

In this paper, we present a framework which enables us to deal with the two kinds of uncertainty in a unified way. On the basis of standard models for modal logic of belief, we examine conditions that are adequate for dealing with incompleteness and inconsistency. This leads us to have the so-called *general uncertain* ones. In the models, incompleteness is represented by plurality of accessibile worlds, while inconsistency is represented by plurality of sets of accessible worlds.

2 Possible-Worlds Semantics for Modal Logic

In this section we briefly describe Kripke and minimal models for modal logic (for details, see Chellas[2]) in the context of belief, because we aim at considering uncertainty in the framework of an agent's belief formed by available information. Given a set of *atomic* sentences $\mathcal{P}$, let $\mathcal{L}_{\mathcal{P}}$ be a language, i.e., a set of *sentences*, formed from $\mathcal{P}$ in the usual way of combining ordinary truth-functional logical operators such as $\top$(tautology), $\bot$(inconsistency), $\neg$(negation), $\wedge$(conjunction), $\vee$(disjunction), $\rightarrow$(material implication), $\leftrightarrow$(equivalence), and a modal operator $\mathbf{B}$(an agent's belief). A sentence in $\mathcal{L}_{\mathcal{P}}$ is called *compound* if it is not atomic, and it is called *modal* if it contains the modal operator.

2.1 Kripke Models

A *Kripke* model $\mathcal{M}$ for modal logic is defined as a structure

$$< W, R, v >,$$

where W is a non-empty set of *possible worlds*, or simply *worlds*, R is a relation on W called an *accessibility* relation, and v is a truth-assignment function that assigns exactly one of truth-values, T(true) and F(false), to each atomic sentence at each world, i.e.,

$$v : \mathcal{P} \times W \longrightarrow \{T, F\}.$$

A Kripke model is said to be *finite* just in case its set of possible worlds are finite. A truth-assignment function can be extended for compound sentences in the usual way. In particular, recall that it is extended for modal sentences in the following way:

Definition 1 (*Truth condition for modal sentences*)

$$v(\mathbf{B}p, w) \stackrel{\mathrm{def}}{=} \begin{cases} T, & \text{if } \forall w' \in W \ (wRw' \Rightarrow v(p, w') = T) \\ F, & \text{otherwise} \end{cases} . \quad \square$$

For any sentence p, its *proposition*, or *truth set*, is defined by

$$\|p\| \stackrel{\mathrm{def}}{=} \{w \in W \mid v(p, w) = T\}.$$

It is well-known that various conditions of an accessibility relation have their corresponding axiom schemata in modal logic. A system of modal logic being both sound and complete with respect to the class of all Kripke models is the smallest *normal* system K, which is axiomatized by a rule

RN.   From p infer $\mathbf{B}p$

and an axiom schema

K. $\mathbf{B}(p \to p') \to (\mathbf{B}p \to \mathbf{B}p')$

as well as rules and axiom schemata in a system of classical propositional logic.

Besides there are some ways of formulating Kripke models. First, we can define a set of worlds accessible from a world w using an accessibility relation R :

$$\mathcal{W}_w \stackrel{\mathrm{def}}{=} \{w' \in W \mid wRw'\},$$

whose elements are called *accessible worlds* of the world w in this paper. Then, Definition 1 can be rewritten using $\mathcal{W}_w$:

Lemma 2 (*The second form of truth condition for modal sentences*)

$$v(\mathbf{B}p, w) = \begin{cases} T, & \text{if } \mathcal{W}_w \subseteq \|p\| \\ F, & \text{otherwise} \end{cases} . \quad \square$$

Hence we may reformulate a Kripke model using the sets of accessible worlds as

$$< W, \{\mathcal{W}_w\}_{w \in W}, v > .$$

In the paper, a Kripke model is said to be *uniform* just in case $\mathcal{W}_w = \mathcal{W}_{w'}$ for any w, w' in W.

Next we can define a function

$$N : W \longrightarrow 2^{2^W}$$

using $\mathcal{W}_w (w \in W)$ by

$$N(w) \overset{\text{def}}{=} \{X \in 2^W \mid \mathcal{W}_w \subseteq X\}.$$

$N(w)$ is abbreviated to N_w following Chellas[2]. We can easily see that the function N satisfies

$$
\begin{array}{ll}
[\text{m}] & \forall w \in W \; \forall X, Y \subseteq W((X \in N_w \text{ and } X \subseteq Y) \Rightarrow Y \in N_w), \\
[\text{a}] & \forall w \in W \; (\cap N_w (= \cap_{X \in N_w} X) \in N_w), \\
[\text{n}] & \forall w \in W \; (W \in N_w),
\end{array}
$$

and $\mathcal{W}_w = \cap N_w$. By means of N, Lemma 2 can be rewritten again:

Lemma 3 (*The third form of truth condition for modal sentences*)

$$v(\mathbf{B}p, w) = \left\{ \begin{array}{ll} T, & \text{if } \|p\| \in N_w \\ F, & \text{otherwise} \end{array} \right. \quad . \; \Box$$

Thus we have another reformulation of Kripke models as

$$< W, N, v >$$

or equivalently

$$< W, \{N_w\}_{w \in W}, v > .$$

2.2   Minimal Models

A *minimal*, or *Scott-Montague*, model $\mathcal{M}$ for modal logic is just a triple

$$< W, N, v >,$$

where W and v are the same as in Kripke models, and N is a function, i.e.,

$$N : W \longrightarrow 2^{2^W},$$

on which, however in general, conditions are not necessarily required. Of course, a minimal model can also be reformulated as

$$< W, \{N_w\}_{w \in W}, v > .$$

A minimal model is said to be *finite* just in case its set of possible worlds are finite. In the paper, a minimal model is said to be *uniform* just in case $N_w = N_{w'}$ for any w, w' in W. Minimal models obviously can be found to be a generalization of Kripke ones. A truth-assignment function can be extended for compound sentences in the usual way and, in particular, we can use the formula in Lemma 3 for modal sentences.

It is known that various conditions of N have their corresponding axiom schemata in modal logic, some of them have been already shown in Section 2.1. The system of modal logic being both sound and complete with respect to the class of all minimal models is the smallest *classical* system E, which is axiomatized by a rule

$$\mathbf{RE}. \quad \text{From } p \leftrightarrow p' \text{ infer } \mathbf{B}p \leftrightarrow \mathbf{B}p'$$

as well as rules and axiom schemata in classical propositional logic. Do not confuse the above use of 'classical' in modal logic(cf.[2], p.231) with the ordinary use such as in 'classical' propositional logic.

Let us consider conditions [m], [n] mentioned in Section 2.1, and

$$[c] \quad \forall w \in W \ \forall X, Y \subseteq W((X \in N_w \text{ and } Y \in N_w) \Rightarrow X \cap Y \in N_w).$$

They correspond to the following axiom schemata

$$
\begin{aligned}
&\textbf{M.} \quad \mathrm{B}(p \wedge p') \to (\mathrm{B}p \wedge \mathrm{B}p'), \\
&\textbf{N.} \quad \mathrm{B}\top, \\
&\textbf{C.} \quad (\mathrm{B}p \wedge \mathrm{B}p') \to \mathrm{B}(p \wedge p'),
\end{aligned}
$$

respectively. With respect to the class of minimal models satisfying [m], [c], and [n], a system *EMCN* is shown to be sound and complete, where *EMCN* is obtained by adding to *E* the above three schemata **M**, **C**, and **N**. Note that *EMCN=K* as a set of theorems(cf.[2], p.236).

If a minimal model in the class further satisfies [a], then we can define an accessibility relation:

$$wRw' \overset{\text{def.}}{\iff} w' \in \cap N_w.$$

Hence, the class of Kripke models can be identified with the class of minimal ones satisfying [m], [a], and [n], and we can strictly see that Kripke models are a special case of minimal models. In the rest of this paper, we make discussion mainly in terms of minimal models.

Lastly we note that [c] is equivalent to [a] in any finite model, although in general [c] is only implied by [a]. We can always define accessibility relations in finite minimal models satisfying [m], [n], and [c].

3 General Uncertain Models

In this section, we examine an agent's belief state described by a standard model for modal logic of belief and then draw conditions which should be further required on or removed from the models in order to manipulate both incompleteness and inconsistency in a unified way.

3.1 Rational Belief Models

A normal system *KD45* is widely accepted as a standard system of modal logic for belief of an ideal rational agent(cf.[5]). It is obtained by adding to *K* the following three axiom schemata

$$
\begin{aligned}
&\textbf{D.} \quad \mathrm{B}p \to \neg \mathrm{B}\neg p, \\
&\textbf{4.} \quad \mathrm{B}p \to \mathrm{BB}p, \\
&\textbf{5.} \quad \neg \mathrm{B}p \to \mathrm{B}\neg \mathrm{B}p.
\end{aligned}
$$

By noticing *KD45=EMCND45*, where the latter system is obtained by adding **D**, **4**, and **5** to *EMCN*, the standard system is sound and complete with respect to the class of minimal models satisfying

$$
\begin{aligned}
&[d] \quad \forall w \in W \ \forall X \subseteq W \ (X \in N_w \Rightarrow X^\mathrm{C} \notin N_w), \\
&[iv] \quad \forall w \in W \ \forall X \subseteq W \ (X \in N_w \Rightarrow \{w' \in W \mid X \in N_{w'}\} \in N_w), \\
&[v] \quad \forall w \in W \ \forall X \subseteq W \ (X \notin N_w \Rightarrow \{w' \in W \mid X \notin N_{w'}\} \in N_w),
\end{aligned}
$$

as well as [m], [a], and [n]. We call a minimal model in the class a *rational belief* model in this paper.

There are two basic kinds of an agent's belief in a sentence p

$$\mathbf{B}p : \ p \text{ is believed to be true,}$$
$$\mathbf{B}\neg p : \ p \text{ is believed to be false.}$$

Then, combining them, we can generate the four belief states in the sentence p:

Definition 4

$$
\begin{array}{lll}
(\mathrm{b_1}) \ \text{A positive state:} & \mathbf{B}p \wedge \neg\mathbf{B}\neg p, \\
(\mathrm{b_2}) \ \text{A negative state:} & \neg\mathbf{B}p \wedge \ \mathbf{B}\neg p, \\
(\mathrm{b_3}) \ \text{An ambiguous state:} & \neg\mathbf{B}p \wedge \neg\mathbf{B}\neg p, \\
(\mathrm{b_4}) \ \text{An inconsistent state:} & \mathbf{B}p \wedge \ \mathbf{B}\neg p. \quad \square
\end{array}
$$

Next, we examine what state holds in rational belief models and what does not.

3.2 Incompleteness

In general, we can deal with *incompleteness* just in rational belief models because the third state $(\mathrm{b_3})$ may happen in them. That is, in the models, a schema

$$\mathbf{D_c}. \quad \neg\mathbf{B}\neg p \to \mathbf{B}p$$

does not necessarily hold. Note that the schema is equivalent to

$$\neg(\neg\mathbf{B}p \wedge \neg\mathbf{B}\neg p),$$

which denies the third state.

Conversely speaking, if the schema $\mathbf{D_c}$ holds, there is no incompleteness. Its corresponding condition in minimal models is

$$[\mathrm{d_c}] \quad \forall w \in W \ \forall X \subseteq W(X^c \notin N_w \Rightarrow X \in N_w).$$

Thus we have the following theorem:

Theorem 5

1. In general, there may be incompleteness in rational belief models.

2. There is no incompleteness in a rational belief model if it further satisfies $[\mathrm{d_c}]$. $\square$

Note that the counterpart of $[\mathrm{d_c}]$ in Kripke models is

$$(\mathrm{d_c}) \ \text{partial functionality:} \ \forall w, w', w'' \in W \ ((wRw' \wedge wRw'') \Rightarrow w' = w''),$$

by which we have the following lemma:

Lemma 6 There are at most one possible world accessible from each world in complete cases. $\square$

3.3 Inconsistency

The above fourth state (b_4) cannot happen in rational belief models, because a schema **D** guarantees that

$$\neg(\mathbf{B}p \wedge \mathbf{B}\neg p)$$

holds for any sentence p.

Then which conditions should be removed from rational belief models to deal with inconsistency? An immediate answer is a removal of [d] corresponding to **D**, which means that there exists at least one sentence p such that

$$v(\mathbf{B}p \wedge \mathbf{B}\neg p, w) = T,$$

that is, $\|p\|, \|\neg p\| \in N_w$. But this leads us to an awful consequence: by condition [c], we have

$$\phi = \|p\| \cap \|\neg p\| \in N_w$$

and then, noticing both [m] and the truth condition given in Lemma 3, we must conclude that

$$(\phi \subseteq)\|p'\| \in N_w$$

for any sentence p', which means that an agent believes all sentences in $\mathcal{L}_\mathcal{P}$ at the world w. This is a modal logical counterpart of the fact that *any conclusion can be deduced from inconsistency* in classical logic and, of course, should be avoided.

Thus what we should defend in order not to suffer belief in all sentences is not [d] but

$$[\mathrm{p}] \quad \forall w \in W \ (\phi \notin N_w)$$

which corresponds to a schema

$$\mathbf{P}. \quad \neg\mathbf{B}\bot.$$

But, to our regret, conditions [p] and [d] are equivalnet to each other in any rational belief model.

Then can't we remove [d] without sufferring belief in all sentences?—Our answer here is YES. Notice that condition [c] plays the essential role in the above treatment of inconsistecy. If we remove [c] from rational belief models, then, in general, the two conditions [d] and [p] are no longer equivalent in the models and only the implication

$$[\mathrm{d}] \ \Rightarrow \ [\mathrm{p}]$$

holds. Thus we can also remove [d] without losing [p]. Consequently it is sufficient to remove both [c] and [d] from and add [p] to a rational belief model in order to deal with inconcsistency without sufferring belief in all sentences:

Theorem 7

1. There is no inconsistency in rational belief models.

2. There may be belief in two inconsistent sentences, which does not conclude belief in all sentences as its consequence, in a minimal model if it satisfies [m], [n], [p], [iv], and [v]. $\square$

3.4   General Uncertain Models

By the above consideration, we define the following class of minimal models:

Definition 8 A minimal model is called a *general uncertain* model (GU-model, for short) iff it satisfies [m], [n], [p], [iv], and [v]. □

A GU-model is a model of a system *EMNP45*. We can show the following remarkable property of a finite GU-model (FGU-model, for short), which suggests a unified way of representation of both incompleteness and inconsistency to us:

Theorem 9 Let $\mathcal{M} = <W, \{N_w\}_{w\in W}, v>$ be a FGU-model. Then, for each world w, there exist k_w non-empty minimal elements $\mathcal{W}_w^1, \cdots, \mathcal{W}_w^{k_w}$ of N_w with respect to set inclusion such that

$$\forall X \in N_w \; \exists i \in \{1, \cdots, k_w\}, \; \mathcal{W}_w^i \subseteq X. \quad \square$$

By the theorem, a FGU-model can be reformulated as

$$< W, \{\{\mathcal{W}_w^i\}_{1 \leq i \leq k_w}\}_{w \in W}, v >,$$

and a truth-condition for modal sentences is given by the following lemma:

Lemma 10 (*Truth condition for modal sentences in a FGU-model*)

$$v(\mathbf{B}p, w) = \begin{cases} T, & \text{if } \mathcal{W}_w^i \subseteq \|p\| \text{ for } \exists i \in \{1, \cdots, k_w\} \\ F, & \text{otherwise} \end{cases} \quad . \quad \square$$

The condition means that *an agent may have plural sets of accessible worlds* at each world if there is inconsistency.

    Then, noticing Lemmas 6 and 10, we have the following characterization of incompleteness and inconsistency in a FGU-model:

Theorem 11 Let $\mathcal{M}$ be a FGU-model.

1. There is *incompleteness* at a world w in $\mathcal{M}$ iff $|\mathcal{W}_w^i| \geq 2$ for some $i \in \{1, \cdots, k_w\}$.

2. There is *inconsistency* at a world w in $\mathcal{M}$ iff $k_w \geq 2$. □

4   Concluding Remarks

In this paper, we have specified the class of general uncertainty models in which we can represent both incompleteness and inconsistency in information in a unified way. The key idea is plurality of sets of accessible worlds which may collapse a single set of accessible worlds in case that information is consistent, and which may further collapse a singleton of an accessible world in case that information is complete. Thus incompleteness is represented by plurality of accessibile worlds, while inconsistency is represented by plurality of sets of accessible worlds.

Acknowledgements.
The first author was partially supported by Grant-in-Aid No.12878054 for Exploratory Research of the Japan Society for the Promotion of Science of Japan.

References

[1] Carnap, R. (1947): *"Meaning and Necessity : A Study in Semantics and Modal Logic"*. The University of Chicago Press.

[2] Chellas, B.F. (1980): *"Modal Logic : An Introduction"*. Cambridge University Press.

[3] Dubois, D., Lang, J. and Prade, H. (1992): "Inconsistency in Possibilistic Knowledge Bases: To Live with It or not Live with It". L.A.Zadeh et al(eds.), *Fuzzy Logic for the Management of Uncertainty*. John Wiley and Sons, pp.335-351.

[4] Dubois, D. and Prade, H. (1985): *"Théorie des Possibilités : Applications à la Représentation des Connaissances en Informatique"*. Masson, 1985.

[5] Fagin, R. and Halpern, J.Y. (1988): "Belief, Awareness, and Limited Reasoning. Artificial Intelligence", Vol.34, pp.39-76.

Applications of EVALP Based Reasoning

Kazumi NAKAMATSU [1], Jair Minoro ABE [2]
and Atsuyuki SUZUKI [3]

[1] Himeji Institute of Technology, Shinzaike 1-1-12, HIMEJI 670-0092 JAPAN
`nakamatu@hept.himeji-tech.ac.jp`
[2] SENAC College of Computer Science and Technology,
Rua Galvao Bueno 430, 01506-000, Sao Paulo-SP-BRAZIL
`jairabe@uol.com.br`
[3] Shizuoka University, Johoku 3-5-1, HAMAMATSU 432-8011 JAPAN
`suzuki@cs.inf.shizuoka.ac.jp`

Abstract. We have already proposed an annotated logic program called
an EVALPSN (Extended Vector Annotated Logic Program with Strong
Negation) to deal with defeasible deontic reasoning. The aim of this
paper is to show two practical applications of EVALPSN based defeasible
deontic reasoning, a defeasible deontic action control system for a virtual
beetle robot and an automated safety verification system for railway
interlocking.
keywords :   defeasible deontic reasoning, extended vector annotated
logic program, railway interlocking, robot action control

1   Introduction

There are various situations in which agents have to choose one thing among some
conflicting things in our world. Defeasible logic is a formalization that can deal with such
decision making. We have also a case in which an agent reasons the next action to be
done based on norms (eg. laws, ethics, instructions etc.). We, human being, can reason
something based on some deontic notions (eg. obligation, permission, forbiddance etc.).
This kind of reasoning is formalized as deontic reasoning. The combination of these
reasonings provides defeasible deontic reasoning. Suppose that you are wondering if you
should go back to your home or drink beer in front of a pub. You might be forbidden
from drinking by your wife or a doctor, however you love to drink. Then you must have
a conflict in your mind. But you have to make a decision with some reasons after all.
Such reasoning is very common in our everyday life. We have formalized such defeasible
deontic reasoning from the viewpoint of paraconsistent(annotated) logic programming.

We have already proposed three kinds of annotated logic programs, ALPSN (Annotated Logic Program with Strong Negation) that can deal with default reasoning
[5], VALPSN(Vector Annotated Logic Program with Strong Negation) that can deal
with defeasible reasoning [6, 7, 8] and EVALPSN(Extended Vector Annotated Logic
Program with Strong Negation), an extended version of VALPSN, that can deal with
defeasible deontic reasoning [9].

In this paper, we introduce two major applications of EVALPSN, robot action control and automatic safety verification. In these applications, basically some kinds of
constraints(eg. policies, regulations, laws, etc.) are formalized in EVALPSN and permission in a form of EVALP clauses for those constrains can be derived from the

EVALPSN. If more than two permitted literals are derived, more appropriate literals are chosen by defeasible reasoning.

We introduce two applications of EVALPSN :

- a defeasible deontic action control for an autonomous robot. We assume a beetle robot that is supposed to travel in a maze having obstacles, the wall of the maze, pit halls and alcohol(he is forbidden from alcohol), to the goal. He has three different kinds of sensors to detect the obstacles. Depending on the sensor values, he has to deceid his next action by defeasible deontic reasoning. This robot control can be implemented as EVALPSN programming such that the inputs of the EVALPSN are different kinds of sensor values.

- an automatic verification system for railway interlocking safety. The interlocking safety verification is carried out by checking whether a route interlocking request and a sub-route release request contradict the safety properties that assures safe interlocking or not. In this system, EVALP with no strong negation is used as its inference engine, those requests and the safety properties are expressed in EVALP, and the verification can be executed as EVALP enquiry.

Generally, as EVALPSN contains strong negations, it has stable model semantics[3] and the computation of the stable models takes long time. Therefore it is not so appropriate for real time processing. However, if an EVALPSN is a stratified program, it has a well-founded model[2] and the strong negation in the EVALPSN can be treated as the Negation as Failure in PROLOG. Since the EVALPSN in the robot control is stratified, it can be good for real time control.

This paper is organized as follows : first, EVALPSN is reviewed, next, EVALPSN based robot action control and the safety verification for railway interlocking based on EVALP are introduced, and last, we provide the conclusion.

2   EVALPSN

We have formally defined VALPSN in [7]. After reviewing VALPSN briefly, we extend it to EVALPSN.

Generally, a truth value called an annotation is explicitly attached to each atom of annotated logic. For example, let p be an atom, μ an annotation, then $p\!:\!\mu$ is called an annotated atom. A partially ordered relation is defined on the set of annotations that has a complete lattice structure. An annotation in VALPSN is a 2-dimensional vector such that each component is a non-negative integer. We assume the complete lattice of vector annotations as follows : for each integer m, let

$$\mathcal{T}_v = \{(x,y)|\ 0 \leq x \leq m,\ 0 \leq y \leq m,\ x,y \text{ are integers }\}.$$

The ordering of $\mathcal{T}_v$ is denoted in the usual fashion by a symbol $\preceq_v$. Let $\vec{v_1} = (x_1, y_1)$ and $\vec{v_2} = (x_2, y_2)$. Then the ordering is defined as :

$$\vec{v_1} \preceq_v \vec{v_2} \ \texttt{iff} \ \ x_1 \leq x_2 \text{ and } y_1 \leq y_2.$$

In a vector annotated literal $p:(i,j)$, the first component i of the vector annotation (i,j) indicates the degree of positive information to support the literal p and the second one j indicates the degree of negative information. For example, a vector annotated literal $p:(3,2)$ can be informally interpreted that p is known to be true of strength 3 and false of strength 2.

Annotated logics have two kinds of negations, an epistemic negation($\neg$) and an ontological negation($\sim$). The epistemic negation followed by an annotated atom is interpreted as a mapping between annotations and the ontological negation is a strong negation that appears in classical logics. The epistemic negation of vector annotated logic is defined as the following exchange between the components of vector annotations,

$$\neg(p : (i,j)) = p : \neg(i,j) = p : (j,i).$$

Therefore, the epistemic negation followed by an annotated atomic formula can be eliminated by the above syntactic operation. On the other hand, the epistemic one followed by a non-atom is interpreted as a strong negation [1].

Definition 1(Strong Negation$\sim$)   Let A be an arbitrary formula in VALPSN.

$$\sim A =_{def} A \rightarrow (\neg(A \rightarrow A) \wedge (A \rightarrow A)).$$

Definition 2 (well vector annotated literal)   Let p be a literal. $p : (m, 0)$ is called a *well annotated literal*, where m is a non-negative integer.

Definition 3 (VALPSN)   If $L_0, \cdots, L_n$ are well vector annotated literals,

$$L_1 \wedge \cdots \wedge L_i \wedge \sim L_{i+1} \wedge \cdots \wedge \sim L_n \rightarrow L_0$$

is called a *vector annotated logic program clause with strong negation*(VALPSN clause). A VALPSN is a finite set of VALPSN clauses.

The main difference between VALPSN and EVALPSN is annotation and the lattice structure of annotations. An annotation in EVALPSN has a form of $[(i,j), \mu]$ such that the first component (i,j) is a 2-dimentional vector as same as an annotation of VALPSN and the second one μ is an index that expresses fact(α), obligation(β), and so on. An annotation in EVALPSN is called an extended vector annotation. We assume the complete lattice $\mathcal{T}$ for extended vector annotations as follows : $\mathcal{T} = \mathcal{T}_v \times \mathcal{T}_d$,

$$\mathcal{T}_v = \{(x,y)\mid 0 \leq x \leq 3,\ 0 \leq y \leq 3\} \quad \text{and} \quad \mathcal{T}_d = \{\perp, \alpha, \beta, \gamma, *_1, *_2, *_3, \top\},$$

where x and y are integers. The ordering of $\mathcal{T}_d$ is denoted by a symbol $\preceq_d$ and described by the Hasse's diagram(cube) in Figure 1. The intuitive meanings of the members of $\mathcal{T}_d$ are : $\perp$ (unknown), α (fact), β (obligation), γ (not obligation), $*_1$ (both fact and obligation), $*_2$ (both obligation and not obligation), $*_3$ (both fact and not obligation) and $\top$ (inconsistent). The diagram $\mathcal{T}_d$ shows that $\mathcal{T}_d$ is a trilattice in which the direction of a line

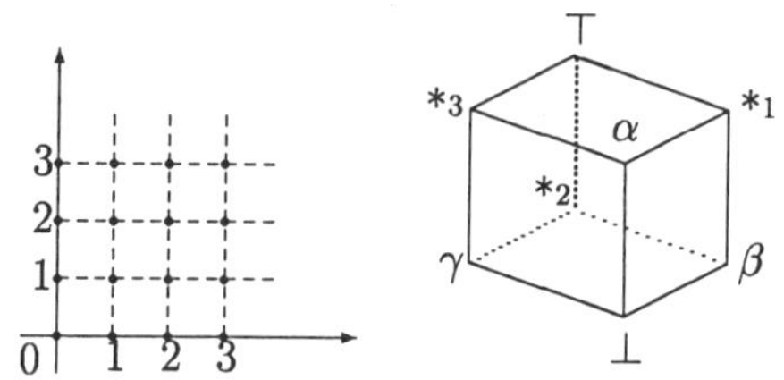

Fig. 1.   Lattice $\mathcal{T}_v$ and Lattice $\mathcal{T}_d$

$\overline{\gamma\beta}$ indicates the direction of a line $\overline{\gamma\beta}$ indicates *deontic truth*, the direction of a line $\overline{\perp *_2}$ indicates *deontic knowledge* and the direction of a line $\overline{\perp\alpha}$ indicates *actuality*. The ordering of $\mathcal{T}$ is denoted by a symbol $\preceq$ and defined as follows : let $[(i_1, j_1), \mu_1]$ and $[(i_2, j_2), \mu_2]$ be extended vector annotations,

$$[(i_1, j_1), \mu_1] \preceq [(i_2, j_2), \mu_2] \quad \text{iff} \quad (i_1, j_1) \preceq_v (i_2, j_2) \text{ and } \mu_1 \preceq_d \mu_2.$$

We provide an intuitive interpretation for some members of $\mathcal{T}$. For example, an extended vector annotated literal $p : [(3,0), \alpha]$ can be informally interpreted as "it is known that it is a fact that p is true of strength 3", and $q : [(0,2), \beta]$ can be also informally interpreted as "it is known to be obligatory that q is false of strength 2.

There are two kinds of epistemic negation, $\neg_1$ and $\neg_2$, in the extended vector annotated logic, which are regarded as mappings over $\mathcal{T}_v$ and $\mathcal{T}_d$, respectively.

Definition 4(Epistemic Negation of EVALPSN)

$$\neg_1([(i,j),\mu]) = [(j,i),\mu], \quad \forall \mu \in \mathcal{T}_d$$
$$\neg_2([(i,j),\bot]) = [(i,j),\bot], \quad \neg_2([(i,j),\alpha]) = [(i,j),\alpha],$$
$$\neg_2([(i,j),\beta]) = [(i,j),\gamma], \quad \neg_2([(i,j),\gamma]) = [(i,j),\beta],$$
$$\neg_2([(i,j),*_1]) = [(i,j),*_3], \quad \neg_2([(i,j),*_2]) = [(i,j),*_2],$$
$$\neg_2([(i,j),*_3]) = [(i,j),*_1], \quad \neg_2([(i,j),\top]) = [(i,j),\top].$$

We can eliminate syntactically the epistemic negation followed by annotated atom based on the definition. We can also define the strong (ontological) negation ($\sim$) in EVALPSN by the epistemic negations as well as **Definition 1**. The formal interpretations of the epistemic negations and the strong negation can be defined as well as the case of ALPSN [5, 1].

Definition 5 (well extended vector annotated literal)   Let ϕ be a literal. $\phi : [(m,0), d]$ is called a well extended vector annotated literal, where m is a non-negative integer and $d \in \{ \alpha, \beta, \gamma \}$.

Definition 6 (EVALPSN)   If $L_0, \cdots, L_n$ are well extended vector annotated literals,

$$L_1 \wedge \cdots \wedge L_i \wedge \sim L_{i+1} \wedge \cdots \wedge \sim L_n \rightarrow L_0$$

is called an *extended vector annotated logic program clause with strong negation* (EVALPSN clause). An EVALPSN is a finite set of EVALPSN clauses.

Deontic notions, obligation, permission and forbiddance, and fact can be expressed in EVALPSN as follows :

- fact is expressed by an extended vector annotation, $[(m,0),\alpha]$
- obligation is expressed by an extended vector = annotation, $[(m,0),\beta]$
- forbiddance is expressed by an extended vector = annotation, $[(0,m),\beta]$
- permission is expressed by an extended vector = annotation, $[(0,m),\gamma]$,

where $m(1 \le m \le 3)$ is an integer.

3   Robot Action Control

In this section, we introduce a robot action control system based on EVALPSN.

3.1   Environment of the Robot

We suppose a beetle robot with some sensors called Mr.A travelling in a maze. There are some kinds of obstacles waiting for him in the maze as follows :

- wall — detectable by ultra-sonic sensors and he can not climb over it ;

- pit hall — detectable by floor sensors, although Mr.A might be able to jump over it, he wants to avoid it as possible ;

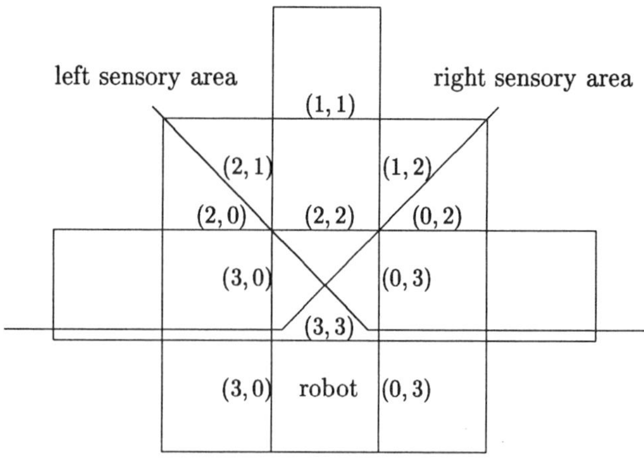

Fig. 2.   Sensory   Areas

- alcohol(Japanese Sake) — detectable by alcohol sensors, and he is forbidden from drinking however he loves Sake.

We also assume that :

- the maze is divided into square blocks, the width of the maze is 2 blocks, and all obstacles and Mr.A's own body can be held in just one block,

- Mr.A can do the three actions, **1.** going straight, **2.** left quarter turn, and **3.** right quarter turn ;

- Mr.A's action is decided at each step based on the discrete time sequence { 0, 1, 2, $\cdots$ }.

3.2   Specification of the Robot Mr. A

The beetle robot Mr.A has the following sensors : touch sensors(an) that are used to detect a special obstacle(eg. the goal of the maze) touching to Mr.A, ultra-sonic sensor(ob) that are used to detect an obstacle(eg. the wall of the maze) in front of Mr.A, floor sensors(fl) that are used to detect a pit hall, and alcohol sensor(al) that are used to detect alcohol(Sake). Mr.A has right and left sensors per each kind. Sensor values can be represented by vector annotations. For example, $ob(t) : [(2,2), \alpha]$ represents that it is a fact that an obstacle has been detected in front of Mr.A at the t-th step. The sensor value is represented as a 2-dimensional vector annotation such that the first component expresses that the belief degree of the sensor value of the left ultra-sonic sensor and the second one does that the belief degree of the sensor value of the right one. The belief degree takes an integer value and ranges from 0 to 3. Each vector (i,j) attached to the block border line in **Fig.2** expresses that if an obstacle is detected in the block, the sensor value is (i,j). The sensory areas of those sensors are also described in **Fig.2**. The sensor values { (3,0), (0,3), (3,3) } out of the sensory areas are only for the touch sensors. For example, (2,1) expresses that the left sensor sensed an obstacle in the strength 2 and the right sensor sensed it in the strength 1.

3.3　the Policy of the Robot's Behavior

Mr.A has the following policies to behave in the maze : **1.** he wants to reach the goal without crashing the wall, **2.** he does not want fall into a pit hall, **3.** he wants to reach the goal with following the right wall, and **4.** he is forbidden from drinking alcohol. We assume that there are superiority relations among the policies **1** to **4** such that **4 < 3 < 2 < 1**. We formalize his behavior in EVALPSN based on the superiority relations.

3.4　EVALPSN Formalization for Robot Action Control

We formalize the robot action control in EVALPSN. Generally, if it is a fact that an obstacle has been detected, forbiddance with different kinds of strength is derived depending on the kind of the obstacle. For example, if the wall is detected in left front of Mr.A, the left turn is strongly forbidden, because he can never climb over the wall. Moreover, if a pit hall is also detected in front and alcohol is detected in right front of him, then there is no way to avoid those obstacles. Each direction with obstacles can be regarded as forbiddance. Then he has to determine the next action. Each forbiddance has differnt degrees of strength, and he choose the weakest forbiddance by alcohol and turn to the right. Then the forbiddance turns to permission.

We formalize the defeasible deontic reasoning in EVALPSN. The strength(0—3) of forbiddance is determined based on the superiority relations shown in the previous subsection. The detection of the wall implies the strongest forbiddance to head the direction, because Mr.A can not climb over it. The strongest forbiddance has the level 3 strength. The level 2 strength means "defeasibly derived", the level 1 strength means "defeasibly defeated", and the level 0 strength means no information to support the reasoning based on forbiddance. We use the following abbreviations to represent the strength of the sensor values that imply the positions of obstacles :

front($\mathbf{f}$) : $(3,3)$	left($\mathbf{l}$) : $(3,0)$
right($\mathbf{r}$) : $(0,3)$	far front($\mathbf{ff}$) : $(2,2)$
left front($\mathbf{lf}$) : $(3,0)$	right front($\mathbf{rf}$) : $(0,3)$
left far front($\mathbf{lff}$) : $(2,1)$	right far front($\mathbf{rff}$) : $(1,2)$
left left front($\mathbf{llf}$) : $(2,0)$	right right front($\mathbf{rrf}$) : $(0,2)$

where the abbreviations $\mathbf{l}$ and $\mathbf{r}$ are only for touch sensors.

We have the following EVALPSN to derive various forbiddance.

- If the wall is detected in far front of Mr.A, then he is forbidden from going straight.

$$ob(t) : [\mathbf{ff}, \alpha] \rightarrow forward(t) : [(0,3), \beta] \tag{1}$$

Similarly, we have the following EVALPSN clauses.

$$ob(t) : [\mathbf{lf}, \alpha] \rightarrow lturn(t) : [(0,3), \beta] \tag{2}$$
$$ob(t) : [\mathbf{rf}, \alpha] \rightarrow rturn(t) : [(0,3), \beta] \tag{3}$$
$$fl(t) : [\mathbf{ff}, \alpha] \rightarrow forward(t) : [(0,2), \beta] \tag{4}$$
$$fl(t) : [\mathbf{lf}, \alpha] \rightarrow lturn(t) : [(0,2), \beta] \tag{5}$$
$$fl(t) : [\mathbf{rf}, \alpha] \rightarrow rturn(t) : [(0,2), \beta] \tag{6}$$

- If alcohol is detected in far front of Mr.A, the wall, a pit hall or alcohol are detected in left front of him, and the wall, a pit hall or alcohol are detected in right front of him, then he is forbidden on the level 1 strength from going straight. In this case, alcohol derives the weakest forbiddance that turns out to permission.

$$al(t) : [\mathtt{ff}, \alpha] \wedge ob(t) : [\mathtt{lf}, \alpha] \wedge ob(t) : [\mathtt{rf}, \alpha] \rightarrow forward(t) : [(0,1), \beta] \qquad (7)$$

$$al(t) : [\mathtt{ff}, \alpha] \wedge ob(t) : [\mathtt{lf}, \alpha] \wedge fl(t) : [\mathtt{rf}, \alpha] \rightarrow forward(t) : [(0,1), \beta] \qquad (8)$$

$$al(t) : [\mathtt{ff}, \alpha] \wedge ob(t) : [\mathtt{lf}, \alpha] \wedge al(t) : [\mathtt{rf}, \alpha] \rightarrow forward(t) : [(0,1), \beta] \qquad (9)$$

$$al(t) : [\mathtt{ff}, \alpha] \wedge fl(t) : [\mathtt{lf}, \alpha] \wedge ob(t) : [\mathtt{rf}, \alpha] \rightarrow forward(t) : [(0,1), \beta] \qquad (10)$$

$$al(t) : [\mathtt{ff}, \alpha] \wedge fl(t) : [\mathtt{lf}, \alpha] \wedge fl(t) : [\mathtt{rf}, \alpha] \rightarrow forward(t) : [(0,1), \beta] \qquad (11)$$

$$al(t) : [\mathtt{ff}, \alpha] \wedge fl(t) : [\mathtt{lf}, \alpha] \wedge al(t) : [\mathtt{rf}, \alpha] \rightarrow forward(t) : [(0,1), \beta] \qquad (12)$$

$$al(t) : [\mathtt{ff}, \alpha] \wedge al(t) : [\mathtt{lf}, \alpha] \wedge ob(t) : [\mathtt{rf}, \alpha] \rightarrow forward(t) : [(0,1), \beta] \qquad (13)$$

$$al(t) : [\mathtt{ff}, \alpha] \wedge al(t) : [\mathtt{lf}, \alpha] \wedge fl(t) : [\mathtt{rf}, \alpha] \rightarrow forward(t) : [(0,1), \beta] \qquad (14)$$

$$al(t) : [\mathtt{ff}, \alpha] \wedge al(t) : [\mathtt{lf}, \alpha] \wedge al(t) : [\mathtt{rf}, \alpha] \rightarrow forward(t) : [(0,1), \beta] \qquad (15)$$

- If alcohol is detected in left front of Mr.A, the wall, a pit hall or alcohol are detected in right front of him, and the wall, a pit hall or alcohol are detected in far front of him, then he is forbidden on the level 1 strength from turning to the left.

$$al(t) : [\mathtt{lf}, \alpha] \wedge ob(t) : [\mathtt{rf}, \alpha] \wedge ob(t) : [\mathtt{ff}, \alpha] \rightarrow lturn(t) : [(0,1), \beta] \qquad (16)$$

$$al(t) : [\mathtt{lf}, \alpha] \wedge ob(t) : [\mathtt{rf}, \alpha] \wedge fl(t) : [\mathtt{ff}, \alpha] \rightarrow lturn(t) : [(0,1), \beta] \qquad (17)$$

$$al(t) : [\mathtt{lf}, \alpha] \wedge ob(t) : [\mathtt{rf}, \alpha] \wedge al(t) : [\mathtt{ff}, \alpha] \rightarrow lturn(t) : [(0,1), \beta] \qquad (18)$$

$$al(t) : [\mathtt{lf}, \alpha] \wedge fl(t) : [\mathtt{rf}, \alpha] \wedge ob(t) : [\mathtt{ff}, \alpha] \rightarrow lturn(t) : [(0,1), \beta] \qquad (19)$$

$$al(t) : [\mathtt{lf}, \alpha] \wedge fl(t) : [\mathtt{rf}, \alpha] \wedge fl(t) : [\mathtt{ff}, \alpha] \rightarrow lturn(t) : [(0,1), \beta] \qquad (20)$$

$$al(t) : [\mathtt{lf}, \alpha] \wedge fl(t) : [\mathtt{rf}, \alpha] \wedge al(t) : [\mathtt{ff}, \alpha] \rightarrow lturn(t) : [(0,1), \beta] \qquad (21)$$

$$al(t) : [\mathtt{lf}, \alpha] \wedge al(t) : [\mathtt{rf}, \alpha] \wedge ob(t) : [\mathtt{ff}, \alpha] \rightarrow lturn(t) : [(0,1), \beta] \qquad (22)$$

$$al(t) : [\mathtt{lf}, \alpha] \wedge al(t) : [\mathtt{rf}, \alpha] \wedge fl(t) : [\mathtt{ff}, \alpha] \rightarrow lturn(t) : [(0,1), \beta] \qquad (23)$$

$$al(t) : [\mathtt{lf}, \alpha] \wedge al(t) : [\mathtt{rf}, \alpha] \wedge al(t) : [\mathtt{ff}, \alpha] \rightarrow lturn(t) : [(0,1), \beta] \qquad (24)$$

- If alcohol is detected in right front of Mr.A, the wall, a pit hall or alcohol are detected in far front of him, and the wall, a pit hall or alcohol are detected in left front of him, then he is forbidden on the level 1 strength from turning to the right.

$$al(t) : [\mathtt{rf}, \alpha] \wedge ob(t) : [\mathtt{ff}, \alpha] \wedge ob(t) : [\mathtt{lf}, \alpha] \rightarrow rturn(t) : [(0,1), \beta] \qquad (25)$$

$$al(t) : [\mathtt{rf}, \alpha] \wedge ob(t) : [\mathtt{ff}, \alpha] \wedge fl(t) : [\mathtt{lf}, \alpha] \rightarrow rturn(t) : [(0,1), \beta] \qquad (26)$$

$$al(t) : [\mathtt{rf}, \alpha] \wedge ob(t) : [\mathtt{ff}, \alpha] \wedge al(t) : [\mathtt{lf}, \alpha] \rightarrow rturn(t) : [(0,1), \beta] \qquad (27)$$

$$al(t) : [\mathtt{rf}, \alpha] \wedge fl(t) : [\mathtt{ff}, \alpha] \wedge ob(t) : [\mathtt{lf}, \alpha] \rightarrow rturn(t) : [(0,1), \beta] \qquad (28)$$

$$al(t) : [\mathtt{rf}, \alpha] \wedge fl(t) : [\mathtt{ff}, \alpha] \wedge fl(t) : [\mathtt{lf}, \alpha] \rightarrow rturn(t) : [(0,1), \beta] \qquad (29)$$

$$al(t) : [\mathtt{rf}, \alpha] \wedge fl(t) : [\mathtt{ff}, \alpha] \wedge al(t) : [\mathtt{lf}, \alpha] \rightarrow rturn(t) : [(0,1), \beta] \qquad (30)$$

$$al(t) : [\mathtt{rf}, \alpha] \wedge al(t) : [\mathtt{ff}, \alpha] \wedge ob(t) : [\mathtt{lf}, \alpha] \rightarrow rturn(t) : [(0,1), \beta] \qquad (31)$$

$$al(t) : [\mathtt{rf}, \alpha] \wedge al(t) : [\mathtt{ff}, \alpha] \wedge fl(t) : [\mathtt{lf}, \alpha] \rightarrow rturn(t) : [(0,1), \beta] \qquad (32)$$

$$al(t) : [\mathtt{rf}, \alpha] \wedge al(t) : [\mathtt{ff}, \alpha] \wedge al(t) : [\mathtt{lf}, \alpha] \rightarrow rturn(t) : [(0,1), \beta] \qquad (33)$$

- If alcohol is detected in far,left or right front of Mr.A, and there is at least one direction with no obstacle, then he is forbidden on the level 2 strength from drinking alcohol. These conditions can be formulated in EVALPSN :

$$al(t) : [\mathtt{ff}, \alpha] \wedge \sim al(t) : [\mathtt{lf}, \alpha] \wedge \sim fl(t) : [\mathtt{lf}, \alpha] \wedge \sim ob(t) : [\mathtt{lf}, \alpha]$$

$$\to forward(t) : [(0,2), \beta] \tag{34}$$

$$al(t) : [\text{ff}, \alpha] \wedge \sim al(t) : [\text{rf}, \alpha] \wedge \sim fl(t) : [\text{rf}, \alpha] \wedge \sim ob(t) : [\text{rf}, \alpha]$$
$$\to forward(t) : [(0,2), \beta] \tag{35}$$

$$al(t) : [\text{1f}, \alpha] \wedge \sim al(t) : [\text{ff}, \alpha] \wedge \sim fl(t) : [\text{ff}, \alpha] \wedge \sim ob(t) : [\text{ff}, \alpha]$$
$$\to lturn(t) : [(0,2), \beta] \tag{36}$$

$$al(t) : [\text{1f}, \alpha] \wedge \sim al(t) : [\text{rf}, \alpha] \wedge \sim fl(t) : [\text{rf}, \alpha] \wedge \sim ob(t) : [\text{rf}, \alpha]$$
$$\to lturn(t) : [(0,2), \beta] \tag{37}$$

$$al(t) : [\text{rf}, \alpha] \wedge \sim al(t) : [\text{ff}, \alpha] \wedge \sim fl(t) : [\text{ff}, \alpha] \wedge \sim ob(t) : [\text{ff}, \alpha]$$
$$\to rturn(t) : [(0,2), \beta] \tag{38}$$

$$al(t) : [\text{rf}, \alpha] \wedge \sim al(t) : [\text{1f}, \alpha] \wedge \sim fl(t) : [\text{1f}, \alpha] \wedge \sim ob(t) : [\text{1f}, \alpha]$$
$$\to rturn(t) : [(0,2), \beta] \tag{39}$$

If Mr.A travels with following the right wall, he has to consider the superiority relation, left turn < going straight < right turn. Then, we have the following EVALPSN to derive an action that he should do at the next step.

• If the left turn is permitted, and going straight and the right turn are forbidden, he should turn to the left.

$$\sim lturn(t) : [(0,2), \beta] \wedge forward(t) : [(0,2), \beta] \wedge rturn(t) : [(0,2), \beta]$$
$$\to lturn(t+1) : [(3,0), \beta] \tag{40}$$

Similary, we have the following EVALPSN :

$$lturn(t) : [(0,2), \beta] \wedge \sim forward(t) : [(0,2), \beta] \wedge rturn(t) : [(0,2), \beta]$$
$$\to forward(t+1) : [(3,0), \beta] \tag{41}$$

$$lturn(t) : [(0,2), \beta] \wedge forward(t) : [(0,2), \beta] \wedge \sim rturn(t) : [(0,2), \beta]$$
$$\to rturn(t+1) : [(3,0), \beta] \tag{42}$$

$$\sim lturn(t) : [(0,2), \beta] \wedge \sim forward(t) : [(0,2), \beta] \wedge rturn(t) : [(0,2), \beta]$$
$$\to forward(t+1) : [(3,0), \beta] \tag{43}$$

$$lturn(t) : [(0,2), \beta] \wedge \sim forward(t) : [(0,2), \beta] \wedge \sim rturn(t) : [(0,2), \beta]$$
$$\to rturn(t+1) : [(3,0), \beta] \tag{44}$$

$$\sim lturn(t) : [(0,2), \beta] \wedge forward(t) : [(0,2), \beta] \wedge \sim rturn(t) : [(0,2), \beta]$$
$$\to rturn(t+1) : [(3,0), \beta] \tag{45}$$

$$\sim lturn(t) : [(0,2), \beta] \wedge \sim forward(t) : [(0,2), \beta] \wedge \sim rturn(t) : [(0,2), \beta]$$
$$\to rturn(t+1) : [(3,0), \beta] \tag{46}$$

3.4.1 A Sinple Example for the Action Control

We take an example of the action control(**Fig.3**). Initially, suppose there is a pit hall in far front of Mr.A and the wall of the maze in right front of him. Then the sensor values, $fl(0) : [\text{ff}, \alpha]$ and $ob(0) : [\text{rf}, \alpha]$, are input and the EVALP clause, $lturn(1) : [(3,0), \beta]$ is derived by the EVALPSN clauses, (3), (4) and (40). Nextly, there is the wall in far front of Mr.A after the left turn. Then the sensor value, $ob(1) : [\text{ff}, \alpha]$, is input and the EVALP clause, $rturn(2) : [(3,0), \beta]$ is derived by the EVALPSN clauses, (1) and (45). Lastly, there is the wall in left front of Mr.A and a pit hall in right front of him after the right turn. Then the sensor values, $ob(2) : [\mathtt{lf}, \alpha]$ and $fl(2) : [\mathtt{rf}, \alpha]$, are input and the EVALP clause, $forward(3) : [(3, 0), \beta]$ is derived by the EVALPSN clauses, (2), (6) and (41).

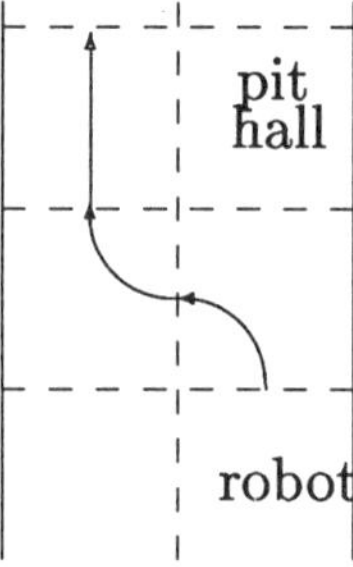

Fig. 3. Eg. Mr.A's Actions

3.5　Automatic Safety Verification for Railway Interlocking

Verification for railway interlocking safety is a crucial issue. One method to resolve the issue has been proposed and explained in details in Morley's Ph.d thesis[4]. The railway interlocking safety verification is carried out by checking whether a route interlocking request and a sub-route release request contradict the safety properties or not. The safety properties are :

MX It is never the case that two or more of the sub-routes over a given track section are simultaneously locked ;

RT Whenever a route is set, all its component sub-routes are locked ;

PT Whenever a sub-route over a track section containing points is locked, the points are controlled in alignment with that sub-route.

The basic idea to prove the interlocking safety based on EVALP is :
• since those safety properties can be regarded as regulations that imply deontic notions, obligation, forbiddance and permission, they can be interpreted deontically and represented in EVALP ;
• both the requests consist of conditional parts(if-parts) and conclusion parts(then-parts), and if the conditional part is assumed and the conclusion part does not contradict the safety properties, then the safety of the request is assured.

We regard this verification process as that if the conditional part holds as obligation and the conclusion part is permitted against the safety properties, then the safety of the request is assured ; therefore, those requests can be also represented and checked the safety in EVALP programming system.

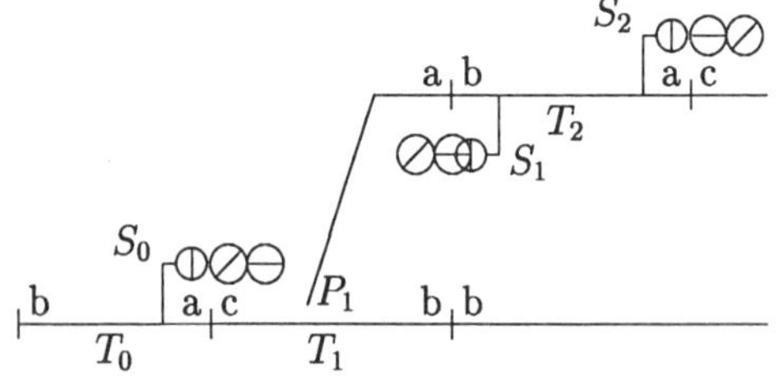

Fig. 4. Signalling Scheme Plan

We take a concrete example of the safety verification based on a signaling schema plan (**Fig.4**) from [4] and show how to construct the EVALP. The entities declared in this schema are : **Track Circuits** $\mathcal{T} = \{T_0, T_1, T_2\}$, **Points** $\mathcal{P} = \{P_1\}$ and **Signals** $\mathcal{S} = \{S_0, S_1, S_2\}$ that represent physical entities in the network, and **Routes** $\mathcal{R} = \{R_{02}, R_1\}$

and **Sub-routes** $\mathcal{U} = \{T_0^{ab}, T_0^{ba}, T_1^{ca}, \ldots, T_2^{ba}\}$ that represent logical control entities. For example, T_0^{ab} denotes the sub-route from a to b in the track section T_0, R_{02} denotes the route from the signal S_0 to the signal S_2, and R_{02} contains the sub-routes T_0^{ba}, T_1^{ca} and T_2^{ba}. $\mathbf{1}$ and $\mathbf{f}$ denote the two states, *locked* and *free*, of a sub-route. $\mathbf{s}$ and $\mathbf{xs}$ denote the *set* or *unset* states of the route. $\mathbf{o}$ and $\mathbf{c}$ denote track circuits *occupied* or *clear*. Occupied denotes there is a train in the track circuit. Points have 4 states : $\mathbf{cn}$(*controlled normal*), (eg. P_1 $\mathbf{cn}$ denotes the point P_1 is controlled normal(ca or ac) direction), $\mathbf{cr}$ (*controlled reverse*), $\mathbf{cnf}$ (*controlled normal* or *free to move*) (eg. P_1 $\mathbf{cnf}$ denotes the point P_1 is controlled normal, or if it is not, P_1 can be moved to reverse position if the normal sub-routes are free), and $\mathbf{crf}$ (*controlled reverse* or *free to move*). Then some panel route requests are also declared. For example,

$$*\text{Q}02 \quad \text{if } P_1 \ \mathbf{crf}, \ T_1^{ac} \ \mathbf{f}, \ T_2^{ab} \ \mathbf{f}$$
$$\text{then} \quad R_{02} \ \mathbf{s}, \ P_1 \ \mathbf{cr}, \ T_1^{ca} \ \mathbf{1}, \ T_2^{ba} \ \mathbf{1} \tag{47}$$

is a route request for the route R_{02}.

We translate the safety properties, **MX**, **RT**, **PT** into an EVALP. We use the denotations $\mathbf{1}$, $\mathbf{f}$, $\mathbf{s}$, $\mathbf{xs}$, $\mathbf{cn}$, $\mathbf{cr}$, etc., as the first component of an extended vector annotation in this EVALP instead of a 2-dimensional vector annotation.

MX

We assume that sub-route's states locked and free are represented by the annotations, $\mathbf{1}$ and $\mathbf{f}$, in EVALP such that $\neg_1([\mathbf{1}, \mu]) = [\mathbf{f}, \mu]$, $\neg_1([\mathbf{f}, \mu]) = [\mathbf{1}, \mu]$. We also assume the vector annotations constitute the lattice structure $\mathcal{T}_1 = \{ \perp, \mathbf{f}, \mathbf{1}, \top \}$ that are translated into the lattice $\{ (0,0), (0,1), (1,0), (1,1) \}$ of vector annotations when programming actually. As well as the sub-route's states, we assume the lattice structure $\mathcal{T}_2 = \{ \perp, \mathbf{c}, \mathbf{o}, \top \}$ of the vector annotations for the track section's states, clear(c), occupied(o) and undefined($\perp$). Generally, **MX** states that it is forbidden that two or more of the sub-routes over a given track section are simultaneously locked. We interpret the forbiddance as the following permission : if one of the sub-routes should be free, the other sub-route can be locked. Then $\mathbf{MX}[T_0^{ab}, T_0^{ba}]$ is translated into the following EVALP clauses :

$$T(0, ab) : [\mathbf{f}, \beta] \rightarrow T(0, ba) : [\mathbf{f}, \gamma] \tag{48}$$
$$T(0, ba) : [\mathbf{f}, \beta] \rightarrow T(0, ab) : [\mathbf{f}, \gamma] \tag{49}$$

As well as the above translation, we obtain the following EVALP for the properties, $\mathbf{MX}[T_2^{ab}, T_2^{ba}]$ and $\mathbf{MX}[T_1^{ac}, T_1^{bc}, T_1^{ca}, T_1^{cb}]$:

$$T(2, ab) : [\mathbf{f}, \beta] \rightarrow T(2, ba) : [\mathbf{f}, \gamma] \tag{50}$$
$$T(2, ba) : [\mathbf{f}, \beta] \rightarrow T(2, ab) : [\mathbf{f}, \gamma] \tag{51}$$
$$T(1, ac) : [\mathbf{f}, \beta] \wedge P(1) : [\mathbf{cn}, \gamma] \rightarrow T(1, ca) : [\mathbf{f}, \gamma] \tag{52}$$
$$T(1, ca) : [\mathbf{f}, \beta] \wedge P(1) : [\mathbf{cn}, \gamma] \rightarrow T(1, ac) : [\mathbf{f}, \gamma] \tag{53}$$
$$T(1, cb) : [\mathbf{f}, \beta] \wedge P(1) : [\mathbf{cr}, \gamma] \rightarrow T(1, bc) : [\mathbf{f}, \gamma] \tag{54}$$
$$T(1, bc) : [\mathbf{f}, \beta] \wedge P(1) : [\mathbf{cr}, \gamma] \rightarrow T(1, cb) : [\mathbf{f}, \gamma] \tag{55}$$

RT

We assume the states, set and unset, of routes are represented by the vector annotations, $\mathbf{s}$ and $\mathbf{xs}$, which constitutes the lattice structure $\mathcal{T}_3 = \{ \perp, \mathbf{xs}, \mathbf{s}, \top \}$, and the same assumption as well as the case of **MX**. This **RT** indicates if all the sub-routes contained in a route can be locked, the route can be set. We interpret $\mathbf{RT}(R_{02}, [T_1^{ca}, T_2^{ba}])$ as if the sub-routes T_1^{ca} and T_2^{ba} can be locked then the route R_{02} can be set. We obtain the following EVALP :

$$T(1, ca) : [\mathbf{f}, \gamma] \wedge T(2, ba) : [\mathbf{f}, \gamma] \to R(02) : [\mathbf{xs}, \gamma] \tag{56}$$

PT

This **PT** indicates that if one of the sub-routes, T_1^{bc}, T_1^{cb}, should be free and the other one can be locked, the point P_1 can be controlled normal. We represent the safety properties $\mathbf{PTcn}(P_1, [T_1^{bc}, T_1^{cb}])$ and $\mathbf{PTcr}(P_1, [T_1^{ac}, T_1^{ca}])$ in the following EVALP :

$$T(1, bc) : [\mathbf{f}, \gamma] \wedge T(1, cb) : [\mathbf{f}, \beta] \to P(1) : [\mathbf{cr}, \gamma] \tag{57}$$
$$T(1, bc) : [\mathbf{f}, \beta] \wedge T(1, cb) : [\mathbf{f}, \gamma] \to P(1) : [\mathbf{cr}, \gamma] \tag{58}$$
$$T(1, ac) : [\mathbf{f}, \gamma] \wedge T(1, ca) : [\mathbf{f}, \beta] \to P(1) : [\mathbf{cn}, \gamma] \tag{59}$$
$$T(1, ac) : [\mathbf{f}, \beta] \wedge T(1, ca) : [\mathbf{f}, \gamma] \to P(1) : [\mathbf{cn}, \gamma] \tag{60}$$

Here we represent the information about the point states $\mathbf{cnf}$ and $\mathbf{crf}$ in EVALP. The point state $\mathbf{cnf}(\mathbf{crf})$ represents the permission to move to the state $\mathbf{cr}(\mathbf{cn})$ if not $\mathbf{cn}(\mathbf{cr})$. The state $\mathbf{cnf}(\mathbf{crf})$ also represents that the point has already been controlled normal(reverse), or if the reverse(normal) side sub-routes should be free and no train(cleared) in the track section, the point can move to be controlled reverse(normal). The states of points are represented by the lattice $\{ \perp(0, 0), \mathbf{cnf}(1, 0), \mathbf{crf}(0, 1), \mathbf{cn}(2, 0), \mathbf{cr}(0, 2), \cdots, \top(2, 2) \}$ of vector annotations. It can be regarded that if one of the sub-routes T_1^{ac} and T_1^{ca} should be free, the point P_1 can be controlled normal($\mathbf{cn}$) under the condition the point P_1 should be $\mathbf{crf}$. Moreover, if both the sub-routes T_1^{bc} and T_1^{cb} should be free and no train in the track section T_1, the point P_1 can be controlled normal. Therefore, we obtain the first three EVALP clauses for P_1 $\mathbf{crf}$, and similarly the second three EVALP clauses for P_1 $\mathbf{cnf}$:

$$P(1) : [\mathbf{crf}, \beta] \wedge T(1, ac) : [\mathbf{f}, \beta] \to P(1) : [\mathbf{cn}, \gamma] \tag{61}$$
$$P(1) : [\mathbf{crf}, \beta] \wedge T(1, ca) : [\mathbf{f}, \beta] \to P(1) : [\mathbf{cn}, \gamma] \tag{62}$$
$$P(1) : [\mathbf{crf}, \beta] \wedge T(1, bc) : [\mathbf{f}, \beta] \wedge T(1, cb) : [\mathbf{f}, \beta] \wedge T(1) : [\mathbf{c}, \beta]$$
$$\to P(1) : [\mathbf{cr}, \gamma] \tag{63}$$
$$P(1) : [\mathbf{cnf}, \beta] \wedge T(1, bc) : [\mathbf{f}, \beta] \to P(1) : [\mathbf{cr}, \gamma] \tag{64}$$
$$P(1) : [\mathbf{cnf}, \beta] \wedge T(1, cb) : [\mathbf{f}, \beta] \to P(1) : [\mathbf{cr}, \gamma] \tag{65}$$
$$P(1) : [\mathbf{cnf}, \beta] \wedge T(1, ac) : [\mathbf{f}, \beta] \wedge T(1, ca) : [\mathbf{f}, \beta] \wedge T(1) : [\mathbf{c}, \beta]$$
$$\to P(1) : [\mathbf{cn}, \gamma]. \tag{66}$$

3.5.1 Safety Verification Example

Now we take the route request Q02 (47) as an example of the safety verification. It is verified as follows :

1. translate some definitions and the safety properties into EVALP EP_1;

2. translate all terms in the conditional part as obligation into EVALP EP_2;

3. translate each term in the conclusion part as permission into an EVALP clause and inquire it against an EVALP $EP = EP_1 \cup EP_2$, then if **yes** is returned, the safety is assured, and if **no** is returned, not assured.

Then $EP = \{(48), \cdots, (66), P(1) : [\text{crf}, \beta], T(1, ac) : [\text{f}, \beta], T(2, ab) : [\text{f}, \beta]\}$ and the inquired EVALP clauses are $\{R(02) : [\text{xs}, \gamma], P(1) : [\text{cn}, \gamma], T(1, ca) : [\text{f}, \gamma], T(2, ba) : [\text{f}, \gamma]\}$. We obtain the answer **yes** from EP for all the inquired EVALP clauses.

4 Conclusion

In this paper, we proposed two types of practical applications of EVALP based defeasible deontic reasoning, a defeasible deontic action control for a robot and an automated safety verification for railway interlocking. The robot action control is for a virtual robot and not so practical. However, it shows the guidelines for more practical applications (eg. control systems for moving agents) of EVALP(SN) based defeasible deontic reasoning. On the other hand, the safety verification for railway interlocking is a very useful application and it is expected to be applied to actual systems. There are various applications to safety verification. Some of them require faster real time processing, (eg. air traffic control). Here what we want to say is that EVALP(SN) can be implemented on a microchip and we have already implemented the robot control EVALPSN on a microchip aiming faster processing, although we did not describe it in this paper. The existence of such a microchip must extend the applicatory area of EVALP(SN) in the viewpoint of defeasible deontic reasoning.

References

[1] Da Costa,N.C.A., Subrahmanian,V.S., and Vago,C. : The Paraconsistent Logics $P\mathcal{T}$. Zeitschrift für Mathematische Logic und Grundlangen der Mathematik **37** (1989) 139–148

[2] Gelder,A.V., Ross,K.A. and Schlipf,J.S. : The Well-Founded Semantics for General Logic Programs. J.the Association for Computing Machinery, **38** (1991) 620–650

[3] Gelfond,M. and Lifschitz,V. : The Stable Model Semantics for Logic Programming. Proc. 5th International Conference and Symposium on Logic Programming (1989) 1070–1080

[4] Morley,J.M. : Safety Assurance in Interlocking Design. Ph.D Thesis, University of Edinburgh, 1996

[5] Nakamatsu,K. and Suzuki, A. : Annotated Semantics for Default Reasoning. Dai,R.(ed.) Proc. 3rd Pacific Rim International Conference on Artificial Intelligence, International Academic Publishers (1994) 180–186

[6] Nakamatsu,K.and Abe, J.M. : Reasonings Based on Vector Annotated Logic Programs. Computational Intelligence for Modelling, Control & Automation, Concurrent Systems Engineering Series **55**. IOS Press (1999) 396–403

[7] Nakamatsu,K., Abe,J.M. and Suzuki,A. : Defeasible Reasoning Between Conflicting Agents Based on VALPSN. Proc. AAAI Workshop Agents' Conflicts, AAAI Press (1999) 20–27

[8] Nakamatsu,K., Abe,J.M. and Suzuki,A. : Defeasible Reasoning Based on VALPSN and its Application. Proc. The Third Australian Commonsense Reasoning Workshop, (1999) 114–130

[9] Nakamatsu,K., Abe,J.M. and Suzuki,A. : Annotated Semantics for Defeasible Deontic Reasoning. Proc. the Second International Conference on Rough Sets and Current Trends in Computing (to appear as an LNAI volume, Springer-Verlag), (2000) 432–440

Logic, Artificial Intelligence and Robotics
J.M. Abe & J.I. da Silva Filho (Eds.)
IOS Press, 2001

Logic–A Key of New Information Technology

Setsuo Ohsuga

Waseda University,
Department of Information and Computer Science,
3-4-1 Ohkubo Shinnjyuku-ku Tokyo, 169-8555,   JAPAN
Ohsuga@ohsuga.info.waseda.ac.jp

Abstract. This paper discusses the role of logic as a key of new information technology in near future.   First, a way of developing an intelligent information system is discussed. A system arcitecture and its major component technologies are considered. These are all based on the decralative language and its processing.   Among them a way of integrating different methods of information processing is discussed in more details because it nees especially the feature of declarative method of representation and its processing.   So far the term 'integration' has been thought as a special method to put two specific programs together in an ad hoc manner. There has been no idea of defining a general way of integrating any pair of information processing methods.   To find such a general way is the purpose of integration discussed in this paper.   By classifying the various information-processing methods a possible way of integration is explored.   Among various methods of information processing, only the declarative knowledge-based system is suited for being used as an integrator. This is an example but logic is becoming more and more important in information technology in the future as the basis of declarative representation and processing in near future.   First, a way of developing an intelligent information system is discussed.   A system arcitecture and its major component technologies are considered. These are all based on the decralative language and its processing.   Among them a way of integrating different   methods of information processing is discussed in more details because   it requires especially the feature of declarative method of representation and its processing.   So far the term 'integration' has been thought as a special method to put two   specific programs together in an ad hoc manner.   There has been no idea of defining a general way of integrating any pair of information processing methods.   To find such a general way is the purpose of integration discussed in this paper. By classifying the various information-processing methods a possible way of integration is explored. Among various   methods of information processing, only   the declarative knowledge-based system is suited for being used as an integrator.   This is an example but logic is becoming more and more important in information technology in the future as the basis of declarative representation and processing.

1. Introduction

As the social activity expands problems arising there grow large and complex. It is worried that these problems go beyond the human capability for managing them and it may bring many troubles in human society [3]. If human being cannot resolve this problem, then what can do it in place of them? A key to resolve this problem is computer. But it is not the computers as have been so far and are still being today. Future computers must become more intelligent. Generally speaking, a problem solving is composed of a number of stages such as problem generation, problem representation, problem understanding, solution planning, searching and deciding solution method, execution of the method, and displaying the solution. Currently computers join only partly in this process. Persons have to do most parts, i.e. from the beginning until deciding solution method and, in addition, making programs thereafter in order for using computers. With these computers resolving the scale and complexity problems are still left to human. It is required to computers in near future to undertake the larger part of problem solving. Automating problem solving to a large extent is necessary for the purpose. Thus autonomy is one of the most important characteristics of future information systems.

But, what is autonomy? Autonomy is defined in [1] as an operation without direct intervention of human. But this is a too broad definition for those who are going to realize autonomy actually. Every problem solving method is different by the problems to be solved. How can computers find a proper method of problem solving for a specific problem? Moreover it is difficult for human to specify a large and complex problem. Autonomy must be exhibited not merely in processing problem but also in human-computer interaction such that computers can understand user's intention and aid users to specify his/her problem easily.

Various novel techniques have to be developed in achieving this goal. As the key of the techniques logic plays an important role. It is shown in the sequel of this paper.

2. Problem Solving Structure for Enabling Autonomy and Easy Specification

Instead of the broad definition as mentioned above, autonomy in processing is defined in this paper as the capability of a computer to represent and process the problem solving structure required by the problem. The problem-solving structure means a realization of methodology of problem solving and is the structure of the operations for arriving at the solution of a given problem. It is different by each specific problem but the same type problems have the same skeletal problem solving structure. For example design as a non-deterministic problem solving is represented as the repetition of model analysis and model modification (Figure 1). Its structure is not algorithm-based as in the ordinary computerized methods. The detail of the operations for design is different by the domain of the problem and the extent to which the problem is matured. But the difference in the detail can be absorbed by the use of the domain specific knowledge base and the method of using the domain knowledge base for problem solving is made the same even for the different problems. Then a unified problem solving structure is built for wide class of problems of this type. Conversely, the problem type is defined as a class of problems with the same problem solving structure. The necessary condition for a computer to be autonomous for wide classes of problems therefore is that the computer is provided with such a method as to generate the proper problem-solving structures dynamically for the different types of problems. If any problem solving structure is not made clear for a certain type problem, the computer system cannot accept problems of this type. But it is not the defect of this approach because, if human cannot define problem solving structure, then the computer cannot solve the class of problems in any way.

Figure 1 shows a few simple examples of problem solving structure. By combining these structures with the domain knowledge variety of problems can be represented. Actually these structures are made more elaborate. For example, a finer structure is defined by considering the characteristics of problem domains. In this case problem-solving structure is no more independent from domains. But it can be more comprehensive for the users in this domain.

When a problem is large, the problem decomposition is necessary before going into the detailed problem solving procedure for finding solution. The problem decomposition is problem dependent to a large extent but the decomposition method can be made common to many types of problems and be included in problem solving structure. The more complex problem-solving structure must be defined. The method of autonomous problem decomposition to cover wide range of problems is one of the key issues of large-scale problem solving.

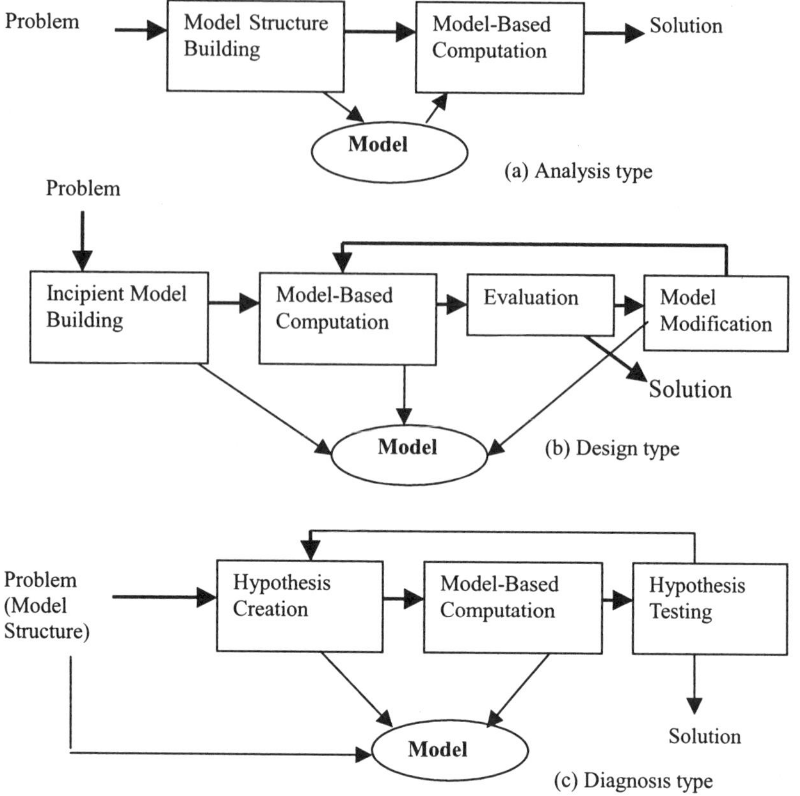

Figure 1 Problem solving structure

3. System Architecture to Satisfy Basic Requirements

A necessary condition for a computer to represent these problem-solving structures is that the computer is provided with the capabilities for solving various types of problems. Basically problems are classified into two types; the deterministic type and the non-deterministic type. For example design is represented as a non-deterministic problem solving to explore a solution in an open space. Thus the computer should be provided with the capability for solving not only deterministic but also non-deterministic problems.

From the viewpoint of information processing technology, the characteristic of deterministic problem solving is that providing a problem-solving method and executing it for obtaining a final solution can be separated as the independent operations. That of non-deterministic problem solving is that these two operations cannot be separated but are closely related to each other. This requires trial-and-error approach. Current computer's method, i.e. to make a program and to execute it, can correspond only to deterministic problem solving. This is an intrinsic nature of computers at present that has come from very the principle of computers built up on the basis of procedural language.

Trial-and-error operations can be realized if modularized representations of operations are available in computers because then different combinations of operations can be generated in the computer to correspond to the different trials. For realizing this idea a computer system based on the completely modularized declarative language must be developed. But, since the conventional CPU developed on the basis of procedural language is the only available hardware system today, an information system that uses both kinds of languages has to be developed as software architecture. Figure 2 illustrates a conceptual architecture for achieving this idea. A CPU is the processor of procedural language. An inference engine is necessary as the processor for declarative language. It is implemented as a special procedural program.

Then the procedural language and declarative language share the all activities that the future information systems must be able to do. The procedural language is used for representing operations closely related to CPU like OS but it is not suited for user's problem solving as has been discussed above. On the other hand, the declarative language is used for representing applications. This architecture gives computers potentiality for autonomy for various problems.

4. Some Important Concepts for Realizing New Information Systems

What is discussed in the sequel is on a research project that the author's group is making in order to achieve the goal of realizing this idea, in particular part B in Figure 2. Its conceptual architecture is shown in Figure 3. Its major component technologies are as follows.
1. New modeling method
2. Aiding human externalization
3. Large knowledge base management and problem solving system generation
4. Autonomous problem decomposition and solving
5. Generation of program
6. Integration of different Information Processing Units (IPUs) of different styles
7. Knowledge gathering, acquisition and discovery

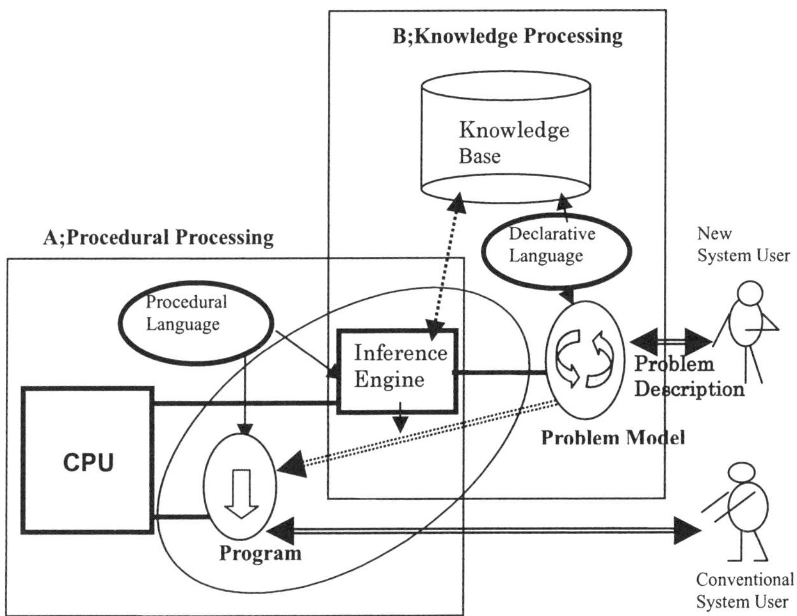

Figure 2   Conceptual architecture of new information systems

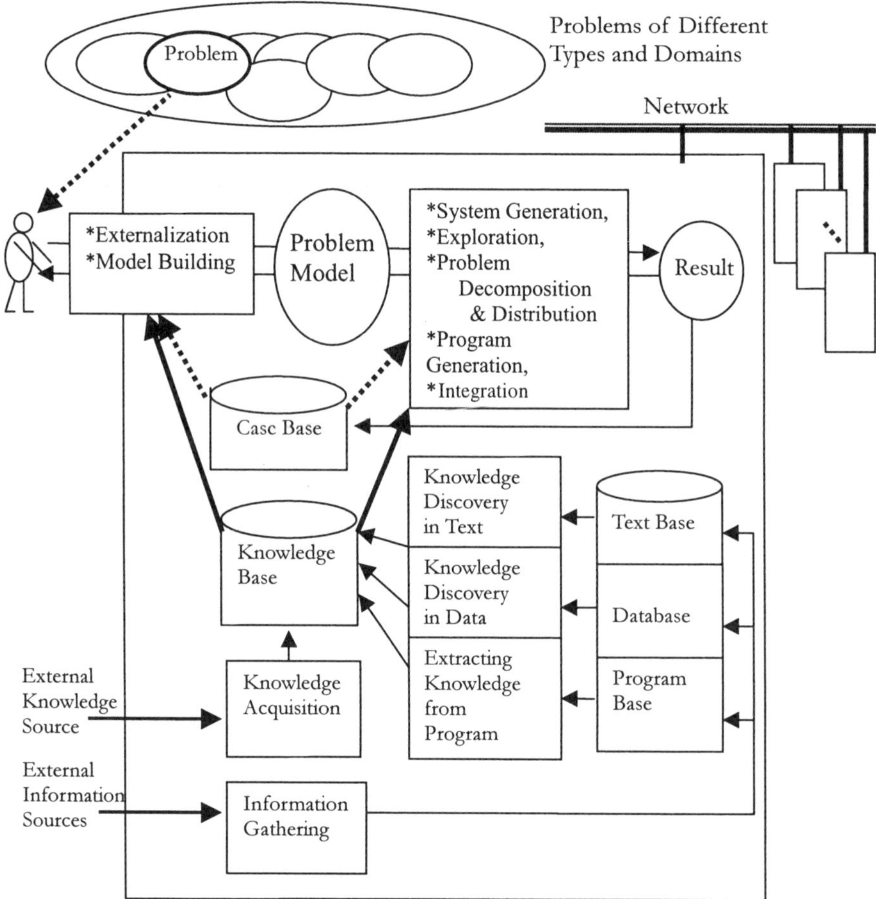

Figure 3   A system architecture

The system is organized using a declarative language as the kernel representation scheme. These items therefore are all defined on the basis of logical expression. Only the brief summaries of these items are introduced in this paper because of the lack of the space except the sixth item. Integration of different information processing methods requires especially the characteristic of logic. For the other items see [12].

4.1   New Modeling Method

The difference between the conventional style and the new style of problem solving is in the location of a problem being transferred from human to computer in the problem solving process mentioned before and accordingly its representation. In the old style the problem is represented in the form of program but in the new style the problems must be represented in the form close to those generated in human brains, be transferred to computers as soon as they are created and be processed there. The formal representation of the problem is called here a problem model. A problem is a concept created in a person's brain and a model is its explicit representation. The principle of problem modeling is to represent everything that relates problem solving explicitly so that no information remains in human brain in the form invisible to the others. A method of problem model representation has been decided such that variety of problems can be represented in the same framework.

It is assumed in this paper that every problem is created in relation with some object in which a user has interest. If a representation of an object is not complete but some part lacked, then problems arise therefrom. Therefore to represent a problem is to represent the object with some part lacking, and to solve problem is to fill the lacked part. This is named an object model. The basis of the object model is to represent the relation between the structure of the object being constructed from a set of components and functionality of every conceptual object (the object itself and its components).

Actually only the limited aspects of object within the scope of user's interest can be represented explicitly.

Then different problems can arise from the same object depending on the different views from human. It means that representations not only of an object but also of human view to the object must be included in the problem model in order to represent the problem correctly. A person may have interests in everything in the world, even in the other person's activity. This latter person being interested by the person may have an interest in still the other person. Let the persons be called the subjects of problem solving. It implies that if the persons are represented explicitly as subjects in a problem model, then the model forms a nest structure of the subject's interests. Thus a modeling method including conceptual object and conceptual subject is defined. It is called a Multi-Strata Model [7,9,10]. It reveals a large expressive power to represent various problems of different types and domains in the same framework.

Various types of problems are defined depending on the lacked part in a model. For example, if some functionality of an entity in a pure object model is lacked, an analytic problem is generated. If some functionality is given as the requirement but the structure of entities in an object model is lacked, then a design type problem arises. If a structure of activities by subjects is wanted, then scheduling type problem arises. By representing problems explicitly in this form information is made visible.

4.2　Human Interface for Aiding Human Externalization

Externalization is a computer's supports for (1) human cognitive process to help user for clarifying his/her idea and (2) model building in order to represent the idea in the form of model. It is a human-computer interface in the much broader sense than the ordinary ones.

Very often novice users are not accustomed to represent their ideas formally. Sometimes, their ideas are quite nebulous and they cannot represent the ideas in the correct sentences. How can the system help these users? This issue belongs to cognitive science. What the system can do is to stimulate the users to notice what they are intending. Some researches are made on this issue [5,16] but these are not discussed in this paper any more. In the following it is assumed that the users have clear ideas on their problems. The system aids them to build models to represent the ideas. Easy problem specification, computer-led modeling and shared modeling are considered. A key word to identify a problem solving structure is provided in the system and is shown to users. If the user can points one of the keyword in the list presented by the system, the system identifies the corresponding problem solving structure. As well as the key word an explanation of the problem that can be dealt with by the problem solving structure is provided. Users can select the correct keyword referring to the explanation [13].

The task of building a problem model for a specific problem may be shared by number of persons. It facilitates model building because usually a person can be careful enough to not-so-many number of aspects at a time but a complex model must be made taking very large number of aspects in consideration. The multi-strata model is useful for this purpose.

4.3   Large Knowledge Base and Problem Solving System Generation

To deal with a multi-disciplinary problem automatically means that the system is provided with multi-domains knowledge. Since a specific problem uses only a specific knowledge that concerns the related domains, a method to extract dynamically only the relevant knowledge for the given problem in a large knowledge base must be developed. This means that the real problem solving system has to be generated automatically for every specific problem. The large knowledge base must be conveniently organized in order to facilitate this knowledge extraction and system generation. This knowledge base is managed by a Knowledge Base Management System.

In order to enable the system to generate a specific problem solving system for the given problem it is desirable to divide a large knowledge base into a set of proper knowledge chunks and organize them in a structure so that a substractures to form a specific problem solving system for the given problem can be easily extracted. A knowledge base is divided into chunks depending on the division of problems domains and the hierachy of these knowledge chunks is formed corresponding to the domain hierarchy. Assume a domain, say Domain1, is divided into (Domain11, Domain12,--, Domain1N). Correspondingly knowledge is divided. A part of knowledge that relate specifically to Domain1i but not to the other domains is KnowledgeChunk1i. If a part of knowledge relates more than two domains in the lower domain sets, then the part remain in the knowledge chunk corresponding to Doamin1. This is the specific knowledge to Domain1. In this way knowledge structure is formed [12].

4.4   Autonomous problem decomposition and solving

If the problem is large, it is decomposed to a set of smaller problems before the execution of the problem solving structure. This problem decomposition and sharing is also achieved (semi-) automatically based on the problem model. By the combination of the decomposition methods with the multi-strata modeling, various interesting problems are represented and solved. The autonomous problem solving follows the decomposition. In principle, it is to execute the problem solving structure for the given type of problem. The detail is abbreviated here. Refer to [8,14,17].

In decomposition a conceptual object is defined first and is decomposed top-down forming a structure. Corresponding to the object decomposition a new subject is assigned to each new lower level object resulting in a new subject structure. Thus created pair of new object and new subject forms a sub-task coped with in the system. It forms an agent in the system. That is, this problem decomposition creates a new multi-agent system. Each agent shares a part of the workload given to the system. The way for decomposing the object model is different by the problem. By the way of decomposition problem definition is different. In the design type problem the object model is built in such a way that the required functionality of the object can be satisfied and a new person is assigned to a new object. In the other case the user at the top specifies the other activity. For example the user specifies an evolutionary rule to decompose an object automatically as an activity to a subject. This rule is to decompose an object toward adapting the environment given by the user. Then an evolutionary system to adapt to the environment is created.

4.5   Generation of program

Every procedural program is a representation of a compound of subject activities in the programming language. This compound is represented as a structure of the related activities. Programming is to specify a structure of the activities and then to translate it to a program. An object to be programmed, i.e. the structure of the activities, is obtained by the exploratory operation to find the route to reach the goal from the given problem. That is, the programming is the posterior process of normal problem solving.

4.6   Knowledge gathering, acquisition and discovery

Different from the items 1 through 6 concerning the user's problem solving directly, knowledge gathering, acquisition and discovery are for preparing knowledge to be used for problem solving. At present knowledge is mostly prepared manually by person. An automatic knowledge acquisition including knowledge discovery from data, texts and programs are being made research. But these topics are rather the long-term research target [2,18].

5.   Integration of two or more information processors of different styles

5.1 Significance of integration

There are various information-processing methods based on the different principles of representation and information processing. Ordinary procedural programming method, declarative knowledge-based method, neural network method are the typical examples. Each of these methods represents a framework in which various actual systems are developed as instances. For example a neural network is developed in the general framework of the network-style method of information processing as a general concept.

At the implementation level, every method is approximately represented in a procedural form. But the difference of the methods based on the different principles is interested in this paper. Both procedural programming method and declarative knowledge-based method are based on the symbolic representation while the neural network method is non-symbolic one. There is a large difference between them [11]. Every method has its own characteristics and its own limitation when used in the real environment. When someone wishes to solve a problem in computer he/she needs a specific processor for the problem. Person selects a method that is best suited for one's purpose of problem solving. To select one method however means to lose chances to use advantages that the other methods provide.

If an information processor that satisfies a user's requirement could be generated by integrating two or more existing processors, it is often easier than to develop a new processor starting from scratch. Moreover when problems that persons wish to solve get large, they cannot be coped with by a single method. Then there arises the need for combining different methods. What methods should be combined and how are these combined in order to satisfy the requirement of solving large-scale problem? Researches are being made on the way of integrating them [4, 6]. But most of the research works are in the conceptual level.

What are actually combined are not the methods themselves but the instance processors derived from these methods. By combining the instances however the combination of the methods is achieved. In the following the term integration is used to mean to put the instances together in a processor. Integration of the different systems is not easy but is possible via knowledge-based systems [15].

A term information-processing unit (hereafter IP-Unit for short) is used in order for denoting such an instance processor. Two or more IP-Units can be integrated to form another (integrated) IP-Unit. In an integrated IP-Unit, the methods based on which each original IP-Unit is defined are combined.

Every method as a framework has its own representation scheme and a processing mechanism. These representation scheme and processing mechanism are inherited to every IP-Unit in this framework. In addition, each IP-Unit is endowed its own characteristics. For example, every IP-Unit has its own scope of variables only in which the IP-Unit is valid. These characteristics specify the IP-Unit and decide the condition of integration with the other IP-Units. Not necessarily every combination of IP-Units is possible for integration but there are some conditions. What is the condition of more-than-two IP-Units being able to be integrated? What is the measure of the effectiveness of the integration? These are the main issues that are discussed in the sequel.

In the following integration is defined formally as an operation to integrate the different IP-Units in order to obtain a new IP-Unit with a larger scope than any of the old IP-Units and to construct a processing mechanism in the new scope. There are two approaches of integration. The first is to merge two or more different IP-Units specifically in order to make a new IP-Unit that has a larger functional capability including both functions of the original IP-Units but without considering any specific problem to solve at the time of integration. This integration is called the static integration. The second is to integrate IP-Units in real time in order to satisfy the necessity for solving a given problem as an application from outside. This is called the dynamic integration. The former is for providing the more powerful set of IP-Units than those currently available while the latter is for solving given problem. The former is usually an ad hoc operation that has been performed manually by person. The person sometimes resolves the original IP-Units into parts and rebuilds a new IP-Unit. This is almost the same as making a new IP-Unit from scratch. For example, two procedural programs with small scopes can be integrated to acquire the large scope such as the case of integrating image-processing program with acoustic processing program in order to acquire the program to deal with image and acoustic information at the same time. It is excluded from the following discussion. For the latter, on the other hand, it is required to perform integration automatically in a computer system. It is difficult to take such a procedure as to resolve the IP-Units into parts and rebuild them as taken by persons. In the sequel therefore, integration is limited only to using the original IP-Units without resolving into parts. It is however allowed that some extra function(s) is added from outside to paste the IP-Units. If the IP-Unit is the one that has been obtained by integration of the smaller IP-Units however, it can be resolved to the level of the original IP-Units. As will be shown below, a processor that is described explicitly is identified as an independent IP-Unit and is used for creating the other IP-Units by integration.

5.2   *Definition and scope of IP-Unit*

Every IP-Unit is built on the basis of a specific representation and its processing mechanism inherited from the method it originates. Every IP-Unit has its own scope of representation as an inherent characteristic of the IP-Unit. If this scope cannot cover the requirement for information processing came from outside as problem, then the IP-Unit alone cannot solve this problem but needs to expand its scope by augmenting the capability. If the expansion of the scope is difficult by modifying the original IP-Unit, then a new IP-Unit has to be developed. This was often the case in the ordinary software method. Integration of different IP-Units is another way to expand the scope so that the integrated IP-Unit has a larger scope and function to respond to what is required. Thus from the viewpoint of integration, the scope of an IP-Unit, i.e. what information it can accept and process, is the primary importance. Integration of IP-Units is not restricted to the case of IP-Units with the different methods but those of the same method, for example two procedural programs, can also be the objects of integration.

Every IP-Unit receives input from the outside and generates output. Let the input and the output be specified by a set of variables, $x = x1,x2,--,xm$ and $y = y1,y2,--,yn$ respectively. The scope of a variable is defined by 'type $\times$ domain'. Let the domains and ranges of the input variable xi and the output variables yj be Xi and Yj respectively. Then the scope of the IP-Unit is defined as $X1 \times X2 \times -- \times Xm \times Y1 \times Y2 \times --\times Yn$, i.e. a sub-space $\prod_i Xi \times \prod_j Yj$ in a large open universe. The function that the IP-Unit can achieve on the other hand is a mapping P from $X1 \times X2 \times -- \times Xm$ to $Y1 \times Y2 \times --\times Yn$. This is the external definition of the IP-Unit that is necessary and sufficient for discussing integration. The internal structure of this IP-

Unit has to be specified to the detail for the purpose of implementation. But it is not necessary for the discussion of the possibility of integration. Thus an IP-Unit is represented formally as P(X1, X2,--, Xm ; Y1, Y2, --, Yn).

In some simple methods of information processing like a neural network, the processing mechanism defines the scope directly. In the more complex scheme, especially the information processing methods based on symbolic representation, syntax and semantics are defined first as the representation scheme and then a processing mechanism is defined based on the scheme. Procedural program processing, database processing and declarative knowledge processing are all in this class, but the actual representation schemes and processing mechanisms, and accordingly the possibility of integration of an IP-Unit with the other, are different by the methods.

The scope of an IP-Unit is decomposed as follows.

(1) Type ; There are types of different nature. Every variable has its own type and is classified by the type as follows. (i) Numerical variable with such types as 'real', 'integer' and so on, (ii) Object variable to represent some entity in the universe, for example a variable xi represents some person of a type 'Person' as a set of abstract persons. This is dealt with as a type because it represents entities with a nature that is different from the similar but the other type, say 'Table', 'TV set' and so on, and   (iii) Logical variable to represent a truth value of a proposition. Every variable has its own type. Let Txi and Tyj represent the types of variables xi and yj respectively. Then the type of the IP-Unit is $\prod_i$ Txi $\times \prod_j$ Tyj.   The type (ii) above is defined by a classification structure of conceptual entities. In reality the type (i) is a part of the type (ii) because a numerical entity is also a conceptual entity.   Figure 4 shows an example of the classification structure. Since this is a hierarchy, there can be a close relation between two types. For example, Person > Boy where > denotes that the left side concept is upper to the right side. It is read 'greater-than' in the sequel.

(2) Domain ; every variable is specified a set of values that the variable can take in a type. It is called a domain. In case of numerical variable as (i) above, let the type be ' real number'. Then the domain may be a set of real value, for example, an interval like [a, b]. Similarly, in case of logical variable as (ii), the domain is specified by a set of the individual entities, like (Ohsuga, Smith, McDermot) in Person. Sometimes the domain can be the set of all possible individuals and the same as the type. In case of truth value as (iii) above, the domain is either (0, 1) in the simple case or an interval [0, 1] including incompleteness of truth value.

(3) Granularity ; every variable has its own minimum distance between adjacent values in its domain depending on the syntax of language as representation method or the fineness of processing mechanism. In principle, IP-Units with the different granularity cannot be integrated.

Every IP-Unit P(X1, X2,--, Xm ; Y1, Y2, --, Yn) is endowed a logical value implicitly. It is either logically true if this relation holds or false otherwise. Namely it is represented by a logical variable. Then the IP-Unit works as an evaluation mechanism of this variable. An operation to obtain an output for the given input is to keeps it true. It is possible to define an IP-Unit that includes a logical variable representing a still another IP-Uint. In this way high order IP-Unit can be defined.

In a neural network, the truth value of the relation between input and output is not always exactly one but can be any value between zero and one. This causes a difficulty in integrating symbolic system and non-symbolic system. This issue is discussed in section 5 and it is assumed for a while that every neural network is evaluated either True or False.

The type of an IP-Unit represents a skeletal structure of its scope. The domain is specified as the detailed information on the scope.

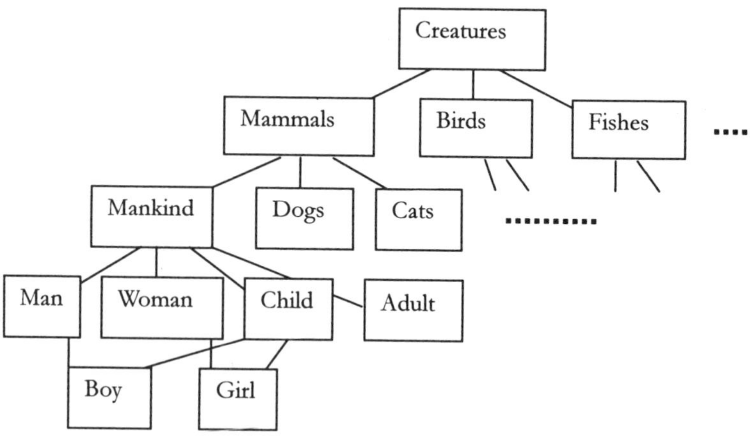

Figure 4   Hierarchy of conceptual entities

5.3 Possibility of integration

To find a transformation between different representation schemes as a general method for merging them was not easy but in many cases it has been performed manually in an ad hoc way for each specific pairs of IP-Units to be merged. It has been discussed in [15] however that the integration of logic-based IP-Unit with the other IP-Unit is possible. In the following its automation is intended. For this purpose the integration problem is considered in two steps.

In the first step integration is defined as an operation of an IP-Unit to take the other IP-Unit (s) in itself and to use its function in order to realize a new function as a whole. The former is called the master IP-Unit and the latter is the slave IP- Unit. In order to take in the slave IP-Units operation, the master IP-Unit must be able to represent the feature of the slave IP-Units in its own representation scheme. The master is also required to transform the format of some data in the master to the input of the slave and the output of the slave to the master in such a way that the operation of the slave can be activated properly in the master IP- Unit. Some additional operations other than those existing in the master IP-Unit are necessary. These are added to the master IP-Unit from outside. Thus the master IP-Unit has to be able to accept this additional information and expand its scope without disturbing its own processing. The expanded master IP-Unit can decide the way to evoke the slave and translate the data to send to the slave. Receiving the output from the slave and translating it into the master scheme, the master IP-Unit can continue the operation.

There are number of information-processing styles and for some of them it is difficult to meet the condition to be the master IP-Unit. As the typical examples, non-symbolic system, procedural processing and declarative processing are considered. A non-symbolic system has a fixed mechanism of processing and it is difficult to accept the representation of the slave IP-Unit from outside without changing the whole structure. A procedural processing should be more flexible than the non-symbolic system but in reality it is still difficult to add such information as above from outside without changing the main body of the program. That is, the time and way of evoking the slave and transformation routines must be programmed and embedded in the main program in advance. It is to rebuild the program and automation of integration adapting to the changing environment is difficult. Finally a declarative processing, for example a rule-based system using 'If-Then' rule, adding information from outside does not require the modification of the main part of the system but the system can expand the scope of processing by combining the added rules with the existing ones. This is an important characteristic of the declarative processing system. It makes the declarative processing potentially the best candidate of the master IP-Unit. It can integrate various IP-Units of the different styles. The descriptive information on the slave to be added to the master IP-Unit must exist at integration and is added to the knowledge base of declarative processing system. This information on the slave is assumed created when the slave IP-Unit was first developed. Thus, even if the master system cannot represent the processing mechanism of the slave system to the detail the master IP-Unit could pass the processing to the slave IP-Unit temporarily and use the results. That is, the mechanism of the slave could be a black box to the master but the master is only required to describe the result of the operation to be achieved in the slave as the input-output relation. It is reason why the information processing of the different styles can be integrated.

In the second step this idea is generalized. Let two IP-Units, say (I1, M1) and (I2, M2), of any information processing styles, S1 and S2 respectively, be to be merged. Ix and Mx represent an input scope and a mapping from the input to output of the processor x. respectively. If at least one of them is a declarative IP-Unit, then the integration as discussed above is possible by making this IP-Unit the master. If both IP-Unit1 and IP-Unit2 are not the declarative ones, their direct integration is difficult. But their integration becomes still possible by introducing a new declarative IP-Unit. In this case this third IP-Unit is the master to both IP-Unit1 and IP-Unit2, and the descriptions of both IP-Unit1 and IP-Unit2 are added thereto. Thus non-declarative IP-Units can be integrated indirectly via the third declarative IP-Unit. It is possible to prepare such a declarative IP-Unit for the purpose of integrating arbitrary IP-Units.

This form of integration is possible only when the master IP-Unit accepts information from outside, i.e. input, directly. Even if a slave IP-Unit accepts input directly before being integrated, the system is changed so that a master accepts it first and then it is passed to the slave. After then any integration becomes possible via this master system.

This capability of integration depends on the flexibility of an IP-Unit to expand the scope of information processing as well as its expressive power of representation of one or both of IP-Unit to be integrated. The expandability depends on representation method. Some method has a large expandability but the others have not. Thus it is desirable from integration point of view to make a representation method that has the largest expandability as well as the enough expressive power a central scheme of information processing style. Then, integration is defined between representations of the different methods. Among all methods that are used today, only purely declarative representation scheme has such expandability. A typical example is predicate logic.

Example; As a simple example, assume that two IP-Units, IP-Unit 1(x1, y1) and IP-Unit 2(x2, y2), are to be integrated. These are developed as to achieve the specific activities represented primitiveActivity1(x1, y1) and primitiveActivity2(x2, y2) respectively. The inputs x1 and x2 are from the sensors, sensor1 and sensor2 respectively. On the other hand, the master IP-Unit has an objective of executing an activity based on some function of the results of these primitive activities. For example, the activity is defined as depending on the difference of the results of the primitive activities, i.e. z = y1 − y2. But the master IP-Unit has not any function for executing the primitive activities. Therefore it integrates the IP-Unit 1 and IP-Unit 2 for the purpose of using their functions. The sensor signals are first accepted by a master IP-Unit and sent to the IP-Unit 1 and IP-Unit 2. The format of the sensor signals accepted by the master IP-Unit and the inputs to the IP-Unit 1 and IP-Unit 2 can be different. Let an activity(u1, u2, w) be executed to generate w as a function of the sensor inputs. Then the following predicate is provided to the master IP-Unit as the definition of the required activity.

activity(u1, u2, w):- sensor1(u1), sensor2(u2), primitiveActivity1(u1, v1),
                      primitiveActivity2(u2, v2), subtract(v1, v2, z), function(z, w).

For the purpose of integration the following definitions of IP-Unit 1 and IP-Unit 2 are added to the master IP-Unit. The symbol # denotes that a calling procedure including the conversions of inputs and outputs are necessary in the master IP-Unit because of the difference of the formats.

primitiveActivity1(x1, y1) :- #IP-Unit 1(x1, y1),
primitiveActivity1(x2, y2) :- #IP-Unit 1(x2, y2),

The IP-Unit 1 and IP-Unit 2 are executed in the rule of activity(u1, u2, w) via the above rules and the result is obtained. After then the result, w, can be used in whatever the ways in the master IP-Unit. In this example the function(z, w) is assumed being defined in the master IP-Unit. It can be still the other processor, IP-Unit 3. In this case, IP-Unit 1, IP--Unit 2 and IP-Unit 3 are integrated via the master IP-Unit.

6. Conclusion

As the social activity accelerates problems arising there grow large and complex. It is worried that these problems go beyond the human capability for managing them. As a probably only possibility of solving this problem development of intelligent information system becomes more and more urgent. Automating problem solving to a large extent is necessary for the purpose. Autonomy is one of the most important characteristics of future information systems.

This paper has discussed a way of developing an intelligent information system to deal with large and complex problems. First a system architecture and its major component technologies including integration of different methods of information processing were mentioned. These are all based on the declarative representation and processing.

Among them a way of integrating different methods of information processing was discussed in more details because it requires especially the characteristic of declarative method of representation and processing. Very often the term 'integration' meant a special method to put two specific programs together in an ad hoc manner. There has been no view to define a general way of integrating any pair of information processing methods. To find such a general way was the main issue in this paper.

By classifying the various information-processing methods a possible way of integration and a way of realizing this integration were discussed. Among various methods of information processing, only the declarative knowledge-based system is suited for being used as an integrator. As the basis of declarative representation and processing logic is becoming more and more important in information technology in the future.

References

[1]   C. Castelfranchi; Intelligence Agents: Thories, Architectures, and Languages, in Guarantees for autonomy in cognitive agent architecture, (edited by M.Wooldridge and N.R.Jennings), Springer, 1995
[2]   J.Z.Dong, N.Zhong, and S.Ohsuga; Rule Discovery by Probabilistic Rough Introduction, J.of Japanese Society for Artificial Intelligence, Vol. 15, No. 2,   2000
[3]   W. Wayt Gibbs; Software Chronic Crisis, Scientific American, Volume 18, No.2, 1994
[4]   V. Honavar and L. Uhr (eds.) ; Artificial Intelligence and Neural Networks, Steps toward Principled Integration, Academic press, 1994.
[5]   K.Hori, A system for aiding creative concept formation, IEEE Transactions on Systems, Man and Cybernetics, Vol.24, No.6, 1994
[6]   L. Monostori, Cs. Egresits and B. Kadar ; Hybrid AI Solution and their Application in Manufacturing, Artificial Intelligence and Expert   Systems,   Gordon and Breach Publishers,   1996.
[7] S. Ohsuga; Multi-Strata Modeling to Automate Problem Solving Including Human Activity, Proc.Sixth European-Japanese Seminar on Information Modelling and Knowledge Bases,1996
[8]   S. Ohsuga ; Toward Truly Intelligent Information Systems - From Expert Systems To Automatic Programming, Knowledge Based Systems, Vol. 10, 1998
[9]   S. Ohsuga ; A Modeling Scheme for New Information Systems -An Application to Enterprise

Modeling and Program Specification,　　IEEE International Conference on Systems, Man and Cybernetics, 1999

[10]　S. Ohsuga; New Modeling Scheme To Let Computers Closer to Persons, International Workshop on Strategic Knowledge and Concept Formation, 1999

[11] S. Ohsuga; The Gap between Symbol and Non-Symbol Processing -An Attempt to Represent a Database by Predicate Formulae-, Proc. Pacific Rim International Conference on Artificial Intelligence (PRICAI-2000), 2000

[12] S. Ohsuga and H. Ohshima; A Practical Approach to Intelligent Multi-Task Systems- Structuring Knowledge Base and Generation of Problem Solving System, 11th European-Japanese Conference on Information Modelling and Knowledge Bases, 2001

[13] S. Ohsuga, H. Matsuura and T. Nishiguchi ; Toward More Friendly Human-Computer Interaction, Proc. The 5th World Multi-Conference on Systemics, Cybernetics and Informatics, SCI, 2001

[14] S. Ohsuga ; How Can AI Systems Deal with Large and Complex Problems? Internationa Journal of Pattern Recognition and Artificial Intelligence, Vol.15, No.3, 2001__

[15] S. Ohsuga and N. Ueda; Dealing with Information in the Different Styles Together—Is Skill Inheritance and Integration of Information Possible? In (L. Monostori, et. al.) Engineering of Intelligent Systems, 14th International Conference on Industrial and Engineering Applications of Artificial Intelligence and Expert Systems, Springer-Verlag, 2001

[16] Y.Sumi, et. al., Computer Aided Communications by Visualizing Thought Space Structure, Electronics and Communications in Japan, Part 3, Vol. 79. No.10, 11- 22, 1996.

[17] K. Tanaka and S. Ohsuga; Problem Decomposition and Multi-Agent System Creation for Distributed Problem Solving, (To appear in) Proc.The Second Asia-Pacific Conference on Intelligent Agent Technology (IAT-2001), Maebashi City, Japan, October 23-26, 2001

[18] N. Zhong, Juzhen Dong & S. Ohsuga ; Rule Discovery by Soft Induction Techniques, Neurocomputing, 36,　2001,

Logic, Artificial Intelligence and Robotics
J.M. Abe & J.I. da Silva Filho (Eds.)
IOS Press, 2001

Non Local Computation by Semantic Field
(Semantic Web)

Germano RESCONI

Mathematical & Physical department, Catholic University, Via Trieste 17, 25121 Brescia, Italy

E-mail resconi@numerica.it

Tetsuya MURAI

Division of Systems and Information Engineering, Graduate School of Engineering,
Hokkaido University, Sapporo 060-8628, Japan
E-mail: murahiko@main.eng.hokudai.ac.jp

Abstract. A computational space integrates and links applications and data sources to create heterogeneous computational systems. Any application has local components for controlling data flow that use classic logic variables and expressions. In this paper inside the computational space, we introduce a set of fields that transport the values of the logic variables or expressions from one application to another. This field transform a local computation in a non-local computation. Computation originate from contribute of all the logic variables in the computational space. With external logic variable, we control flows of data in the application G by a sentence of the form $\Box$ A – necessarily A – that is true when A is true in G. and in any other application connected by the field with G. We can also use a sentence of the form $\Diamond$ A – possibly A – that is true when A is true in almost one application connected with G. Expressions as $\Box$ A and $\Diamond$ A are modal logic expressions that extend classical logic. In the modal logic, any application or context is a world. Inside the computational space, we have two types of connections. The first is the classical local wire connection that links the output of one application with the input of another application. The second is a high parallel and non-local connection by a field whose intensity instructs synchronically target applications. In the non-local computation, the field that connect the applications is denoted "semantic field". By semantic field, the computational space becomes an entity where we cannot separate the single applications. In fact, any application G is not only connected with the other by input output wire structure, but is connected in a more deeper way by semantic field. The structure of Web can be used as physical support for Semantic Field.

1. Modal Logic

There is a wide spread acceptance that Kripke models[1,2] offer the fundamental semantics for modal logic. In this section, we give a brief description of Kripke

1.1 Kripke Modal Framework

Let ATOMS, be elements of a denumerable set of *atomic sentences*. Let L_{ML} be a language for modal logic formed from ATOMS using connectives such as $\neg$ (negation), $\wedge$ (conjunction), $\vee$ (disjunction), $\rightarrow$ (material implication), and $\leftrightarrow$ (equivalence) and modal operators such as $\Box$ (necessarily) and $\Diamond$ (possibility).

A *Kripke model* is defined as the three-tuple M = < W, R, V >, where W is a non-empty set of possible worlds, R $\subseteq$ W × W is an accessibility relation on W, and V is a valuation operator for atomic sentences at each possible world, w in W : V: ATOM × W → {T, F}. The valuation operator is extended in the usual way and, in particular, for modal operators, $\Box$ - necessarily and $\Diamond$ - possibility:

$$V(\Box p, w) = T \Leftrightarrow \forall\, w'(wRw' \Rightarrow V(p, w')) = T \Leftrightarrow \Xi_w \subseteq \|p\|^M,$$
$$V(\Diamond p, w) = T \Leftrightarrow \exists w'\, (wRw' \text{ and } V(p, w')) = T \Leftrightarrow \Xi_w \cap \|p\|^M \neq \varnothing$$

where $\Xi_w = \{w' | wRw'\}$. A sentence is necessarily true in one world when is true in any accessible world. A sentence is possibility true when is true in almost one accessible world. Note that $\|p\|^M$ is called the truth set associated with the proposition of an atomic sentence p in a model M. In what follows, we omit the superscript $^{'M'}$ unless confusion arises. It is well known that different conditions on accessibility relations correspond to different axiom schemas.

2. Semantic Field

Definition. The semantic field[3] is formally defined as the tuple $\mathsf{SF} = <H, W, \delta, F, \eta, \mathsf{S}>$, where H is the computational space, W is the set of worlds, δ: W → H, F is the value of the semantic field, η: W → F, S is the set of sources of the semantic field. The sources S are worlds' where the field is generated. In a given computational space H, we locate possible worlds. In Fig.1, we show the relation that exists between the accessible relation and the semantic field.

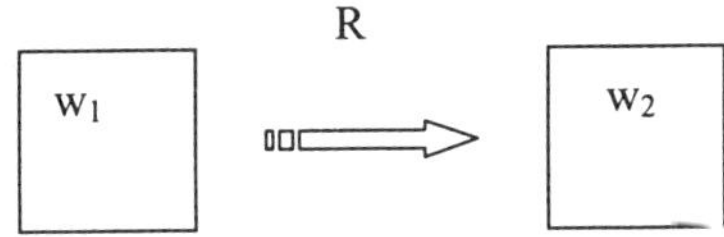

Accessible relation
w₁ R w₂ , w₁ can see the
values of the sentences in w₂

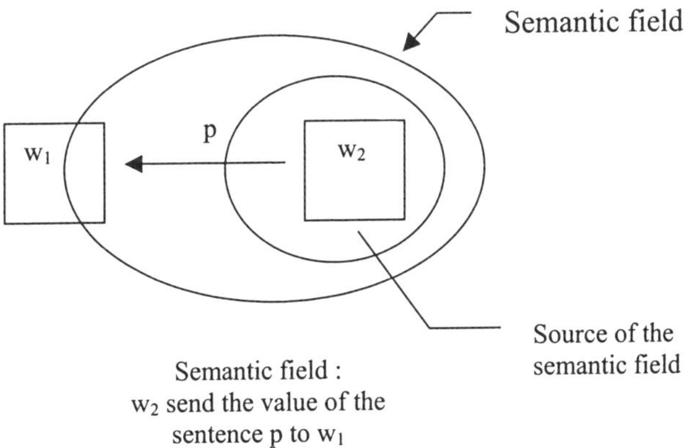

Semantic field :
w₂ send the value of the
sentence p to w₁

Fig.1 : The semantic field is the carrier that transports the values of the proposition p from one world to another. One world can see the logic values of another world by the accessibility relation.

3. Non Local Computation by Semantic Field

3.1. Computational Space, Worlds, Applications and Projects

Computational space is an environment for integrating and linking applications and data sources to create heterogeneous computational systems. MathConnex, Excel or other software products are concrete realisations of the semantic field. *Computational space can be also the Web where the* worlds are storage units called entities. Example of computation space is given in Fig.1, where any box is a world or context associated to an application or a source of data.

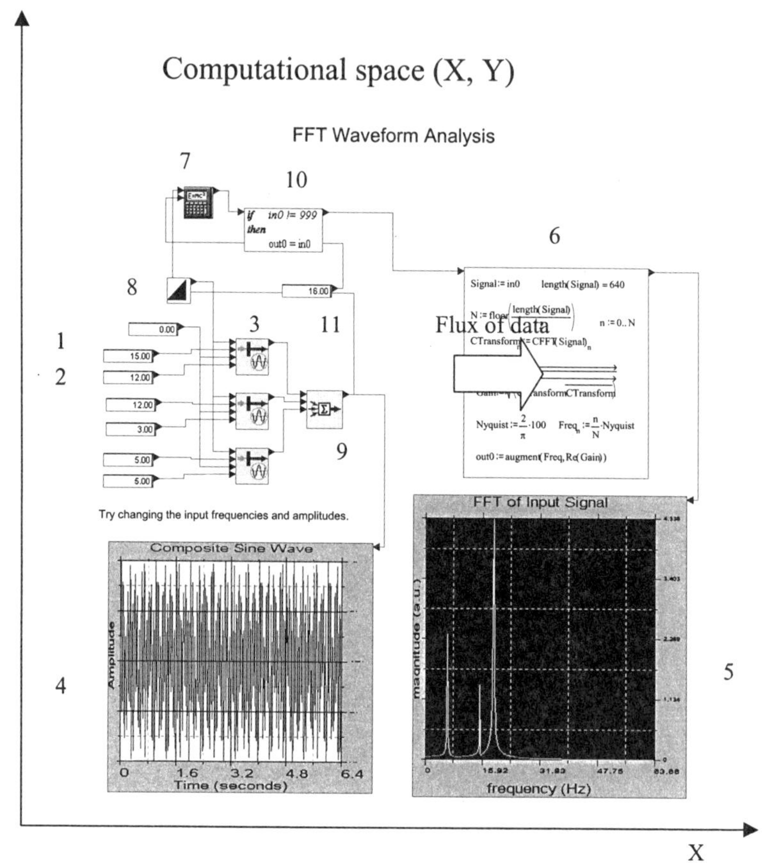

Fig.2   The environment or computational space is shown to obtain the FFT waveform analysis. In the box 1, there is the value of the frequency. In the box 2, there is the amplitude. In the box 3, there is the sin wave generator. In the box 4, there are the composite sin waves. In the box 5 there is the fast Fourier transform FFT. In the box 6, there is the Mathcad FFT program. In the box 7, there is the accumulator. In the box 8, there is the counter. In the box 9, there is the sum operator. In the box 10 there is the conditional box. In the box 11, there is the FFT update frequency

Any box in a position of the space X, Y is an application that we denote world or context. In the computational space, we realise a project.

A project is included inside the computational space. Any computation begins from the source of data at the left and end at the box that show the results.

3.2. Computational Space with Semantic Field

In the computational space, any box is a world, *where logic values of the variables are obtained by* data *that we introduce in the box.* Fluxes of data inside the computational space define the true and false values of the propositions in the worlds or boxes. Inside at any box we can compose the different logic variables in a way to realise logic expressions as implications, conjunctions, disjunctions and so on. Inside the computational space, any box that has a sentence p generates a semantic field H_p that send to the other worlds the value of p. All the fields generated from the different boxes give the total semantic field. The structure of the semantic field connects all the boxes or worlds. We use modal logic operators to control the flux of the data in all the boxes.

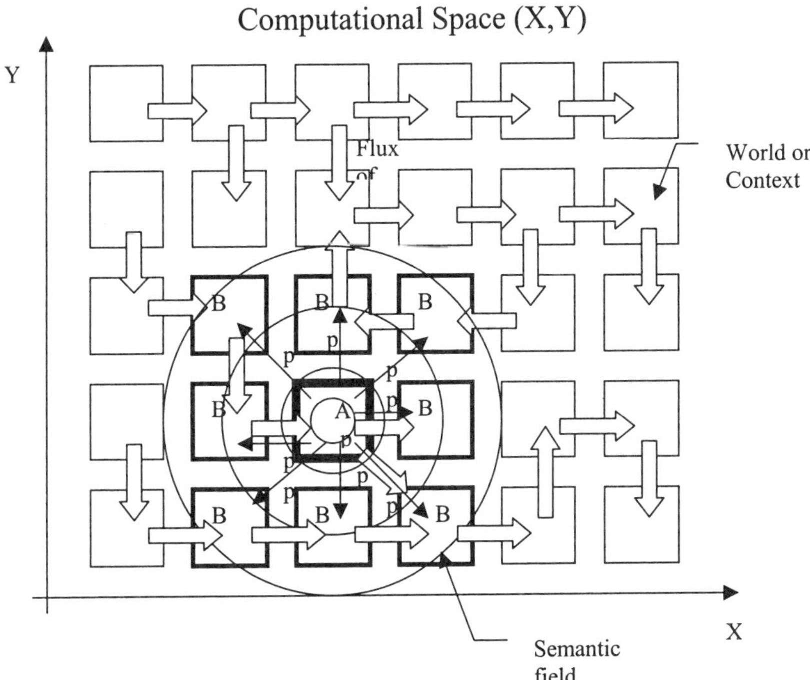

Fig.3 A computational space with boxes is shown, where data are computed; any box is a world or context. For the sentence p, we show the associate semantic field. In A, we have a source of the semantic field. Boxes B receive the logic value of p from the source box A.

With the input/output connection, we obtain a sequential computation. With the semantic field, we go beyond the sequential computation and we generate a non-sequential or non-local connection that generates non-local computation.

Game example as a computational space: To take decision in projects for games we use the modal logic operators. In fact, when a player wins he eliminates any contact with the other players. The semantic field is put at zero. The player cannot receive information on the logic value of the sentence "is possibly to lose" and continue to play. However, when he loses, he generates the semantic field to receive the logic value of the sentence "is possibly to win." In fact, there exists almost one player that wins. Because it is possible to win he continues to play. In this way with creation or elimination of the semantic field, the player can convince himself that is good to play in any case.

3.3. Relation between Set of Worlds and Semantic Field in a Computational Space

In Fig.4, we show the set of worlds P where the sentence P is true and Q where the sentence Q is true. As we can see in Fig.4 when we put the sources in different positions, by modal logic "possibly","necessarily" we can see in part one of Fig.4 that Q is necessarily false in P, in part two of Fig.4   P is possibly true in Q, in part three of Fig.4.   Q is always necessarily false and in part four of Fig.4 Q is always possibly true.

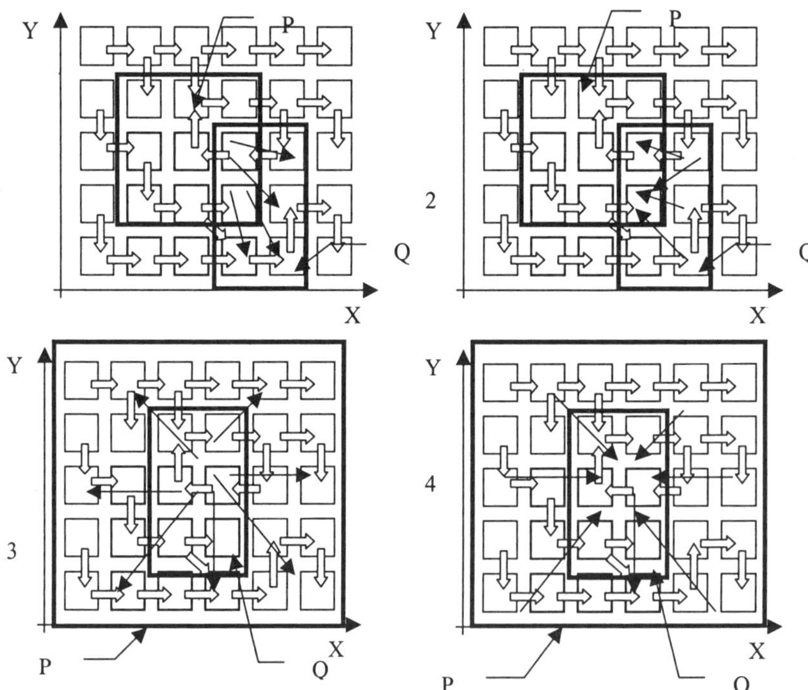

Fig.4 In one Q is necessarily false in P, in two P is possibly true in Q, in three Q is always necessarily false and in four Q is always possibly true. In three and four Q is included in P that covers the entire semantic field.

3.4. Non-Additive Computational Space

In Fig.5 the number of worlds inside P∪Q in three is N[◊(p∨q)]. In one, the semantic field send the logic value of p from P to all the worlds in K. Therefore, the set of worlds where ◊ p is true or N(◊p) is |P∪K|. In two, the semantic field send the logic value of q from Q to all the worlds in K. Therefore, the set of worlds where ◊q is true or N(◊q) is |Q∪K|. As we can see in Fig.5, we have

$$N[\,\Diamond\,(\,p \vee q\,)\,] \le N(\,\Diamond\,p\,) + N\,(\Diamond\,q)$$

Therefore, we have a non-additive computational space

$$N[\,\Diamond\,(\,p \vee q\,)\,] \ne N(\,\Diamond\,p\,) + N\,(\Diamond\,q)$$

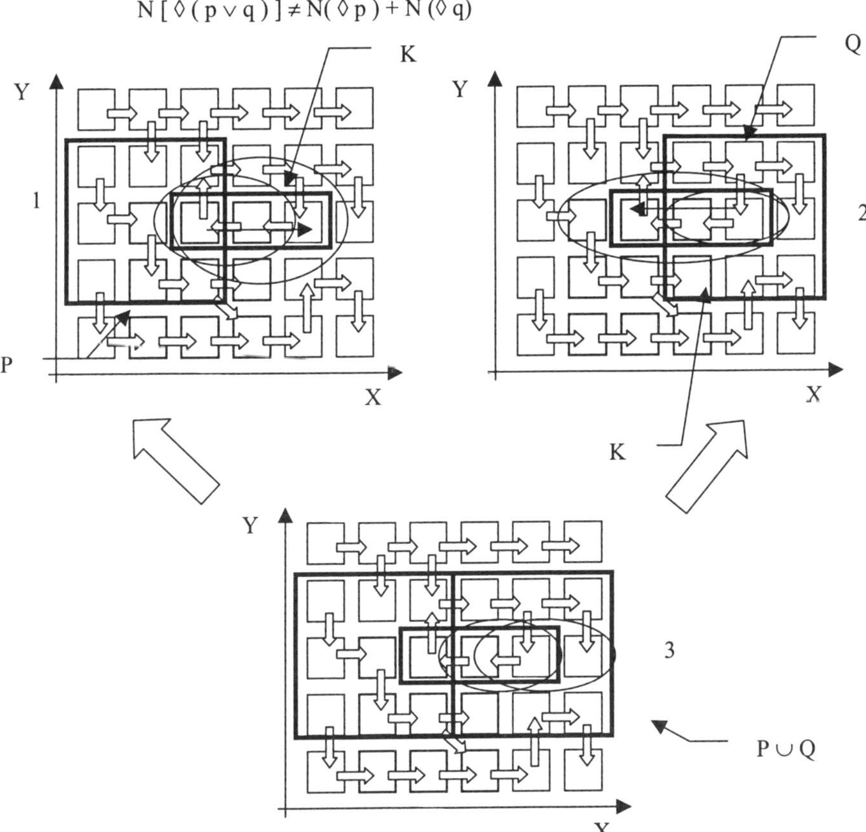

Fig.5 In this figure at the point, one in the set K the sentence p true in the set P is true in one part and false in another part of K. In the part of K where p is false p is possibly true. At the point, two in the set K the sentence q true in Q is true in one part and false in another part. In the part of K where q is false q is possibly true. In the set P ∪ Q, the proposition p ∨ q is true in all the set K. So is necessary true. In fact, K is included in P ∪ Q. For P and Q separate, the number of worlds where p or q are possibly true is greater of the number of worlds in P ∪ Q where p ∨ q is possibly true.

The non-additive system can be found also in the evidence theory[3]

3.5. Expert Systems with Local and non-Local Rules

As we know, expert system is based on the local inferential rules as

$$\textbf{If A then B else C} \tag{1}$$

Where A is a logic sentence in one possible world, and B, C are two different computations.
With the semantic field, we can create the non-local inferential rules

$$\textbf{If Possibly A then B else C} \tag{2}$$

$$\textbf{If Necessarily A then B else C} \tag{3}$$

The sentences (1) and (2) are evaluated in one world by the semantic field that transports the logic value of A from all the other worlds. We cannot isolate the worlds but only the communications among words can give us the value of the sentence "possibly A," and "necessarily A." When almost one world sends a message to the world w that A is true, then A is possibly true in w also if other worlds send the message that A is false. When all the worlds that are in contact with w send a message that A is true, in w A is necessarily true. When in w A is necessarily true can exist worlds that are not in contact with w where A is false. When we put in contact these worlds with w, in w A is not necessarily true. So the necessary operator is relative to the connection system or semantic field. We remark that Web net can be a useful instrument to transport logic values and generate semantic fields.

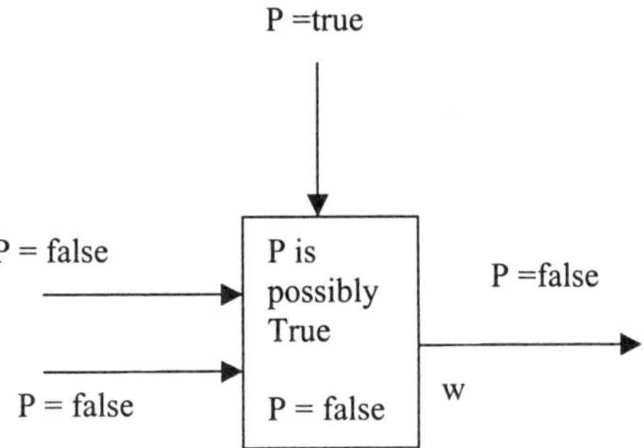

Fig.6 The world w, where the sentence P is false, receives three messages from the other worlds. One message says that the sentence P is true; the others say that the sentence P is false. In the world w, the sentence P is false but possibly true.

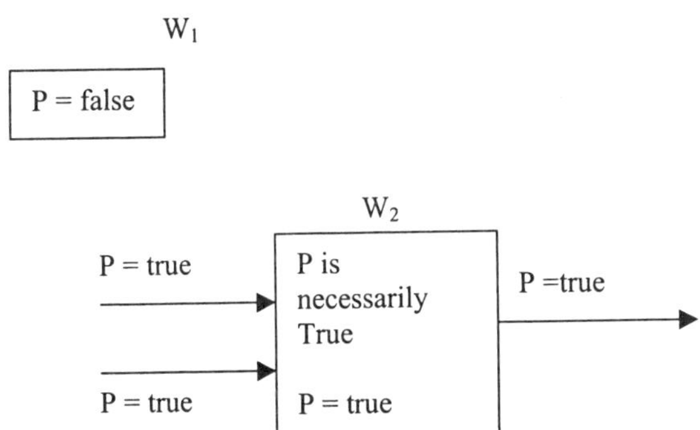

Fig.7 :The world W_2, where the sentence P is true, receives two messages from the other worlds. For the two messages, P is true. So P is necessarily true also if exist one world W_1 where P is false. In fact does not exist any communication between W_1 and W_2 so W_2 cannot know that exist one world where P is false.

References

[1] B.F.Chellas, *Modal Logic: An Introduction.* Cambridge University Press, 1980.

[2] G.E.Hughes and M.J.Cresswell, *An Introduction to Modal Logic.* Methuen, 1968.

[3] G.Resconi, T.Murai, and M.Shimbo, Field theory and modal logic by semantic fields to make uncertainty emerge from information. Int. J. General Systems, Vol.29(5), pp.737-782

Logic, Artificial Intelligence and Robotics
J.M. Abe & J.I. da Silva Filho (Eds.)
IOS Press, 2001

Ehrenfeucht Theorem for the Annotated Logics Q_τ

Alexandre SCALZITTI
Jair M. ABE

Laboratory of Artificial Intelligence and Robotics – LIAR
SENAC College of Computer Science and Technology
Rua Galvão Bueno, 430 – CEP 01506-000 - São Paulo – SP – Brazil

Abstract. Annotated Logics belong to the class of non-classical logics and has found many applications in AI, Robotics and other areas of Mathematics and Computer Science. After being studied from a foundational point of view by N. C. A. da Costa, J. M. Abe, S. Akama, and others, "annotated versions" of classical results have been obtained. One example is the development of Annotated Model Theory. In this paper we extend the well-known Ehrenfeucht Theorem of (Classical) Model Theory to the Annotated Logics.

1. Introduction

Annotate Logics are a category of non-classical logics (paraconsistent, paracomplete and non-alethic logics) introduced by V. S. Subrahmanian [8] in the area of paraconsistent logic programming. Due to a large number of applications of these logics, it became essential to study them from a foundational point of view. This was done in [3] by N. C. A. da Costa, J. M. Abe and V. S. Subrahmanian and in [2] by J. M. Abe. Many theoretical advances were made in the area of Annotated Logics and among them we find the results in Annotated Model Theory which was studied and developed by J. M. Abe in [1], [2], [9] and [10].

One of the most important results in the Classical Model Theory is the Ehrenfeucht Theorem which is directly related to the Ehrenfeucht Game. The Ehrenfeucht Game is, roughly speaking, a back-and-forth technique in which we use two players, Spoiler and Duplicator, performing over two structures, say A and B. The goal of this game is to produce a partial isomorphism between A and B. The Ehrenfeucht Theorem gives us a characterization of when two structures, in some special sense, are equivalent using the Ehrenfeucht Game.

In this paper, we extend the Ehrenfeucht Theorem for the Annotated Logics Q_τ. Our paper is divided as follows: in section 2 we present the basic notions of Annotated Model Theory (Model Theory for Q_τ) such as structures, satisfatibility, etc.; in section 3, we present the logics Q_τ which are the annotated first-order predicate calculi and which will also be our framework; in section 4 we present the basic ideas of Ehrenfeucht Games as well as other concepts involved: τ-isomorphisms, partial τ-isomorphisms, winning strategies for the Game as well as some preliminary lemmas; in section 5, we present the m-Hintikka formulas which characterize the Ehrenfeucht Game along its moves and play an important role in the proof of the Ehrenfeucht Game. Moreover, we present some preliminary lemmas which will be crucial for the proof of the "annotated version" of the Ehrenfeucht Theorem; in section 6 we finally write the proof of the Ehrenfeucht Theorem for $Q\tau$. In section 7, we present comments on some aspects of the lattices involved in our discussion. In section 8, we present some final comments and concluding remarks. At last, we find the bibliographical references.

2. The logics Q_τ

Q_τ is a family of first-order logics, called annotated first-order predicate calculi. Let us fix some arbitrary, finite lattice of truth values $\tau = (dom(\tau), \leq)$. The assumption that τ is finite will be crucial to our purposes as we shall see later. The least element of τ is denoted by $\bot$ and the greatest element is denoted by T. We also assume that there is a fixed unary operator $\sim : dom(\tau) \to dom(\tau)$ which constitutes the "meaning" of our negation. The symbols $\vee$ and $\wedge$ denote respectively the the least upper bound (supremum) and the greatest lower bound (infimum) operators of τ. For details, see [2] and [10].

The language $\mathcal{L}_\tau$ of Q_τ is a first-order language (with equality) whose alphabet is composed by:

1. Variables: enumerable infinite set of variable symbols;
2. Boolean conectives: $\neg$ (negation), $\wedge$ (conjunction), $\vee$ (disjunction) and $\to$ (implication);
3. Quantifiers: $\forall$ (for all) and $\exists$ (there exists);
4. Equality symbol: $=$;
5. Annotated constants: each element of τ is called an *annotational constant*;
6. Auxiliary symbols: parenthesis and commas;
7. Symbols of L.

In the sequel, we assume that L has at least one relation symbol. A *term of Q_τ* is a variable or a constant. We use t_1, t_2, t_3, ... to represent terms. A *formula of Q_τ* is defined in the following way:

1. If t_1 and t_2 are terms then $t_1 = t_2$ is a formula. We will call this sort of formulas, *atomic formulas*;
2. If R is a n-ary relation symbol of L, λ is an annotational constant and t_1, t_2, ... t_n are terms, then $R_\lambda (t_1, t_2, ... t_n)$ is a formula. This formula is called an *annotated atom* and intuitively can be read: "It is believed that the truth value for $R_\lambda (t_1, t_2, ... t_n)$ is at least λ;
3. If $R_\lambda (t_1, t_2, ... t_n)$ is an annotated atom, $\neg^k R_\lambda (t_1, t_2, ... t_n)$, for $k \geq 0$, is a formula and it is called *hyper-literal*. The "$\neg^k$" means a sequence of k negations in front of $R_\lambda (t_1, t_2, ... t_n)$. If k is 0, then we define that $\neg^0 R_\lambda (t_1, t_2, ... t_n)$ is $R_\lambda (t_1, t_2, ... t_n)$. If k is 1, we define that $\neg^1 R_\lambda (t_1, t_2, ... t_n)$ is $R_\lambda (t_1, t_2, ... t_n)$. We say that the hyper-literal $\neg^k R_\lambda (t_1, t_2, ... t_n)$, for $k \geq 0$, *has negation rank k*. A formula other than a hyper-literal is called a *complex formula*;
4. If $\varphi \vee \psi$, and ψ are formulas then $\neg\varphi$, $\varphi \vee \psi$, $\varphi \wedge \psi$, $\varphi \to \psi$ are formulas;
5. If φ is a formula then $\exists x\, \varphi$ and $\forall x\, \varphi$ are formulas.

3. Annotated Model Theory

A *vocabulary L* is a non-empty finite set which consist of relation symbols P, Q, R, ... and of constant symbols c, d, e ... Every relation symbol is equiped with a natural number, its *arity*. We say that a vocabulary is *relational* if it contains no constants. A *structure A of vocabulary L for Q_τ* or simply a *L-structure for Q_τ* consists of : (1) a non-empty set called the *domain of A* and denoted $dom(A)$; (2) a n-ary function $R^A : dom(A) \to dom(\tau)$ for each n-ary relation symbol R in L; (3) a distinguished element c^A of *dom(A)* for every constant symbol c in L. From now on, when we say structure, we mean an L-structure for Q_τ, for some fixed vocabulary L.

Let us fix a structure A. An *evaluation function* or simply an *evaluation* α is a function with domain on the set of the terms such that the images of the variable symbols are elements of *dom(A)* and the image of the constants of L are the constants of A. Let us denote by $\alpha_{a/x}$ the evaluation which agrees with α except that $\alpha(x)=a$, for $a \in dom(A)$. Fixed a sentence φ and an evaluation α, we precisely define the notion of $A \models \varphi[\alpha]$ which is read "φ *is satisfied in A through* α" by the following clauses:

1. if φ is $t_1=t_2$, then $A \models \varphi[\alpha]$ iff $\alpha(t_1)= \alpha(t_2)$;
2. if φ is $R_\lambda (t_1, t_2, \ldots t_n)$, then $A \models \varphi[\alpha]$ iff $R^A (\alpha(t_1), \alpha(t_2), \ldots, \alpha(t_n)) \geq \lambda$:
3. if φ is $\psi \wedge \phi$, or $\psi \vee \phi$, or $\psi \to \phi$, then $A \models \psi \wedge \phi [\alpha]$ iff $A \models \psi[\alpha]$ and $A \models \phi[\alpha]$, $A \models \psi \vee \phi [\alpha]$ iff $A \models \psi[\alpha]$ or $A \models \phi[\alpha]$, $A \models \psi \to \phi [\alpha]$ iff it is not true that $A \models \psi[\alpha]$ or $A \models \phi[\alpha]$;
4. if φ is $\neg^k R_\lambda (t_1, t_2, \ldots t_n)$, with $k \geq 0$, then $A \models \varphi[\alpha]$ iff $A \models \neg^{k-1} R_{\sim\lambda} (t_1, t_2, \ldots t_n)$;
5. if φ is $\neg\psi$ but ψ is a complex formula, then $A \models \varphi[\alpha]$ iff it is not true that $A \models \psi[\alpha]$;
6. if φ is $\exists x\psi$, then $A \models \varphi[\alpha]$ iff $A \models \varphi[a/x]$ for some $a \in dom(A)$;
7. if φ is $\forall x\psi$, then $A \models \varphi[\alpha]$ iff $A \models \varphi[a/x]$ for all $a \in dom(A)$.

4. The Ehrenfeucht Game for Q_τ and some basic results

Let us consider two structures A and B. We say that A and B are *τ-isomorphic* if there is a τ-isomorphism from A to B, that is, a one-to-one function $f : A \to B$ such that: (i) for every constant $c \in L$, we have that $f(c^A)=c^B$ and (ii) for all n-ary relation symbol $R \in L$, for all $\lambda \in \tau$, and for all $a_1, a_2,\ldots, a_n \in A$, we have that $A \models R_\lambda (a_1, a_2,\ldots, a_n)$ if and only if $B \models R_\lambda (f(a_1), f(a_2),\ldots, f(a_n))$. Yet for two structures A and B, let us consider a function p such that $Dom_p \subseteq dom(A)$ e $Im_p \subseteq dom(B)$. We call p a *partial τ-isomorphism from A to B* if: (i) p é injective; (ii) for every constant $c \in L$, we have $f(c^A)=c^B$ e (iii) for every n-ary relation symbol $R \in L$, for all $\lambda \in \tau$, and for all $a_1, a_2,\ldots, a_n \in Dom_p$, we have that $R^A_\lambda(a_1, a_2,\ldots, a_n)$ if and only if $R^A_\lambda(f(a_1), f(a_2),\ldots, f(a_n))$. Let us suppose that p in the conditions above is such that $Dom_p = \{a_1, a_2,\ldots, a_s\}$ and $Im_p = \{b_1, b_2,\ldots, b_s\}$ and, besides, $p(a_i)=b_i$, for $i \in \{1, 2, \ldots, s\}$ and $p(c^A)=c^B$ for every $c \in L$. We can denote the partial τ-isomorphism p by a' $\mapsto$ b' with a'$=(a_1, a_2,\ldots, a_s)$ and b'$=(b_1, b_2,\ldots, b_s)$. We define the *quantifier rank of a formula* φ, denoted by $qr(\varphi)$, as being the maximum number of nested quantifiers which occur in φ. Let m be a natural e let us consider two structures A and B. We say that A is *(m,k)-equivalent to B* if A and B satisfy the same Q_τ sentences (Q_τ formulas where all the variables are in the scope of a quantifier) of quantifier rank less or equal m and negation rank less or equal k. We denote this fact by $A \equiv_{m,k} B$. We describe now how the *Ehrenfeucht Game* works. Let us consider two structures A and B (for a previously fixed vocabulary L), a'$\in dom(A)^s$ and b'$\in dom(B)^s$ and a natural m. *The Ehrenfeucht Game $EHR_m (A,a',B,b')$* consists of two players, the *Spoiler* and the *Duplicator* and has m moves. Each *move* consists of two *plays*. In each move, Spoiler is always the first to make a play. A play consists of choosing one of the structures, respectively, A or B, and then of choosing an element of, respectively, *dom(A)* or *dom(B)*. If in the I-th move, Spoiler chooses e_i in A (respectively, f_i in B), Duplicator must choose an element f_i in B (respectively, e_i in A). Let us suppose that after the m moves, $e_1, e_2,\ldots, e_m$ and $f_1, f_2,\ldots,f_m$ have been chosen in A and B, respectively. We say that Duplicator *wins* $EHR_m (A,a',B,b')$ if and only if a'e' $\mapsto$ b'f', with e'$= e_1 e_2\ldots e_m$ and f'$= f_1 f_2\ldots f_m$, is a partial τ-isomorphism. Otherwise, Spoiler wins the game. We say that a player (may he be Spoiler or Duplicator) has a *winning strategy* for $EHR_m(A,a',B,b')$ if for any choices that the opponent makes, he can answer with choices such that, at the end of m moves, we have a partial isomorphism as decribed above. We make a certain abuse of language: if a player has a winning strategy for the game, we simply say that he *wins* the game. We can say that the Spoiler wins $EHR_m(A,a',B,b')$ if, after i moves, $1 \leq i \leq m$, $a'e' |\to$ b'f ', with e'$= e_1 \, e_2... \, e_m$ and f '$= f_1 \, f_2... \, fm$, is not a partial isomorphism. If m is zero, this is, if we have a game with zero moves, we just require that $a' |\to$ b' is a partial τ-isomorphism for Duplicator to win $EHR_0(A,a',B,b')$. In case that s is zero, we denote the game by $EHR_m(A,B)$ and the Duplicator wins the game if $e' |\to f'$ is a partial τ-isomorphism. If m and s is zero, Duplicator wins $EHR_m(A,B)$ because the empty application is a τ-partial isomorphism. We present other two simple facts in the following two lemmas:

Lemma 1 - Let A and B be two structures and a natural m. If A is τ-*isomorphic* to B, then the Duplicator wins $EHR_m(A,B)$.

Proof: Let p the isomorphism between A and B. A winning strategy for Duplicator would be always to choose the image or the inverse image of p depending on the choices of Spoiler. $\square$

Lemma 2 - Let A and B be two structures, $a' \in dom(A)^s$, $b' \in dom(B)^s$ and m a positive integer. The two following statements are equivalent:
1. Duplicator wins $EHR_m(A,a',B,b')$;
2. For all $a \in dom(A)$, there is a $b \in dom(B)$ such as the Duplicator wins $EHR_{m-1}(A,a'a,B,b'b)$ and for all $b \in dom(B)$, there is an $a \in dom(A)$ such as the Duplicator wins $EHR_{m-1}(A,a'a,B,b'b)$.

Proof: Duplicator wins $EHR_m(A,a',B,b')$ if and only if Duplicator wins $EHR_1(A,a',B,b')$ and in the sequel Duplicator wins $EHR_{m-1}(A,a'x,B,b'y)$ for some $x \in dom(A)$ and some $y \in dom(B)$. This is equivalent to say that if the Spoiler chooses $a \in dom(A)$, Duplicator is able to choose $b \in dom(B)$ such that $a'a |\to b'b$ is a partial τ-isomorphism (analogous if Spoiler chooses $b \in dom(B)$) and in the sequel taking $x=a$ and $y=b$, the Duplicator wins $EHR_{m-1}(A,a'a,B,b'b)$. This is equivalent to say that for all $a \in dom(A)$, there is a $b \in dom(B)$ such as the Duplicator wins $EHR_{m-1}(A,a'a,B,b'b)$ and for all $b \in dom(B)$, there is an $a \in dom(A)$ such as the Duplicator wins $EHR_{m-1}(A,a'a,B,b'b)$. $\square$

5. Satisfability of hyper-literals and m-Hintikka formulas for Q_τ

Let us recall the notion of satisfatibility of hyper-literals. Let us consider a structure A and $a' \in dom(A)^s$ along this section. If a sentence φ is of the type $\neg^k R_\lambda \, (t_1, \, t_2, \, ... \, t_n)$, with $k \geq 0$, then $A \models \neg^k R_\lambda \, (t_1, \, t_2, \, ... \, t_n)$ iff $A \models \neg^{k-1} R_{\sim\lambda} \, (t_1, \, t_2, \, ... \, t_n)$. If we continue on this process, we have that $A \models R_{\sim^k\lambda} \, (t_1, \, t_2, \, ..., \, t_n)$. As τ is a finite lattice, we can say that the set $\{ \sim^k\lambda : k \geq 0 \}$ is finite. This means that if we consider to study the satisfatibility in A of each formula of the infinite set $\{ \neg^k R_\lambda \, (t_1, \, t_2, \, ... \, t_n)$, with $k \geq 0 \}$ of hyper-literals is equivalent to study the satisfatibility in A of each formula of the finite set $\{ R_{\sim^k\lambda} \, (t_1, \, t_2, \, ... \, t_n)$, with $k \geq 0 \}$ which is a subset of annotated atoms. With this in mind, let us consider $v'= (v_1, \, v_2 \,,..., \, v_s)$ a vector of variables and the set S of all formulas ϕ which are satisfied in A such that each ϕ is an atomic formula or an annotated atom. We define the formula $\phi^0_{a'}$ as being the conjunction of all formulas of S. More formally :

$$\phi^0_{a'} \, (v') = \wedge \{ \, \phi(v') : A \models \phi[a'] \, \}$$

where each $\phi(v')$ belongs to S. We say that $\phi^0_{a'}$ is the *Hintikka formula of level zero* or the *0-Hintikka* formula. For $m > 0$,

$$\phi^m_{a'} \, (v') = [\, \wedge_{a \in A} \exists v_{s+1} \, \phi^{m-1}_{a'a}(v',v_{s+1}) \,] \, \wedge \, \forall v_{s+1} \, [\, \vee_{a \in A} \, \phi^{m-1}_{a'a}(v',v_{s+1}) \,]$$

which is the *Hintikka formula of level m* or the *m-Hintikka* formula. In the following, we present some basic lemmas about *m*-Hintikka formulas that will help us in the proof of the Ehrenfeucht Theorem.

Lemma 3 - For naturals s, m, the set $\{\phi^m_{a'}\}$ is finite.
Proof: This follows immediately from the fact that the number of all possible vectors a' is finite. For each a' we have m possible m-Hintikka formulas, namely, $\phi^0_{a'}$, $\phi^1_{a'}$, ..., $\phi^m_{a'}$. $\square$

Lemma 4 - Let A be a structure, $a' \in dom(A)^s$ and a natural m. We have that $qr(\phi^m_{a'})=m$.
Proof: We use induction on m. Let us check when m is 0. We don't have any quantifier in $\phi^0_{a'}$ and then we have that $qr(\phi^0_{a'})$ is 0. Let us check for $m > 0$. We have that
$$\phi^m_{a'}(v') = [\wedge_{a\in A} \exists v_{s+1} \phi^{m-1}_{a'a}(v',v_{s+1})] \wedge \forall v_{s+1} [\vee_{a\in A} \phi^{m-1}_{a'a}(v',v_{s+1})]$$
and by induction hypothesis, we have that $qr(\exists v_{s+1} \phi^{m-1}_{a'a}(v',v_{s+1}))=m-1$ and $qr(\vee_{a\in A} \phi^{m-1}_{a'a}(v',v_{s+1}))=m-1$ and therefore we have that $qr(\phi^m_{a'})=m$. $\square$

The next lemma is a first result concerning satisfatibility of *m*-Hintikka formulas.

Lemma 5 - Let A be a structure, $a' \in dom(A)^s$ and a natural m. We have that $A \models \phi^m_{a'} [a']$.
Proof: We use induction on m. Let us suppose that m is 0. By the definition of Hintikka formulas of level zero, we have that $\phi^0_{a'}(v')$ is the conjunction of formulas $\phi(v')$ such that for each $\phi(v')$ we have $A \models \phi[a']$. Then we have that $A \models \phi^0_{a'} [a']$. Now let us suppose that we would like to check if $A \models \phi^m_{a'} [a']$. We know that
$$\phi^m_{a'}(v') = [\wedge_{a\in A} \exists v_{s+1} \phi^{m-1}_{a'a}(v',v_{s+1})] \wedge \forall v_{s+1} [\vee_{a\in A} \phi^{m-1}_{a'a}(v',v_{s+1})]$$
and by induction hypothesis, we have that $A \models \phi^{m-1}_{a'a} [a'a]$ and then $A \models \wedge_{a\in A} \exists v_{s+1} \phi^{m-1}_{a'a}(v',v_{s+1})$ and $A \models \vee_{a\in A} \phi^{m-1}_{a'a}(v',v_{s+1})$. Therefore, we conclude that $A \models \phi^m_{a'} [a']$. $\square$

The next lemma is also a result concerning satisfatibility of *m*-Hintikka formulas but makes a link between the two structures involved.

Lemma 6 - Let A be a structure, $a' \in dom(A)^s$ and a natural m. Then, for every structure B and $b' \in dom(B)^s$, $B \models \phi^0_{a'}[b']$ if and only if $a' \mapsto b'$ is a τ-isomorphism between A and B.
Proof: Let b' and v' be respectively $(b_1,...,b_s)$ and $(v_1,...,v_s)$. By the definition of Hintikka formulas of level zero, we have that $\phi^0_{a'}(v')$ is the conjunction of formulas $\phi(v')$ such that for each $\phi(v')$ we have $A \models \phi[a']$. If $a' \mapsto b'$ is a τ-isomorphism between A and B is equivalent to say that $B \models \phi[b']$ and therefore $B \models \phi^0_{a'}[b']$. $\square$

6. The Ehrenfeucht Theorem for Q_τ

In this section we finally write the proof of the Ehrenfeucht Theorem for Q_τ which has the following statement.

Theorem 7 (Ehrenfeucht) - Let A and B be two structures, $a' \in dom(A)^s$ e $b' \in dom(B)^s$ and a naturals m. The following assertions are equivalent: (i) Duplicator wins $EHR_m(A,a',B,b')$; (ii) $B \models \phi^m_{a'}[b']$ (iii) Structures A and B are m-equivalent.
Proof: First of all, let us consider the evaluation functions α and β such as $\alpha(x_i)=a_i$ and $\beta(x_i)=b_i$, $1 \leq i \leq s$, which are evaluation functions associated respectively to a' and b'.

1. Let us begin with the implication (iii) $\rightarrow$ (ii). We know that A and B are m-equivalent, this is, if $\phi(x_1,...,x_s)$ has quantifier rank less or equal m, then $A \models \phi[a'] \leftrightarrow B \models \phi[b']$. As we know by Lemma 4 that the quantifier rank of $\phi^m_{a'}$ is m and by Lemma 6 that $A \models \phi^m_{a'}[a']$, then $B \models \phi^m_{a'}[b']$.

2. Let us proof the implication (i) $\rightarrow$ (iii). We use induction on m.
Case 1: Let us consider $m=0$. Let us suppose that Duplicator wins $EHR_0(A,a',B,b')$. This means that $p: a' \mapsto b'$ is a partial τ-isomorphism. Moreover, let us suppose that for some sentence $\phi(x_1,...,x_s)$ with quantifier rank less or equal zero, we have that $A \models \phi[a']$. As the quantifier rank of ϕ is less or equal zero, ϕ may be: atomic, annotated atomic or hyper-literal. Let us suppose that ϕ is atomic, this is, ϕ is a formula of the sort $t_0=t_1$. As $A \models \phi[a']$, we have that $\alpha(t_0)= \alpha(t_1)$. As the Duplicator wins $EHR_0(A,a',B,b')$, $\beta(t_0)=p(\alpha(t_0))$ e $\beta(t_1)=p(\alpha(t_1))$ and then $\beta(t_0) = \beta(t_1)$ and then, $B \models \phi[b']$. Let us suppose that ϕ is an annotated atom, this is, ϕ is of the sort $R_\lambda (t_1, t_2, ... t_n)$ with $\lambda \in \tau$, R a s-ary relational symbol and t_1, t_2, ... t_n terms. As $A \models \phi[a']$ and p is a partial τ-isomorphism, we have that $R^A (\alpha(t_1), \alpha(t_2), ... \alpha(t_n))= R^A (a_1, a_2, ...,a_n) \geq \lambda$ implies that $R^B (b_1, b_2, ..., b_n) \geq \lambda$ and then we have that $B \models \phi[b']$. The case that ϕ is a hyper-literal is straightforward since the study of the satisfatibility of a hyper-literal is equivalent to the study of an annotated atom according to what we have already discussed in the beginning of the section 3. If ϕ is not one of the three types discussed before, it is a boolean combination of them. An induction of the complexity of the formula will also give us that $B \models \phi[b']$.
Case 2: Let us consider m positive. Let us suppose that the Duplicator wins $EHR_m(A,a',B,b')$. Moreover we can suppose that $\phi(x') = \exists y\, \psi(x',y)$ and $qr(\phi) \leq m$. Let us assume that $A \models \phi[a']$. So there exists an $a \in A$ such that $A \models \psi [a'a]$. As Duplicator wins $EHR_m(A,a',B,b')$, by Lemma 2, there is a $b \in B$ such that the Duplicator wins $EHR_{m-1}(A,a'a,B,b'b)$. Since $qr(\psi)=m-1$, the induction hypothesis give us that $B \models \psi [a'a]$ and this means that $B \models \phi[b']$.

3. Let us proof the equivalence (i) $\leftrightarrow$ (ii). We use induction on m.
Case 1: Let us consider $m=0$. We have that the Duplicator wins $EHR_0(A,a',B,b')$ if and only if $p: a' \mapsto b'$ is a partial τ-isomorphism if and only if $B \models \phi^0_{a'}[b']$ by Lemma 6.
Case 2: For positive m, Duplicator wins $EHR_m(A,a',B,b')$ if and only if, by Lemma 2, for all $a \in dom(A)$, there is a $b \in dom(B)$ such that the Duplicator wins $EHR_{m-1}(A,a'a,B,b'b)$ and for all $b \in dom(B)$, there is an $a \in dom(A)$ such that the Duplicator wins $EHR_{m-1}(A,a'a,B,b'b)$. By the induction hypothesis, this is equivalent to say that for all $a \in dom(A)$, there is a $b \in dom(B)$ such that the $B \models \phi^{m-1}_{a'a}[b'b]$ and for all $b \in dom(B)$, there is an $a \in dom(A)$ such that $B \models \phi^{m-1}_{a'a}[b'b]$. This is equivalent to say that
$$B \models [\wedge_{a \in A} \exists v_{s+1}\, \phi^{m-1}_{a'a}(v',v_{s+1})] \wedge \forall v_{s+1} [\vee_{a \in A}\, \phi^{m-1}_{a'a}(v',v_{s+1})] [b'b]$$
and consequently that $B \models \phi^m_{a'}[b']$. $\square$

7. Study of some simple cases: particular lattices and operators

In this section we study some special cases for τ and $\sim$. Let us assume throughout this section that A is a structure.

7.1. One-element lattice

Let τ have just one element, that is, $dom(\tau)=\{\lambda\}$. This means that we can annote atoms only with λ. In this case, let $\sim$ be the identity operator, this is, $\sim\lambda=\lambda$, for all $\lambda \in \tau$. In this situation, we have that the study of the satisfatibility of a hyper-literal of positive negation rank is reduced to the same hyper-literal but with negation rank 0. In other words, if we have $\neg^k R_\lambda (t_1, t_2, ... t_n)$, with $k \geq 0$, then according to the discussion in the first paragraph of section 5, the study of $A \models \neg^k R_\lambda$ $(t_1, t_2, \ldots t_n)$ would correspond to study if $A \models R_{\sim^k\lambda}$ $(t_1, t_2, \ldots, t_n)$. But $\sim^k\lambda$ is λ, for $\sim$ is the identity operator. So, it is enough to study $A \models R_\lambda$ $(t_1, t_2, \ldots, t_n)$. For a fixed evaluation α, we have that $A \models R_\lambda$ $(t_1, t_2, \ldots, t_n)$ if and only if R^A $(\alpha(t_1), \alpha(t_2), \ldots, \alpha(t_n)) \geq \lambda$. As the relation R^A takes values from $dom(A)$ to $dom(\tau)$, this means that no matter what values $\alpha(t_1), \alpha(t_2), \ldots, \alpha(t_n)$ may take, R^A $(\alpha(t_1), \alpha(t_2), \ldots, \alpha(t_n))$ is equal λ. This means that any relation with annotation λ is satisfied in the structure A.

7.2. Recovering the Classical Case

Now let us suppose that τ has two elements, this is, $dom(\tau)=\{V,F\}$ where the elements V (true) and F (false) are not comparable. Let us assume also that $\sim V = F$ and $\sim F = V$. Here the role of the operator $\sim$ is intuitively the one of the classical negation. If we have $\neg^k R_\lambda$ $(t_1, t_2, \ldots t_n)$, with $k \geq 0$, then according to the discussion in the first paragraph of section 5, the study of $A \models \neg^k R_\lambda$ $(t_1, t_2, \ldots t_n)$ is equivalent to the study of $A \models R_V (t_1, t_2, \ldots t_n)$ or $A \models R_F (t_1, t_2, \ldots t_n)$ because for $k \geq 0$, we have that $\sim^k\lambda$ is V or F.

8. Concluding remarks

The Ehrenfeucht Theorem provides us a back-and-forth technique to identify "isomorphic pieces" of two structures. In other words, given two structures A and B, the Ehrenfeucht Theorem enables us to search for "pieces of A" which are isomorphic to "pieces of B" (partial isomorphism). The back-and-forth technique used is the Ehrenfeucht Game. The winning strategies used by Duplicator will enable him to win the Games and according to the Ehrenfeucht Theorem, they will be a partial isomorphism. Depending on the structures being studied, we can build winning strategies for Duplicator. A good example of the application of the Ehrenfeucht Theorem is the study of zero-one laws in random structures, especially in Random Graph Theory. An introductory reference for the subject is [7] and a deep reference may be found in [6].

All the above mentioned happens for classical logics but in this paper we have extended the result for the Annotated Logics Q_τ. In other words, we have proved that the idea of *m*-equivalence remains the same if we take sentences of Q_τ, because the Ehrenfeucht Theorem holds for these logics. We have to notice that the assumption that the lattice τ is finite is crucial for our purposes.

References

[1] J. M. Abe, On Annotated Model Theory, Coleção Documentos, Série Lógica e Teoria da Ciência - 13, Institute for Advanced Studies, University of São Paulo, São Paulo, 1993.
[2] J. M. Abe, Fundamentos da Lógica Anotada (Foundations of Annotated Logics), Ph.D. Thesis, University of São Paulo, São Paulo, 1992.
[3] N. C. A. da Costa, J.M. Abe & V. S. Subrahmanian, Remarks on annotated logic, *Zeitschr. f. Math. Logik und Grundlagen d. Math.*, 37: 561-570, 1991.
[4] H.-D. Ebbinghaus and J. Flum, Finite Model Theory, Perspectives in Mathematical Logic, Springer-Verlag, Berlin, 1995.
[5] W. Hodges, A Shorter Model Theory, Cambridge University Press, Cambridge, 1997.
[6] S. Janson, T. Luczak, A. Rucinski, Random Graphs, John Wiley & Sons, Inc. New York, 1998.
[7] A. Scalzitti, Convergence in the Theory of Random Graphs (In Portuguese), Master Degree Dissertation, Institute of Mathematics and Statistics, University of São Paulo, São Paulo, 1999.
[8] V. S. Subrahmanian, On Semantics of Quantitative Logic Programs, Proc. 4th IEEE Symposium on Logic Programming, Computer Society Press, Washington D. C., 1987, 173-182.
[9] J. M. Abe & S. Akama, Annotated logics $Q\tau$ and ultraproducts, *Logique et Analyse* 160, 335-343, 1997.
[10] J. M. Abe, Annotated logics $Q\tau$ and Model Theory, to appear, 2001.

Logic, Artificial Intelligence and Robotics
J.M. Abe & J.I. da Silva Filho (Eds.)
IOS Press, 2001

A Formalization for Signal Analysis of Information in Annotated Paraconsistent Logics

Alexandre SCALZITTI
João Inácio DA SILVA FILHO
Jair Minoro ABE
Artificial Intelligence and Robotics Laboratory - LIAR
SENAC School of Computer Science and Technology
Rua Galvão Bueno, 430 – CEP 01506-000 - São Paulo – SP – Brazil
E-mail: {scal, jinacio, jmabe}@cei.sp.senac.br

Abstract. In this work, we present a set of boolean operations in the unitary square of the cartesian plane (USCP) of the annotated paraconsistent logic with annotation of two values (APL2v). Moreover, we redefine all the presented operations for the lattice of annotations (LA) of APL2v. The method we use to redefine these operations in the LA consists of transformations (functions) between USCP and LA. Besides being used for redefining the operations, these transformations allow us a better flexibility to the control systems based on APL2v which use the USCP or the LA to represent signals: these transformations also allow us to convert easily signals from USCP to LA and vice-versa. The method applied for USCP and LA can be also applied for any other pair of systems where annotations can be represented.

1. Introduction

The control systems used in automation and robotics as well as the expert systems used in AI generally work based on the classical logic, where the world description is considered by only two states, true or false. However, the contradictions and inconsistencies are usual when we describe parts of the real world and these binary systems cannot handle properly the situations where the information are originated from several sources which may be contradictiory or which bring evidences from uncertain knowledge. For projects of analysis systems which are going to take decisions in AI, the control or expert systems are fed with information which should represent the real world in the closest possible way. These analysis systems, in order to be efficient, cannot receive the information in the binary form but in the analogical form and which can be identified with degrees. These information are signals brought through, generally speaking, any signal acquisition system: these acquisition systems are, rougly speaking, sensors (in the case of robotics), cameras (in the case of pattern recognition) or opinions expressed by experts for a certain phenomenon. In fact, since these information come from several sources, they may represent ambiguous or contradicting situations. In order to solve problems of taking decisions face to situations which are not handled by classical logic, new methods of applications of a certain sort of logics, called non-classical, are used: among these logics we find the so-called *paraconsistent logics*.

The paraconsistent logics belong to the class of the non-classical logics and were developed because of the the need to find means of handling contradicting situations. In many studies(among them [1] and [2]), the paraconsistent logics presented results which enable us to consider the inconsistencies in a non-trivial fashion and therefore, they are more proper in the handling of problems generated by contradicting situations which we deal with the real world.

The signals which are brought through any mechanism of signal acquisition are considered *annotations* in the Annotated Paraconsistent Logics. These annotations may be represented in several coordinate systems, depending on the implementation of the control system (or any other sort of system). We present in this work a set of boolean operations on annotations in the *unitary square of the cartesian plane* (USCP). This USCP is a coordinate system for representing annotations of the *annotated paraconsistent logics with two-valued annotations* (APL2v). The fact that we are working with two-valued annotations will be clear later. We redefine these boolean operations for the *lattice of annotations* (LA) which is another coordinate systems for representing annotations. We describe and use transformations (functions) between USCP and LA to redefine these boolean operations on LA. These transformations also give flexibility to the control systems based on APL2v because we can easily convert annotations from USCP into annotations of LA and vice-versa.

Our paper is divided in the following way: in section 2, we slightly present some concepts related to APL2v; in section 3 we present the USCP, the LA and the transformations (mentioned above) between them; in sections 4 and 5, we present the boolean, respectively arithmetical, operations on USCP and on LA; in section 6, we talk about appications. In the sequel, we present in section 7 the conclusions and at last, the bibliographical references.

2. The annotated paraconsistent logics with two-valued annotations-APL2v

The language of the Annotated Paraconsistent Logics (APL), roughly speaking, contains propositional formulas which come together with *annotations*. Intuitively, these annotations represent the signals brought to analysis through any signal acquisition mechanism. Each annotation belongs to a lattice τ which is the traditional lattice of the APL2v (Figure 1a).

We have that annotations in APL2v have two values. Given a propostion p, we have a favorable evidence to p which we generally call *belief degree*. The second value of the annotation is called *disbelief degree*. In the applications and in the literautre, the belief and the disbelief degree are denoted respectively by μ_1 and μ_2.

According to Figure 1(a), we can see that each vertex of τ has a special name: we assign to the superior vertex the symbol **T** (inconsistent), to the inferior vertex, the symbol $\perp$ (indeterminated), to the left one the *f* (false) and to the right one, we assign *t* (true). For practical purposes of τ, we should adopt a cartesian coordinate system to the plane, and then the annotations of a given proposition will be given by points of the plane.

Several examples of coordinate systems adopted can be found in [3]. We call *unitary square of the cartesian plane* (and we denote throughout this paper by USCP) the lattice τ with the coordinate system indicated in Figure 1(b). So, we assign **T** to *(1,1)*, $\perp$ to *(0,0)*, *f* to *(0,1)* and *t* to *(1,0)*. Let us notice that, for each adopted coordinated system, we have that annotations *(μ_1,μ_2)* of τ are identified with different points of the plane. In other words, in the system of Figure 1(b), a certain annotation *(μ_1,μ_2)* can be identified with the point *(1/2;3/4)* and in other system, with *(3/5;4/7)*, for example.

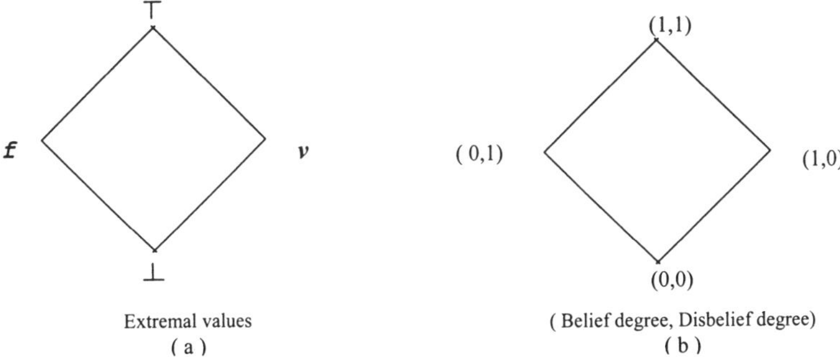

Extremal values   (Belief degree, Disbelief degree)
(a)   (b)

Figure 1 – Representative lattice for the annotated paraconsistent logics.

We present in the following section another coordinate system which can be fixed for τ and very used in applications. We define transformations between USCP and LA which will be the lattice τ together with another coordinate system.

3. The Unit Square of the Cartesian Plane and the Lattice of Annotations

As it was shown in Figure 1(b), let us consider the USCP which is the square which side has measure one, such as in the Figure 2. Let us also consider the lattice of annotations (LA) as being the lattice τ together with the following coordinate system, such as in Figure 3.

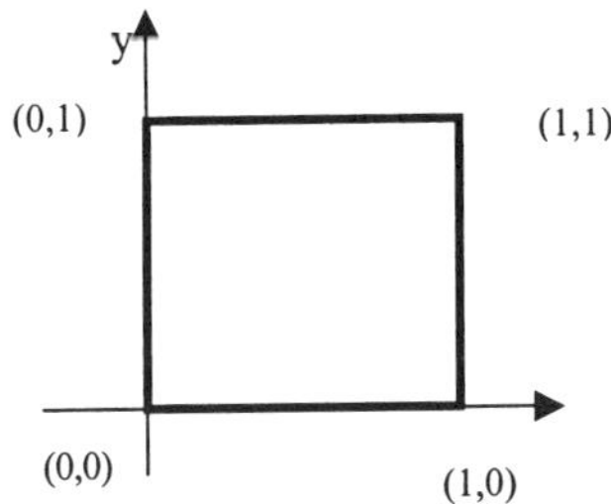

Figure 2- Unitary Square of the Cartesian Plane - USCP.

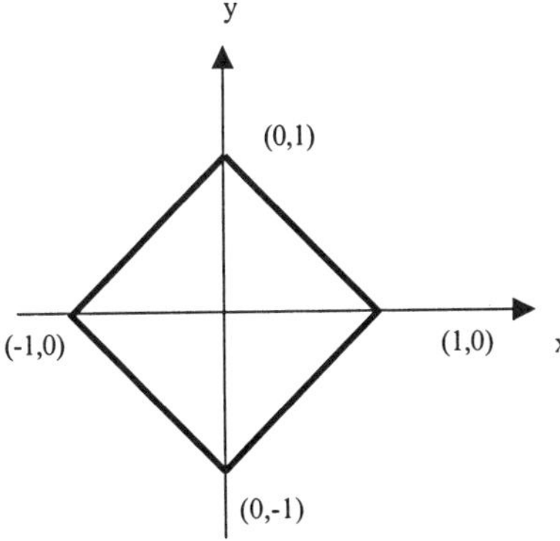

Figure 3- Lattice τ together with a new coordinate system

Similarly to what we have done to USCP, we assign **T** to *(0,1)* , $\perp$ to *(0,-1)* , *f* to *(-1,0)* and *v* to *(1,0)*. We can see that LA can be obtained from USCP through a change of scale, followed by a rotation and, in the sequel, a translation. These transformations begin to be described in Figure 4 where we can see the change (increase) of scale of factor $\sqrt{2}$.

This increase is given by the following linear transformation:

$$T_1(x,y) = (\sqrt{2}\,x,\ \sqrt{2}\,y).$$

whose corresponding matrix is:

$$\begin{bmatrix} \sqrt{2} & 0 \\ 0 & \sqrt{2} \end{bmatrix}$$

In Figure 5, we have a 45° rotation with respect to the origin. This rotation is given by the following linear transformation:

$$T_2(x,y) = \left(\frac{\sqrt{2}}{2}x - \frac{\sqrt{2}}{2}y, \frac{\sqrt{2}}{2}x + \frac{\sqrt{2}}{2}y \right);$$

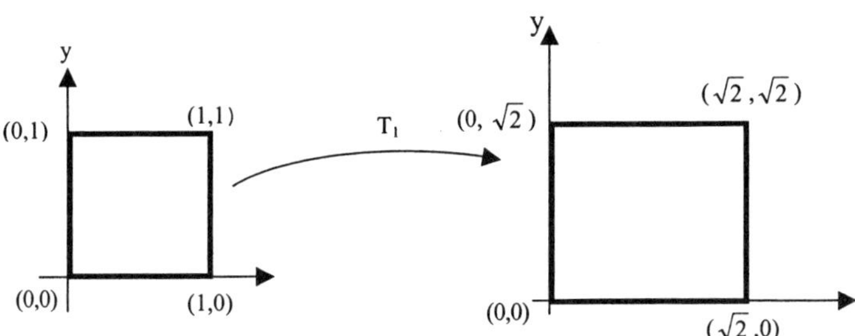

Figure 4 – Increase in scale of factor $\sqrt{2}$.

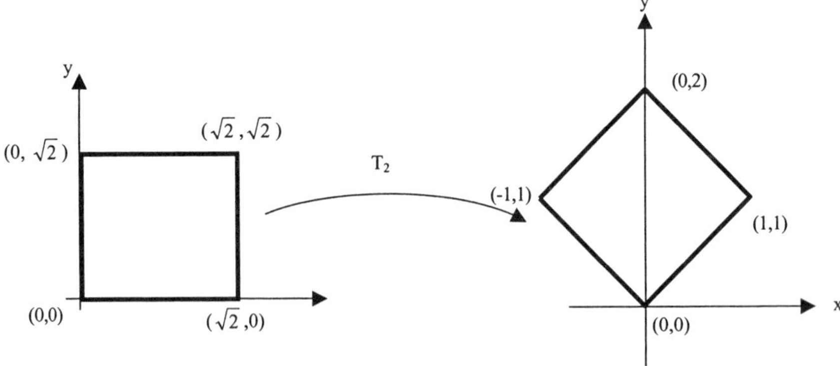

Figure 5 – 45° Rotation with respect to the origin.

whose corresponding matrix is:

$$\begin{bmatrix} \dfrac{\sqrt{2}}{2} & -\dfrac{\sqrt{2}}{2} \\[2ex] \dfrac{\sqrt{2}}{2} & \dfrac{\sqrt{2}}{2} \end{bmatrix}$$

In the following, we have a translation given by the following transformation:

$$T_3(x,y)=(x,y-1)$$

represented in Figure 6. If we do the composition $T_3 \; o \; T_2 \; o \; T_1$ we obtain the transformation

$$T(x,y)=(x-y, \; x+y-1).$$

Let us notice that with the transformation T, we can convert annotations represented in USCP to annotations represented in LA. In order to agree with the literature terminology (such as [1] and [4]), for an annotation (μ_1,μ_2), we have that

$$T(\mu_1,\mu_2) = (\mu_1 - \mu_2, \; \mu_1 +\mu_2 - 1)$$

and we say that $\mu_1 - \mu_2$ is the *certainty degree* and it is denoted by G_c and $\mu_1 +\mu_2 - 1$ is the *contradiction degree* and it is denoted by G_{ct}.

The following transformations F_1 , F_2 and F_3 are the inverse transformations of, respectively, T_1, T_2 and T_3 :

1. $F_1(x,y) = \left(\dfrac{\sqrt{2}}{2}x, \dfrac{\sqrt{2}}{2} y \right)$ which is linear and whose corresponding matrix is:

$$\begin{bmatrix} \dfrac{\sqrt{2}}{2} & 0 \\[2ex] 0 & \dfrac{\sqrt{2}}{2} \end{bmatrix}$$

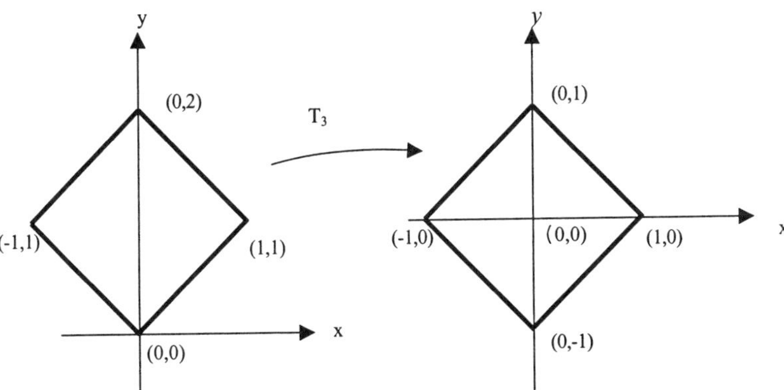

Figure 6 – translation given by the transformation $T_3(x,y)=(x,y-1)$.

2. $F_2(x,y) = \left(\dfrac{\sqrt{2}}{2} x + \dfrac{\sqrt{2}}{2} y, -\dfrac{\sqrt{2}}{2} x + \dfrac{\sqrt{2}}{2} y \right)$ which is also linear and whose corresponding matrix is:

$$\begin{bmatrix} \dfrac{\sqrt{2}}{2} & \dfrac{\sqrt{2}}{2} \\[2mm] -\dfrac{\sqrt{2}}{2} & \dfrac{\sqrt{2}}{2} \end{bmatrix}$$

3. $F_3(x,y) = (x, y+1)$.

If we compute the composition $F_1 \circ F_2 \circ F_3$, we obtain the transformation

$$F(x,y) = \left(\frac{1}{2}x + \frac{1}{2}y + \frac{1}{2}, -\frac{1}{2}x + \frac{1}{2}y + \frac{1}{2} \right).$$

With transformation F, we can convert annotations of the LA into annotations of USCP. With T and F, it is the same thing to represent annotations in the USCP or in the LA because we can easily "move" from this to that and vice-versa, as we show below in Figure 7. This kind of "easy movement" between two different coordinate systems for τ is important for algorithms implemented in control and expert systems based on APL2v. Contradicting information can come from several sources and in different coordinate systems and this fact will not be a problem to the system.

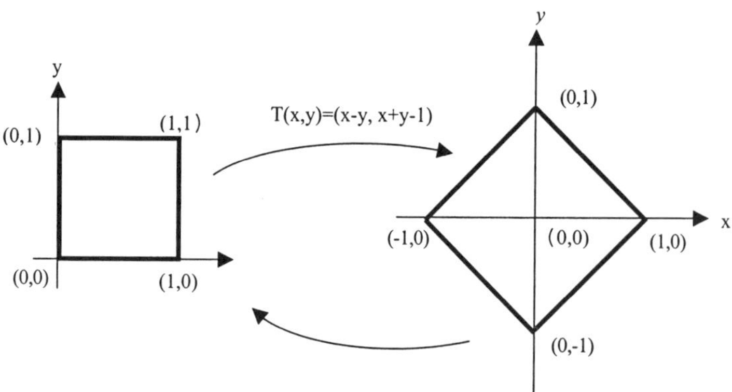

$$F(x,y) = \left(\frac{1}{2}x + \frac{1}{2}y + \frac{1}{2}, \frac{-1}{2}x + \frac{1}{2}y + \frac{1}{2} \right)$$

Figure 7 – Transformations T e F between the USCP and the LA.

4. Boolean Operations in USCP and in LA

In this section, let us define some boolean operations in the USCP and in LA. Given two annotations P_1 and P_2 with coordinates (x_1,y_1) and (x_2,y_2), respectively, we define the following boolean operations in the USCP:

1. Negation $\quad : \neg P_1 = (y_1,x_1)$;
2. Disjunction $: P_1 \vee P_2 = (\ min\ \{\ x_1,\ x_2\}\ ,\ max\ \{\ y_1\ ,\ y_2\}\ )$;
3. Conjunction $: P_1 \wedge P_2 = (\ max\ \{\ x_1,\ x_2\}\ ,\ min\ \{\ y_1\ ,\ y_2\}\ )$.

These operations were defined in [4] and it found many applications. These operations can be redefined in the LA and for this purpose we use the idea expressed in Figure 4 in the following way: let us suppose that given two annotations (x_1,y_1) and (x_2,y_2) in the LA. We can define, for example, the disjunction of these annotations in the following way: let us consider (x_1',y_1') and (x_2',y_2') in USCP such that $(x_1',y_1')=F(x_1,y_1)$ and $(x_2',y_2')=F(x_2,y_2)$. Let us do the disjunction $(x_1',y_1') \vee (x_2',y_2')$ obtaining an annotation (x_3',y_3') and let us compute $(x_3,y_3)=T(x_3',y_3')$. Let us put (x_3,y_3) as being the disjunction of (x_1,y_1) and (x_2,y_2) in the LA. In the sequel, we formalize negation, disjunction and conjunction for the LA. All the operations in the LA were defined with the same above mentioned idea. The negation, disjunction and conjunction in LA are represented respectively by $\neg_{RA}$, $\vee_{RA}$, $\wedge_{RA}$.

1. $\neg_{RA}P_1 := T(\neg F(x_1,y_1))= T(a,b)=(b,a)$.

2. $P_1 \vee_{RA} P_2 := T(F(P_1) \vee F(P_2)) = T(\ (a,b) \vee (c,d)\ ) = T(\ max\{a,c\}\ ,\ min\{b,d\}\ ) = (max\{a,c\} - min\{b,d\}\ ,\ max\{a,c\} + min\{b,d\} - 1)$.

3. $P_1 \wedge_{RA} P_2 := T(F(P_1) \wedge F(P_2)) = T(\ (a,b) \wedge (c,d)\ ) = T(\ min\{a,c\}\ ,\ max\{b,d\}\ ) = (min\{a,c\} - max\{b,d\}\ ,\ min\{a,c\} + max\{b,d\} - 1)$.

with:

$$a = \frac{1}{2}x_1 + \frac{1}{2}y_1 + \frac{1}{2};$$

$$b = \frac{-1}{2}x_1 + \frac{1}{2}y_1 + \frac{1}{2};$$

$$c = \frac{1}{2}x_2 + \frac{1}{2}y_2 + \frac{1}{2};$$

$$d = \frac{-1}{2}x_2 + \frac{1}{2}y_2 + \frac{1}{2}.$$

With these operations, we can work with annotations either in USCP and in LA.

5. Arithmetic Operations in the USCP and in LA

In the following, we define a new operation with annotations. Let us consider a colection of n annotations $a_1, a_2, ..., a_n$ with the respective coordinates
$$(x_1,y_1),\ (x_2,y_2),\ ...,\ (x_n,y_n)$$

in USCP. We define a sum $\oplus$ in USCP in the following way:

$$a_1 \oplus a_2 \oplus ... \oplus a_n := \left(\frac{\sum_{i=1}^{n} x_i}{n}, \frac{\sum_{i=1}^{n} y_i}{n} \right).$$

It is easy to see that $a_1 \oplus a_2 \oplus ... \oplus a_n$ is still an annotation in USCP, this is, the operation is closed in USCP. We can define a new operation $\oplus_{RA}$ for $a_1, a_2, ..., a_n$ (now in the LA) in the following way: let us take a'_i as being $F(a_i)$ for all $i \in \{1,...,n\}$. The sum $\oplus_{RA}$ can be defined in the following way: let us consider $a_1, a_2, ..., a_n$ annotations in the LA. We define:

$$a_1 \oplus_{RA} a_2 \oplus_{RA} ... \oplus_{RA} a_n = T(a'_1 \oplus ... \oplus a'_n)$$

and if we work out on the above formula, we have:

$$a_1 \oplus_{RA} a_2 \oplus_{RA} ... \oplus_{RA} a_n = \left(\frac{\sum_{i=1}^{n} x_i}{n}, \frac{\left(\sum_{i=1}^{n} y_i\right)+1}{n} - 1 \right).$$

We have defined so far a set of operations in USCP and in LA. The operations in LA were naturally defined through the transformations T and F. The arithmetic operations will allow us the computation of annotations. We can then project control systems for taking decisions which is totally based on the concepts of the APL2v whose the results of operations will always remain in USCP or in LA.

6. Application Method – Para-Analyzer Algorithm

The assignment of belief and disbelief degrees has as an objective to solve the problem of contradicting signals collecting evidences and, by means of analysis, to modify the behavior of the system in order to decrease the contradiction "intensity". This is the idea of the para-analyzer algorithm ([4]). The operations that we have defined in the previous section participate in these analyses in the following way: an information (signal) acquisition mechanism collect signals and they are translated into annotations. In order for the system to take a decision, it may be necessary to perform a certain amount of boolean operations with the annotations provided by the sensors (or any other information acquisition system). Through the transformations T and F, which we have defined so far in the previous section, the annotations may be expressed either in the USCP or in the LA for we know how to do the necessary conversions between a coordinate system and the other. Figure 8 shows how the para-analyzer algorithm works:

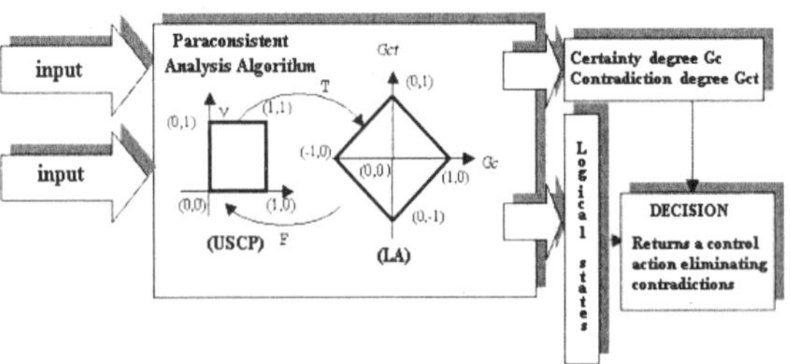

Figure 8 – Paraconsistent analysis with the Para-Analyzer algorithm.

A paraconsistent control system works in the following way:

1-If there is a high degree of contradiction, there is no certainty about the decision, therefore, new evidences must be searched.

2-If there is a low degree of contradiction, the conclusion must be formulated since there is a high degree of certainty.

The Para-Analyzer algorithm is also the central idea of the paraconsistent neural networks ([5]): it is the algorithm of the basic paraconsistent cell.

7. Conclusions

In this paper, we show a formalization to signal analysis using annotations according to the fundamental concepts of the annotated paraconsistent logic which is effective in handling contradictions. When we want to increase the reliability in the information, we must work with signal acquisition systems and this may give rise. to contradictions. The formalization presented through the functions T and F suggests a way to make the control system work with both coordinate systems. The idea can be generalized for any coordinate systems (other than USCP and LA) in which the annotations can be expressed. The signal handling according to the presented formalization allowed us to define boolean operations (NOT, AND, OR) as well as arithmetic operations on annotations. For each new pair of coordinate systems for τ, we can redefine all the presented operations for any other new pair of coordinate systems, using the same idea we used for USCP and LA.

References

[1] Abe, J. M, Fundamentos da Lógica Anotada, (Foundations of Annotated Logics), in Portuguese, Ph D thesis, University of São Paulo, FFLCH/USP - São Paulo, 1992.

[2] Da Costa, N.C.A. & Subrahmanian, V.S. & Vago, C.,The Paraconsistent Logic $P\tau$, Zeitschrift für Mathematische Logik und Grundlagen der Mathematik, Vol.37, pp.139-148,1991.

[3] Da Costa, N. C. A. , Abe, J. M. et alli, Lógica Paraconsistente Anotada, Editora Atlas – São Paulo - 1999

[4] Da Silva Filho, J.I., Métodos de Aplicações da Lógica Paraconsistente Anotada de Anotação com dois valores LPA2v com construção de Algoritmo e Implementação de circuitos Eletrônicos, in Portuguese, *Ph D thesis, EPUSP*, São Paulo, 1999.

[5] Da Silva Filho, J.I., Abe, J.M., Fundamentos das Redes Neurais Artificiais Paraconsistentes, Editora Arte & Ciência, São Paulo, 2001.

Logic, Artificial Intelligence and Robotics
J.M. Abe & J.I. da Silva Filho (Eds.)
IOS Press, 2001

Rule and Agent Oriented Software Architecture for Controlling Automated Manufacturing Systems

Jean Marcelo Simão, Paulo Roberto O. da Silva, Paulo Cézar Stadzisz, Luiz Allan Künzle

Centro Federal de Educação Tecnológica do Paraná
Programa de Pós-Graduação em Engenharia Elétrica e Informática Industrial,
Av. Sete de Setembro, 3165, 80230-901, Curitiba, PR, Brazil
simao@cpgei.cefetpr.brs, stadzisz@cpgei.cefetpr.br

Abstract: This work proposes the use of rule based systems (RBS) in the composition of a generic software architecture for the design and implementation of Control Systems in Automated Manufacturing. The RBS is realised by means of agents encompassing the fact base and the decision and actuation processes (i.e., rules). The use of agents allows an advanced inference, known as RETE. It is also proposed a semi-automatic process for the creation of the rules, to assist the composition of each Control instance. As the design of the Control for Automated Manufacturing Systems is a complex activity (due to the automation level and required flexibility) and due to the risks, costs and time involved, there is a need for tools to aid the analysis and experimentation of control systems. The tests of this architecture are done over a simulation tool called ANALYTICE II. This tool is a discrete event simulator whose primitives are equipments (e.g. lathes, milling machines and conveyors), parts and pallets. The main contribution of this work is the definition of a flexible and comprehensive Control architecture based on artificial intelligence's concepts.

1. INTRODUCTION

The Control of an Automated Manufacturing System (AMS) is responsible for co-ordinating the factory components (e.g. lathes and robots) to carry out predefined production process.

This co-ordination includes: (i) monitoring of the discrete states from each element in the factory; (ii) decision with respect of these states and current process; and (iii) actuation over factory components by means of appropriate commands and respecting specific protocols [6][2].

The generality, automation level, required flexibility and attainable extension, make the Control a complex computational system, whose risks, costs and design time impose rigorous development methods and the use of appropriated support tools for testing and verifying functionalities [2]. The control of discrete event systems, as for the AMS, has been extensively studied in academy and in industry research centres. Many approaches and methods for design and implementation have been proposed [3][9].

This article presents a software architecture for the implementation of the control for AMS. The main goal is to establish a control system design pattern for AMS leading to a quick and robust implementation.

The proposed architecture is flexible allowing the implementation of control systems for different industrial plants and the development of various Control models (e.g. hierarchic, heterarchic and holonic). The proposed architecture embeds artificial intelligence techniques in the maintenance of the production system knowledge base and in the decision process, due to state changes and causal relationships. The architecture has been applied to experiments with the ANALYTICE II simulator [12].

The basis for the proposed AMS Control architecture is grounded in Rule Based Systems (RBS) [10] and on the concept of agents [5][1]. The fact base is composed by agents that represent the components of AMS (e.g. machines and parts), the hierarchy (e.g. station, cell and plant) and abstract elements (e.g. lot of parts, process plan and production plan). The rules, also seen as agents, have conditions based on attributes of agents and actions that modify them. The inference is done by forward chaining, based on the notification principle that was introduced by RETE algorithm [4][10].

This article is organised as follows: section 2 presents the ANALYTICE II simulator, section 3 describes the Control Architecture based on rules and agents and section 5 presents the conclusions of the work.

2. ANALYTICE II

ANALYTICE II is a discrete event simulation tool for AMS that allows experiments for determining qualitative and quantitative parameters through its main modules: equipment model, geometric animation, kernel and simulation monitoring [7].

The modelling primitives are: parts, pallets and equipments. An equipment is represented by the geometric, kinematics and behavioural models. The geometric model can be imported from CAD systems. The kinematics model associates translation and rotation axes to the geometric model, allowing its tridimensional animation. The behavioural model describes the evolution of the states of the equipments and presents an interface to receive command signals, as "open end effector", and send answer signals, as "end of machining". The behavioural models are described by C++ programming language classes, whose instances (i.e. objects) are executed and interact with the simulator kernel.

3. RULE AND AGENT BASED CONTROL ARCHITECTURE

The Control of the AMS is dynamic and event oriented whose controlled variables are changed when discrete events take place [6][8][9]. The Control communicates with the others decision elements (Planning, Scheduling and Supervision) with which form an integrated group for the production process management.

In this work, the control RBS presents the distinguishing characteristic of being accomplished by means of agents which are responsible for the fact base and for the set of rules. The **Fact Base Agent (FBA)** describes discrete states of the AMS components (e.g. stopped robot, robot turning to left, robot end effector opened, etc.) and actuates over these elements. The discrete events are understood as facts by the **Rule Agents (RAs)**, using them as reference in the decision process. An inference in this RBS is accomplished associating FBAs and RAs, following the notification principle from RETE [4][10]. There is another set of agents, called **Formation Agents (FAs)**, responsible for a partial automation and semantic evaluation in the ARs creation process [12].

3.1   Fact Base

The fact base is constituted by a set of FBAs. These agents are derived from the concept of objects, in the software engineering point of view, or from the concept of frames, in the point of view of artificial intelligence. An agent can be defined as a software module or component, with high degree of cohesion, with well defined scope, with autonomy and taking part of a certain context whose changes are perceived by the agent. These perceptions may change the agent behaviour and the agent may promote another changes in the context [5][1].

The FBAs, besides being the self-contained base of facts that capture or instigate changes in facts or states, also command the physical objects of the factory, whenever appropriated.

Each FBA can be specialised in **Active Command Components (ACC)**, in **Control Unity (CU)** or in **Abstract Agent (AB)** to better represent the nature of the modeled or interfaced factory elements.

ACC includes decision and command abilities for an AMS physical component (e.g. equipments, pallets and parts), acting as a specialised supervisor for the physical component. The ACC responsibilities, in respect to the physical component, are: (i) map its behaviour to a state machine; (ii) estimate its states by means of monitoring information; (iii) send commands and receive answers; (iv) storage of information and inference (e.g. what tool to use in an operation). The Control integration to ANALYTICE II is done by means of ACCs that interact, not with real components, but with simulated components. For a better representation of the physical elements, there could be a hierarchy of ACCs (e.g. ACCs for storage, transport and manufacturing equipments).

AB allows representing, maintaining and changing the states of abstract elements that come from other decision elements (e.g. process plan and lot of parts from scheduler).

CU adds the decision and command ability to AMS hierarchical levels (e.g. plant, cell and station). A CU aggregates, at a first level, a set of ACCs, ABs and RAs and at upper levels it aggregates others CUs. Each CU maintains a state machine based on its attributes values and on its aggregated values.

Every **FBA** aggregates two sets of agents called **Attribute Agent (AT)** and **Method Agent (MA)**. An AT is responsible for representing and maintaining the discrete states of an attribute in a FBA (e.g. to know if a given position in a storage has a part or not). A MA is responsible for instigating a FBA to realise some action (e.g. move a robot's end effector).

3.2 Control Rules and Rules Agents

A **Rules Agent (RA)** is a computational representation of a rule (called **Control Rule - CR**), in agent form, that controls the AMS. Each RA is composed by two agents, the **Condition Agent (CA)** and the **Action Agent (AA)** that have a causal relation. CA and AA are respectively the *Condition* and the *Action*. Figure 1 demonstrates one CR. This Control Rule analysis if the CU Station1 instance (i.e. agent) is in a Free state, if the ACC Robot1 instance also is in a Free state, if the ACC Part1 instance has its process plan allowing that the next operation be in the CU Station1 instance and if the ACC Storage1 instance has the ACC Part1 instance.

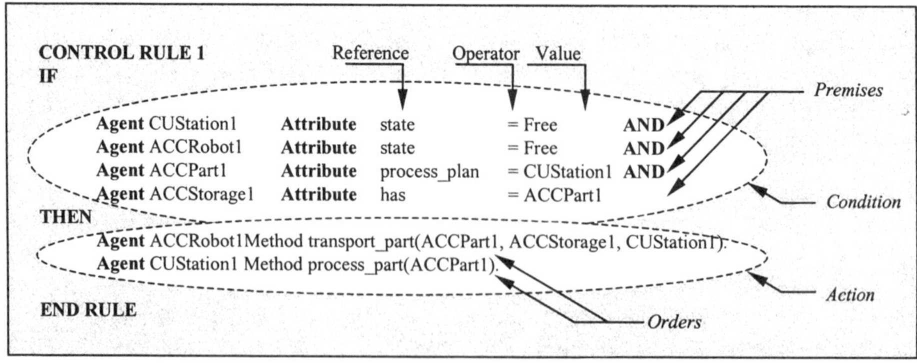

Figure 1- A Control Rule and its constituents.

CA – it does the logic calculus for the RA in which it is contained. The CA is connected to one or more **Premise Agent (PA)**, expressed as a *Premises* in the respective *Condition*. Each **PA** has: (i) a boolean value over itself, (ii) a reference to an unique attribute (AT) of an FBA (called ***Reference***), (iii) a logic operator (called ***Operator***) and (iv) a value or a values' limit (called ***Value***). Each PA does comparisons between *Value* and *Reference* using the *Operator*, that results in its boolean value. Other PA's ability is to compare its *Reference* value with an empty variable, producing an attribution. Therefore, another PA in this same CA can use this attributed variable as its *Value*, creating a connection or correlation between two PAs. The CA does the logic calculus by means of the boolean values conjunction from the connected PAs.

AA – is a sequence of **Orders Agents (OA)**, expressed as *Orders* in the *Action*, that command the FBAs cited in the respective CA in an asynchronous way, through their MAs, to realise their operations. Each AA is linked to a CA and will only be prone to execution if the evaluation produced by the corresponding CA results true.

The general characteristics of each RA are: (i) it may be created by Formation Agent (FA); (ii) it has its knowledge expressed in a CR; (iii) it is destroyed when one of the referenced agents, in its CA, ceases to exist; (iv) it may be activated when the CA generates a true evaluation result. This necessary condition, however, is not enough to have its AA activated, due to the sharing of an *exclusive premise* by two or more RAs (i.e. a conflict); (v) it has scope modularity, that is, the RAs of a CU do not see the RAs of another CU, reducing the communication among agents.

3.3   Formation Rules and Formation Agents

Formation Agents (FA) are computational representations of **Formation Rules (FRs)** and are used to create RAs. A FR is more generic than a CR, therefore its premises only analyse if a FBA is of a given class (e.g. ACC Lathe or CU Station) without considering, at first, the values of the attributes. A rule that specifies if a class of robots may grasp a class of parts is more generic than another one that specifies if a given robot can grasp a given part. A RA derives from a FA, restricting itself to a combination of agents and considering its attributes' restrictions.

Each FA is composed by two agents, the **Condition Formation Agent (CFA)** and the **Action Formation Agent (AFA)**. Figure 2 exhibits a FR that is a mean to express the FA knowledge. The CFA and AFA are, respectively, a *Condition's Formation* and an *Action's Formation*. This Formation Rule filters agents combination from the facts base, where each agent pertains to one of following classes: CU Station, ACC Robot, ACC Part and ACC Storage. Each combination creates a Control Rule following the specified in the Rule Formation elements.

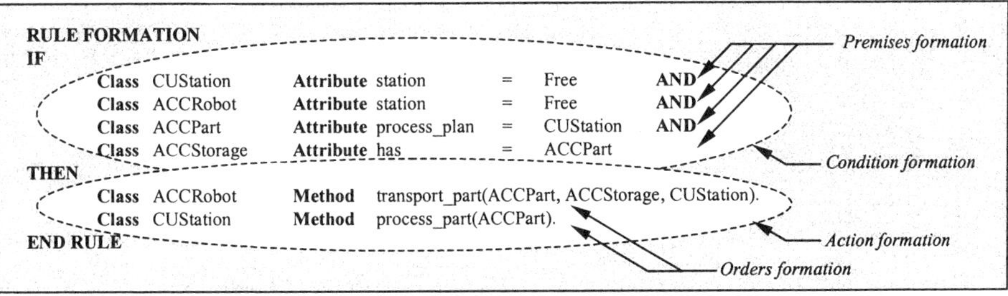

Figure 2 – A Rule Formation and its constituents.

CFA – it is connected to **Premises Formation Agents (PFAs)**, expressed as *Premises Formation* in the *Condition Formation*. The main function of PFA is to filter the agents pertaining to a class. Each PFA already specifies a **Reference** (but to an AT of class), a **Value** and an **Operator**. These information are not used at first, but, may be used as models for PAs creation. The CFA function is to meet the combinations among the PFAs filtered elements.

FAA – is a sequence of **Orders Formation Agents (OFA)**, expressed as *Orders Formation* in the *Action Formation*, that serves only as models for OAs creation.

Each FA obtained combination, by means of CFA, allows a RA instancing. The associates of the FA (i.e. CFA, PFA, AFAs and OFAs) create the RA's associates (respectively CA, PAs, AA and OAs). After the CA creation, the necessary PAs are created and connected. At the PA creation, it receives the *Reference*, the *Value* and the *Operator*, in accordance to the FA specifications. After the creation, the PA executes its logic calculus and auto-attributes a boolean value. Thus a RA, after having its PAs connected, may know its boolean value. As more RAs are created and need already existent PAs, connections between those PAs and the RAs are done, avoiding redundancy and generating a graph of connections. The AA, after being created, is connected to a sequence of OAs, in accordance to the FA's specifications. The AFA's information is used only to create the AAs.

The PFAs will just analyse the restrictions of the FBAs attributes if this has been explicitly determined (what is called *previous logic calculus*). When there is a *previous logic calculus,* a FBA is filtered only if the boolean result is true. In summary, this resource is a pruning in the graph of combination possibilities, avoiding the creation of RAs without semantic meaning or, in other words, RAs that would never reach a 'true' state.

The FRs (FAs) are devices to assist the CRs (RAs) creation process. The architecture, anyway, allows the direct creation of CRs, without the use of FRs. The FRs as well as the CRs need a way (maybe an agent) to extract the knowledge in the respective FAs and RAs.

Yet another important characteristic of FAs, like the RAs, is the ability to share information, they share the PFAs and OFAs, avoiding redundancies in the filtering process.

3.4 Architecture Class Model

Figure 3 exhibits the elements of the architecture, their composition and relationships.

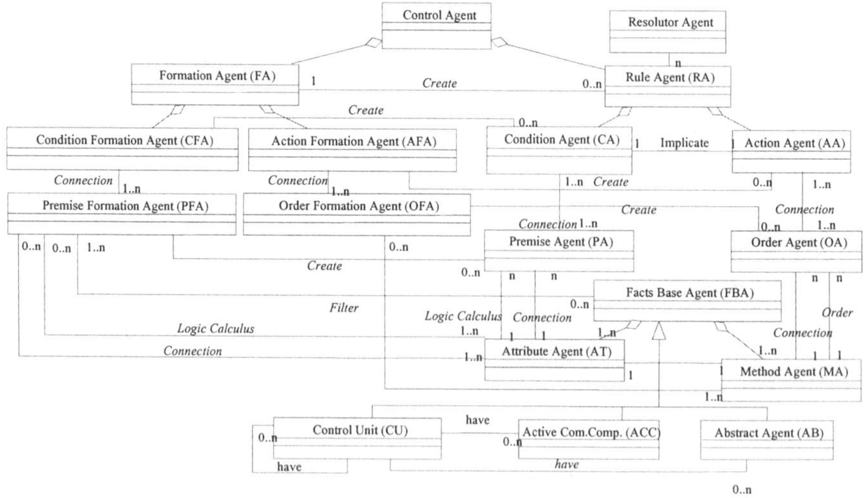

Figure 3- Control Architecture Class Diagram using Unified Modeling Language (UML) [11].

There are, in the Figure 3, two agents not yet mentioned: (i) Resolutor Agent that has as its responsibility the conflicts resolution, and (ii) Control Agent that is the root for all the Control system.

3.5　Inference Process

An usual inference model in RBS is to search through a list like facts base, each time a premise must be evaluated. The complete evaluation, triggered by some event at the facts base, analyses all rules' premises [10]. The RETE algorithm, that inspires the proposed architecture's inference, allows the search process optimisation through the rules notification only at a change in a fact [4].

The inference starts with the RAs and PAs creation that, by this moment, execute the logic calculus. The inference continues through the connections between PAs and ATs of the FBAs. Whenever a change in an AT value occurs, the connected PAs reset their logic calculus. If a state change occurs, it is expanded to the RAs, by means of the connection graph between PAs and CAs, that reappraises its own logic value. The 'true' state for RAs changes their AAs, turning their execution possible. This connection graph, created by the interrelationship among agents, allows a quick inference. In this solution there are not list searches, but messages propagation from ATs, through PAs and RAs, reaching AAs when pertinent. Figure 4 illustrates the communication (notification) chaining by the occurrence of state changes in a factory component.

3.6　Conflict among Rule Agents

To understand the conflict concept, the *exclusive FBA* concept must be understood at first. This concept concerns to FBAs that represent shared resources, for example a robot working for two workstations.33

A conflict occurs when a PA has a *Reference* from an *exclusive ABF* and it is being shared among RAs with true boolean values (eligible RAs). For example: the *Premise* '**Agent** ACCRobot1 **Attribute** state = Free' may collaborate to turn two CRs true. But, if the ACCRobot1 is exclusive, only one of the CRs can be executed, leading to a conflict.

The RAs' conflicts may be from the same origin (when the RAs are created by the same FA) or from a distinct origin (when created by distinct FAs). To solve the conflict, it is chosen only one RA to be activated (the elected RA). This is done based in decision parameters that may arise from many origins (e.g. from a dynamic scheduler or control politics). The decision may also be expressed in an agent form, as the Resolutor Agent in the proposed architecture.

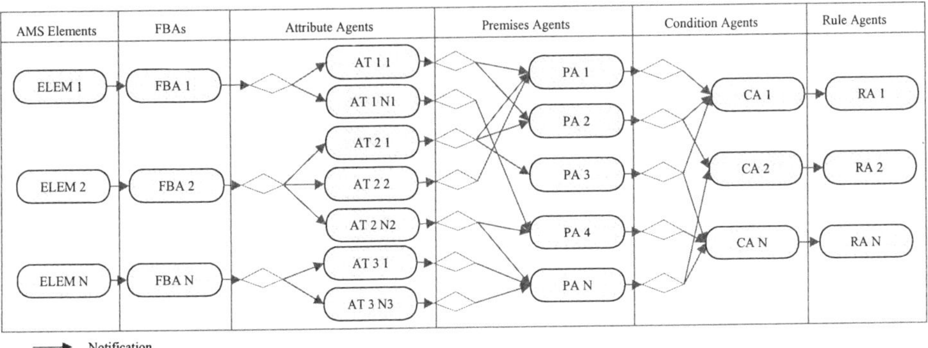

Figure 4 – The Agents notification system represented by an Activity Diagram (UML) [11] .

4. CONCLUSIONS

This article presents an architecture for AMS control whose experiments were carried out over ANALYTICE II simulation's tool. The primitives of the presented Control are modelled as elements of a generic RBS. This primitives are considered agents that communicate through the RETE notification principle, allowing a quick and precise inference. Generically, there are three main agent groups: (i) agents responsible for the Control monitoring and actuation processes, providing the facts base; (ii) agents responsible for the decision and command, providing the set of rules and (iii) agents that allow a certain automation in the conception process of the rules.

The proposed architecture demonstrates the application of artificial intelligence techniques in an important engineering field called discrete control. The use of a generic RBS using causal logic constitute a flexible framework that allows the development of a large number of applications for control systems, either for simulation or concrete factories.

The tests realised over the ANALYTICE II simulation tool demonstrate the easiness of the Control creation under the proposed architecture, and the robustness of the constituted systems. Future works include the expansion of the architecture including distributed processes, and the development of tools to assist the fact base and rule module creation. Others works include the development of methods for the Control synthesis using Petri Nets and mapping techniques from these solutions to the implementation under the proposed architecture.

References

[1] Ávila, Bráulio Coelho & Prado, José Pacheco de Almeida & Abe, Jair Minoro, 1998: "Inteligência artificial distribuída; aspectos". Série Lógica e Teoria da Ciência, Instituto de Estudos Avançados – Universidade de São Paulo.

[2] Bongaerts, Luc, 1998: Integration of Scheduling and Control in Holonic Manufacturing Systems. Ph. D. Thesis PMA / Katholieke Universiteit Leuven.

[3] Chaar, Jarir K. & Teichroew, Daniel & Volz, Richard A., 1993: Developing Manufacturing Control *Software*: A Survey and Critique. The International Journal of Flexible Manufacturing Systems. Kluwer Academic Publishers. Manufactured in The Netherlands, pp. 53-88.

[4] Forgy, C. L., 1982: RETE: A Fast Algorithm for the Many Pattern/Many Object Pattern Match Problem. Artificial Intelligence.

[5] Franklin, Stan & Graesser, Albert, 1996: Is it an Agent, or Just a Program? A Taxonomy for Autonomous Agents, Institute for Intelligent Systems – University of Memphis, Proceedings of the Third International Workshop on Agent Theories, Architectures and Languages, Springer-Verlag.

[6] Künzle, L. A., 1990: Controle de Sistemas Flexíveis de Manufatura – Especificação dos Níveis Equipamento e Estação de Trabalho. Master of Science Thesis, CEFET/PR.

[7] Koscianki, A. & Rosinha, L. F. F. & Stadzisz, P. C. & Künzle, L. A., 1999: FMS Design and Analysis: Developing a Simulation Environment. Proceedings of the 15th Internacional Conference on CAD/CAM Robotics & Factories of the Future CARS & FOF99. Águas de Lindoia – Brazil.

[8] Mendes, R. S., 1995: Modelagem e Controle de Sistemas a Eventos Discretos. In Markus, M. e Pires, P.: Manufatura integrada por computador. Belo Horizonte, Fundação CEFET-MG.

[9] Miyagi, P. E., 1996: Controle Programável – Fundamentos do Controle de Sistemas a Eventos Discretos. Edgard Blücher.

[10] Rich, E. & Knight, K., 1991: Artificial Intelligence. McGraw-Hill.

[11] Rumbaugh, James & Jacobson, Ivar & Booch, Grady, 1999: The Unified Modeling Language Reference Manual. Ed. Addison Wesley Longman.

[12] Simão, Jean Marcelo, 2001: Proposta de uma arquitetura para sistemas flexíveis de manufatura baseada em regras e agentes. Master of Science Thesis. CPGEI, CEFET-PR. Brasil.

Logic, Artificial Intelligence and Robotics
J.M. Abe & J.I. da Silva Filho (Eds.)
IOS Press, 2001

What Can Be Inferred Without Embracive World-Knowledge

Pavel Smrž
Faculty of Informatics, Masaryk University
Botanická 68a, 60200 Brno, Czech Republic
Email: smrz@fi.muni.cz

Abstract. This paper deals with a special type of automatic reasoning—a process of inference of new facts from a knowledge base. The presented type is not based on large embracive common-sense knowledge-base and wide-ranging ontology. Rather, it takes advantage of universal lexical database in the form of semantic network and also universal hypotheses that concern conversational implicatures.

1 Introduction

Human ability to work with information and infere new knowledge presents still untouched goal of artificial intelligence. Despite more than four decades of intensive effort of researchers from all parts of the world today's machines are not able to accomplish human levels of reasoning.

A large knowledge base is necessary for each agent who should act and reason intelligently. Considering limits of nowaday's fully automatic machine learning methods, it seems obvious that at least a part of knowledge about the actual world must be incorporated into the intelligent agent externally. However, the preparation and implementation of large knowledge bases represent an abundance of tedious work.

This article does not disprove the exigency of broad-coverage knowledge bases, but it tries to describe what kind of reasoning can be produced without their integration. Alongside it also shows which properties should not be missing even in the systems that owing to limited domain do not need to work with large knowledge base.

The paper is organized as follows: The following section defines the notion of strict implication—entailment. The third section connects entailment and the most frequent semantic relations included in EuroWordNet—synonymy, hyponymy, meronymy and antonymy. The fourth section deals with presumptive inferences – a special type of reasoning based on generalized, conversationally implicated meaning.

2 Entailment

Implication is one of the most important relations in logic. However, it is well known that the nature of this logical relation does not exactly reflect the relation of implication in natural languages. Always-false propositions implicate anything and anything implicates always-true propositions. Such relations are called **material implications** [2]. Here, we are interested in the complement relation that is called **strict implication**, or **entailment**. Sentence A entails sentence B if and only if in any context where A expresses a true proposition, B also ineluctably expresses a true proposition and A is not a paradoxical sentence or B is not an analytic sentence. (Analytic sentences are sentences that automatically express true propositions in any context, by virtue of the meaning of their constituents and their arrangement.)

Entailment surely plays the crucial role in the area of natural language understanding. However, why is it worthwhile to speak about entailment in the context of semantic networks? In spite of the fact that no semantic network will probably ever encode all the vocabulary of a natural language, semantic nets already capture a significant portion of natural language vocabularies and it is advantageous to be able to infer all relevant facts and relations that hold between them from this database.

Before we start to specify how the entailment is connected to various types of semantic relations, we have to state that entailment (as perhaps all other relations) cannot be understood as a sharp binary relation. There are many "shades" of entailment. It is obvious that it would be very advantageous to consider the degree of entailment in an automatic inference system. This degree should reflect (among others) referential stability of terms [2] and the extra-linguistic features of meaning with full degree of necessity from the true necessity (natural/social [14]), expectedness and possibility, up to impossibility.

3　Correspondence between Linguistic Relations and Entailment

Lexical data organized in the form of semantic networks like WordNet [15] or EuroWordNet [17] find their applications in various areas of natural language processing, e. g. in information retrieval, word sense disambiguation or machine translation. Here, we discuss the applicability of semantic networks in the relation to automatic reasoning, namely the principle of inference through entailment relations.

In EuroWordNet, the position has been taken that the wordnets in particular languages should reflect only the lexicalization patterns [18]. Thus, the main purpose is not to make semantic inferences (the different approach can be exemplified by other projects like Cyc [10]). However, the absence of a universal ontology (that would have to introduce some non-lexicalized concepts) cannot rule out the basic logical inferences from the database. The following subsections present such an approach.

3.1　Synonymy

The basic building blocks of the WordNet semantic network are synsets. Synsets are sets of synonymic word senses. The correspondence between true synonyms and entailment should cause no difficulties because the substitution of a word by its synonym in any sentence should not change the truth-value of the proposition represented by this sentence. It can also be seen as mutual entailment, i. e. equivalence, so that *He is an astronaut* entails *He is a cosmonaut* and vice versa. (Notice that we are assuming co-reference of matching personal pronouns and all other types of reference and deixis in this and all following examples.) There is quite a small set of words connected by the "near-synonym" relation in WordNet. From the perspective of automatic inference system, it is advantageous to extend the "near-synonymy" relation by attributes of denotational distinctions and other types expressing pragmatics—different attitude (pejorative, neutral, favorable, …) , register (field—area of discourse, mode—different language channels, and style—formality/informality), dialectal variation (geographical, temporal, social) [7].

3.2   Hierarchical Relations

Various hierarchical relations in the EuroWordNet database connect synsets, the most important are hyponymy/hyperonymy and meronymy/holonymy (part/whole) relations. The mapping between these relations and entailment is fundamental from the automatic inference system point of view. (Not all factors governing this correspondence are fully understood yet.) To be able to employ the entailment in a transitive manner, it is necessary to distinguish taxonymy from true hyponymy relations (they are mixed in the actual EuroWordNet database). Taxonymy can be exemplified by *horse:animal* relation and true hyponymy by *stallion:horse* [2]. As usual, there are no clear boundaries, however, taxonymy (that has strong relevance to classificatory systems) is not usually defined logically and it is not transitive.

The known example of broken transitivity that has been pointed out [1] is:

- *A (prototypical) car-seat is a type of seat.*

- *A (prototypical) seat is a type of furniture.*

- **A (prototypical) car seat is a type of furniture.*

There is no simple correspondence between meronymy and entailment. The relation is particularly manifested in connection with locative predicates [2]: if X is a meronym of Y, then for an entity A, "A is in X" entails "A is in Y". In order to escape most of the exceptions that are very frequent in the case of the meronymy logical connections, it would be worth to define detailed relations between parts and wholes, especially to identify the position of part (*The wasp is on the steering-wheel* does not entail *The wasp is on the car* [2]), to distinguish ingredients (*grape-juice:wine*), parts from pieces, semi-meronyms (*handle:door* vs. *doors without handles*) etc.

A special type of semantic relation, which is not usually mentioned together with hierarchical relations of hyponymy and meronymy, is called intensity. Of course, the most frequent grammatical category where intensity-related words appear are adjectives (*large:huge*) that form **scales** [9]. But the relation can be found between words from other categories, e. g. nouns (*mist:fog*) The reason why we should deal with intensity here is the similar relation to entailment: *It's huge* entails but is not entailed by *It's large*.

3.3   Incompatibility

Incompatibility plays an essential role in inference processes. Let us start with the simplest case of opposites. In the EuroWordNet database, all types of opposites are joined under the name antonyms, nevertheless, we will use the classification presented in [2] as it has strong influence on the entailment relation. The first type of opposites are complementaries that can be defined as follows: X and Y are complementaries iff F(X) entails and is entailed by not-F(Y) as is exemplified by pairs: *true:false, stationary:moving, possible:impossible*. To emphasize the difference between complementaries and the second type that will be presented in the following paragraph, let us simplify the entailment to the logical implication. Then, the meaning of complementaries can be given as logical equivalence, i.e. $F(X) \Leftrightarrow \neg F(Y)$

or, equivalently, $(F(X) \wedge \neg F(Y)) \vee (\neg F(X) \wedge F(Y))$. The second type of opposites are antonyms (in the narrow sense—see [12]), e. g. the pairs like *high:low, deep:shallow*. X and Y are antonyms iff F(X) entails not-F(Y). In our logical notation: $F(X) \Rightarrow \neg F(Y)$, or, equivalently, $\neg(F(X) \wedge F(Y))$, or, $\neg F(X) \vee \neg F(Y)$, or, finally, $(F(X) \wedge \neg F(Y)) \vee (\neg F(X) \wedge F(Y)) \vee (\neg F(X) \wedge \neg F(Y))$. Therefore, *It's neither high nor low* is not a contradiction. Note that this strict categorization into complementaries and non-complementaries is possible only in the traditional (crisp) logic. In nontraditional—many-valued or even fuzzy logic, we talk about a degree of complementarity.

A special category of opposites is formed by the class of directional opposites—reversives (*up:down, forward:backward, tie:untie,* ...) and converses (*lend:borrow, predator:prey*). To be able to infer knowledge from knowledge bases that take advantage of a wordnet database, these relations should be encoded explicitly, together with a mechanism providing inference.

It is necessary to focus one's attention also to semantic markedness [13], i. e. to such context where the meaning of one term is what is common to the two terms of the opposition (*We saw a group of lions* does not entails *There were no lioness*). Another phenomenon exemplified by:

It's true that it's true. = It's true.   but   *It's false that it's false.* = It's true.

She succeeded in succeeding.   but   *She failed to fail.* (reversal)

is termed logical polarity [2] and must be taken into consideration as well. Note finally that as we are dealing exclusively with logical relations we do not need to distinguish other types of markedness, namely morphological and distributional as well as other types of polarity (morphological, privative and evaluative). These relations of opposition play, however, an important role for the natural language generation systems (see [2] for discussion of these relations). Furthermore, an indication of morphologically markedness can partially replace explicit information about a semantic relation, e. g. about many directional opposites.

4 Presumptive Inference

The aim of this paper is the presentation of the idea that a great portion of relevant inferences can be accomplished even without employment of large and labor-intensive general-purpose ontologies that apart from other things try to capture full common-sense reasoning [10]. Previous section discussed the integration of lexical information from wordnet-like semantic networks. This section deals with an application of presumptive inference in the context of Levinson's **GCI** (Generalized Conversational Implicature) [11].

GCI is based on the notion of a **preferred (default) interpretation**. It continues Grice's approach to the study of meaning and communication [6] articulated especially by two of Grice's **maxims—the maxims of quantity** (make your contribution as informative as is required and do not make it more informative than is required) and **the maxims of manner** (be perspicuous, particularly non-ambiguous and orderly).

Grice's characterization of conversational implicatures as cancelable (or defeasible)—the addition of premises can lead to the cancellation of inferences—directly proposes the use of non-monotonic logic in the implementation of presumptive inference. There are different non-deductive logic systems that can be employed with various planning systems and default logics [5] as demonstrative examples. In his influential book [11], Levinson himself suggests the use of a default logic to model presumptive inference. However, we propose an alternative to the qualitative method of default logic, specifically quantitative **probabilistic reasoning** that allows modeling various degrees of certainty/uncertainty about default assumed consistent conclusions. Probabilities are assigned to particular instantiations of Levinson's inferential heuristics. Let us briefly recapitulate these heuristics together with illustrative examples (the second and the third heuristics are grouped and presented simultaneously what reflects their relation and implicit opposition:

- **The first (Quantity, Q-) heuristic—What isn't said, isn't** – accommodates especially classic scalar implicatures and Gazdar's clausal implicatures [4]. As the included implicatures depend on restriction to a set of salient alternates [11], the interconnection with the lexical semantic modul represented by wordnet-like semantic network (see previous section) plays a crucial role in the application of this heuristic. In addition to scalar and negative scales alternates (some $|\sim$ not all; not all — some) [1] and clausal implicatures (if p then q $|\sim$ Speaker does not know that p, nor that q, p and q uncertain), there are lexical scales (mist $|\sim$ not fog) and sets (red $|\sim$ not blue). Note that the principle "What isn't said, isn't" is very close to the standard logic programming technique known as "negation as failure" (a literal, consisting of an atomic formula preceded by a non-classical negation operator, is considered true if the atomic formula cannot be proven). It directly gives one possible way of implementation in terms of a logic program.

- **The second (Informativeness, I-) heuristic—What is expressed simply is stereotypically exemplified** and **the third (Manner, M-) heuristic—What is said in an abnormal way is not normals**—complementarily stretch a great number of well-known interpretive tendencies. However, only limited portion of such inferences can be applied without an exhaustive knowledge about the actual world because most of them require an immediate common-sense knowledge about typical cases or a typical scenario. It is not easy to find stereotypical interpretation unless a large knowledge base allowing common-sense reasoning will be incorporated. Captured examples of the implicatures under I- and M-heuristic are the replacement of implication by equivalence—"conditional perfection" (*If it rains, I will take my umbrella $|\sim$ Iff it rains, will I take my umbrella*, i. e. also *If it does not rain, I will not take my umbrella*), the enrichment of conjunctions by the assumption of temporal sequence and causality (*p and q $|\sim$ p and then q, p caused q*), and the preference for local co-reference (*Charles returned and he killed the king $|\sim$ Charles = he*). Also a part of "generality-narrowing" principles can be coped with, especially those related to semantic markedness (see Section 4).

Currently, the probabilities associated with particular inferences are combined using quite simple certainty factors in the MYCIN-like style [3]. This simple method will be replaced by a more elaborated representation of uncertain relationships in the next developmental step, presumably by belief (Bayesian) networks [16]. The other problem, that has not find a satisfactory solution yet, concerns the assignment of probabilities to particular cases of implicatures mentioned above. The first approximation gives the same degree of belief to all instances derived from one inference schema supported by a given heuristic. More sophisticated approach would take into consideration not only the involved rule inference schema but also particular expressions that participate in given statements. It would be advantageous especially in the case of conjunctions enrichment where the relationship between the joined parts plays the crucial role in the correct assignment of the inference probability. The question remains how to obtain efficiently the required statistics.

---

[1] We use the symbol '$|\sim$' for a non-monotonic entailment. $P |\sim Q$ means that Q is non-monotonic consequence of P. See [8] for a formulation of general principles characterizing the behavior of '$|\sim$.'

5 Conclusions

We have discussed two main topics in this paper. The first concerns the issue of correspondence between semantic relations and entailment. It has been shown that there is a narrow connection between semantic relations represented in WordNet (synonymy, antonymy, hyponymy, meronymy) and entailment. The second part has touched the possibility of presumptive inference in terms of Levinson's GCI.

Principles mentioned above should find their place in the prototype system SEN (SEmantic Network). It will merge the advantages of wordnet-like networks with the system that will integrate conversational implicatures. The system will form the core of different automatic inference engines specialized in various ontological areas.

References

[1] D.A. Cruse, Lexical Semantics, Cambridge University Press (1986).

[2] D.A. Cruse, Meaning in Language: An Introduction to Semantics and Pragmatics, Oxford University Press (2000).

[3] R. Davis, Production rules as a representation for a knowledge-based consultation program, Artificial Intelligence **8** (1977) 15–45.

[4] G. Gazdar, Pragmatics: Implicature, Presupposition, and Logical Form, Academic Press (1979)

[5] M. Ginsberg (ed.), Readings in Nonmonotonic Reasoning, Morgan Kaufmann, San Mateo, CA (1997).

[6] H.P. Grice, Meaning, Philosophical Review **67** (1957) 377–388.

[7] D.Z. Inkpen and G. Hirst, Experiments on Extracting Knowledge from a Machine-Readable Dictionary. In: Proceedings of CICLing 2001 (The Second International Conference on Computational Linguistics and Intelligent Text Processing), Mexico City, Mexico (2001).

[8] S. Kraus, D. Lehmann, and M. Magidor, Nonmonotonic reasoning, preferential models, and cumulative logics, Artificial Intelligence **44** (1990) 167–207

[9] G.A. Leech, Semantics, Harmondsworth, Penguin (1974).

[10] D. Lenat and R. Guha, Building Large Knowledge-based Systems. Representation and Inference in the Cyc Project, Addison Wesley (1990).

[11] S.C. Levinson, Presumptive Meanings: The Theory of Generalized Conversational Implicature, A Bradford Book, The MIT Press (2000)

[12] J. Lyons, Structural Semantics, Cambridge University Press (1963).

[13] J. Lyons, Semantics, Cambridge University Press (1977).

[14] J. Lyons, Language, Meaning and Context, London, Fontana (1981).

[15] G. Miller, R. Beckwith, C. Fellbaum, D. Gross, K. Miller, Five Papers on WordNet, CSL Report **43**, Cognitive Science Laboratory, Princeton University (1990).

[16] J. Pearl, Probabilistic Reasoning in Intelligent Systems, Morgan Kaufmann, San Mateo, CA (1988)

[17] P. Vossen, Right or Wrong. Combining lexical resources in the EuroWordNet project. In: M. Gellerstam, J. Jarborg, S. Malmgren, K. Noren, L. Rogstrom, C.R. Papmehl (eds.), Proceedings of EURALEX'96, Goeteborg (1996).

[18] P. Vossen and L. Bloksma, Categories and Classifications in EuroWordNet. In: Proceedings of LREC, Granada (1998).

Logic, Artificial Intelligence and Robotics
J.M. Abe & J.I. da Silva Filho (Eds.)
IOS Press, 2001

Semantic Computation by Humans, Computers and Robots

Patrick Suppes
Stanford University

Abstract.

The era of voice recognition by a great variety of technological devices is just beginning. No doubt speech-recognition rates will continue to improve and there will be increasing acceptance of this important channel of communication between human users and devices, from personal computers, PDA's, and private automobiles to kitchen ovens and refrigerators. Much of this communication will be successful. In Section 1, a detailed example from machine learning of robotic natural language will try to make clear why this is so. But the general reason is clear. Well-defined sublanguages of a natural language can be given a restricted and unambiguous semantic interpretation. Their semantic structure is close to that of an elementary formal language. In contrast, open-ended conversations, with the semantic ambiguities characteristic of such language use, present different and much more difficult problems. But progress on these problems will be critical for much wider use of speech in interacting with devices. In Section 2, some conjectures about how the brain handles such language are discussed, together with some detailed data on brain processing of language. Section 3, the final one, is devoted to considering whether device-implementation of conversational language is likely or not to use similar semantic computations.

1 Machine Learning of Regimented Natural Sublanguages

I, together with younger colleagues, have been developing a theory of machine language learning in the context of robotic instruction of elementary assembly actions [1, 2, 3]. More specifically, our situations of language learning concern actions like moving to objects, picking up objects and moving objects in environments. Typical objects are screws, nuts, washers, and sleeves of various colors, sizes, and shapes, related to each other by common spatial relations. Typical commands to be comprehended are *Get a screw., Go to the black washer behind the screw.* and *Put the screw in the left hole.* The system developed so far deals successfully with English, Chinese, and German, as well as several other languages.

We have taken what we believe is a new tack in the approach to machine learning by using in a very explicit way principles of association and generalization derived from classical psychological principles. The principles we used were, however, much more specific and technically developed.

The fundamental role of association as a basis for conditioning is thoroughly recognized in modern neuroscience and is essential to the experimental study of the neuronal activity of a variety of animals. For similar reasons its role is just as central to the learning theory of neural networks, now rapidly developing in many different directions. We have not, however, made explicit use of neural networks, but have worked out our theory of language learning at a higher level of abstraction. In our judgment the difficulties we face need to be solved before a still more detailed theory is developed.

The classical psychological principles of learning used here have been thought by linguists to be wholly inadequate as the basis for a theory of language learning. Nothing could be further from the truth. Skinner's naive formulation of the problems of language learning [4] was rightly attacked by Chomsky [5], but no serious alternative learning theory has been offered by linguists even today.

First, we briefly describe our approach to machine learning of natural language. Second, we focus on the problem of denotation that is important in our use of probabilistic association of words and their meaning. Third, we outline the background cognitive and perceptual assumptions of our machine learning work. Fourth, we formulate explicitly our two general axioms of association and denotation, but do not state the additional axioms describing the full learning process. These may be found in the publications already cited, with some changes being made over time.

1.1 Our Approach to Machine Learning

Without going into all the details, we want to convey a rather clear intuitive sense of the process of learning of natural language in terms of the various events that happen when an utterance is given to a robot. (In this and succeeding sections we shall refer to robots, but it should be understood that the basic program of machine learning would apply without serious modification to other applications. Following standard learning usage, we shall often speak of trials where, of course, we mean that the trial begins with a command in the form of an utterance to be executed by the robot.

The most important way to describe conceptually the learning process our program embodies is in the description of the state of memory of a robot at the beginning of each trial. There are four aspects of this memory that are changed due to learning. The first is the association relation between words of a given language and internal symbols that have as denotations actions, objects, properties and relations in the robot's world. A central problem is to learn in each language what word is properly associated with a given internal symbol. A second aspect of the memory that changes is the denotational value of a given word, which will affect its probability of being associated. The third part that changes is the short-term memory that holds a given verbal command for the period of the trial on which it is effective. This memory content decays and is not available for access after the trial on which a particular command is given. This means that at the beginning of the trial, before a command is given, this short-term buffer is empty. What we have said thus far, could, with some stretching, fit into classical theories of association, but for language learning it is quite evident that the association relation and some simple features of short-term memory are certainly not enough.

The fourth aspect is the important one of learning grammatical forms. Consider the verbal command *Get the nut*. This would be an instance of the grammatical form *A the O*, where *A* is the category of actions and *O* is the category of objects. This form actually represents a mild oversimplification, because we do not have just a single category of actions. There are several subcategories, depending upon the number of arguments required, and certain other natural semantical requirements as well. The example will illustrate how things work, however. The grammatical forms are derived by generalization only from actual instances of verbal commands given to the robot. No prior knowledge of any sort of the grammar of the natural language to be learned is available to the robot. Also important is the fact that associated with each grammatical form as it arises from generalization are the associations of the words which have been the basis for the generalization, along with their internal representations. For example, if *Get the nut* were the occurrence in which the grammatical form just stated was generated, then also stored with that grammatical form would be the associations *get* ∼ $g and *nut* ∼ $n, where $g and $n are internal symbols whose denotations are known to the robot. When incorrect associations are deleted by further learning, the grammatical forms based on such associations are also deleted.

1.2 Problem of Denotation

In the probabilistic theory of machine learning of natural language which we have been developing, we have encountered in a new form a standard problem in the analysis of the semantics of natural language, namely, how to handle words that are nondenoting. We do not mean nondenoting in some absolute sense, but relative to a fixed set of semantic categories. These categories in the robotic case are, roughly speaking, the categories of actions, objects, properties and relations. It may well be that in some elaborate set-theoretical semantics of natural language, nondenoting words like the definite article *the* denote a complicated set-theoretical function, but the relevance of such an elaborate semantics to language learning is doubtful. In the robotic context, we have something simpler and closer to the common man's view of what denotations are. We take as denoting words color and object words, common nouns, familiar concrete action words, *etc.*. We take ordinary prepositions in English and sometimes other devices in other languages to denote relations in most cases.

When a child learning a first language or an older person learning a second language first encounters utterances in that new language, ther is no uniform way in which nondenoting words are marked. There is some evidence that various prosodic features are used in English and other languages to help the child. For example, in many utterances addressed to very young children, the definite or indefinite article is not stressed but rather the common noun it modifies, as in the expression *Hand me the cup*. But such devices do not seem uniform and in any case are not naturally available to us in our machine-learning research, where we use written input of words without additional prosodic notation.

As has already been made clear, a central feature of our approach to machine learning is the probabilistic association between words of the natural language being learned and denoting symbols of the internal language. It is appropriate that at the beginning all words are treated equally, and so the associations are formed from sampling based on a uniform distribution. On the other hand, after many words have been learned and a good deal of language has been acquired by the robot, it is very unnatural, and also inefficient, if the robot is now given, for example, the esoteric command *Get the voltmeter,* to have the internal symbol $vol be associated with equal probability with the definite article *the* and *voltmeter* – we assume here that the association of *get* is already correctly fixed. After much experience, what we want is that there is very little chance of associating the definite article *the* with any denoting symbol.

1.3 Background Cognitive and Perceptual Assumptions

Before explicitly formulating the learning principles of association and denotation we use, we first state informally assumptions we make about the cognitive and perceptual capacities of the class of robots, albeit as yet quite limited, we work with.

Internal language. The robot has a fully developed internal language, which it does not learn. It is technically important, but not conceptually fundamental, that in our case this language is LISP. When we speak here of the internal language we refer only to the language of the internal representation, which is itself a language at a higher level of abstraction, relative to the concrete movements and perceptions of the robot. It is the language of the internal representations held in memory that provides the direct interface to the natural-language learning. In fact, most of the machine learning of a given natural language can take place through simulation of the robot's behavior by using just the language of the internal representation. The first associations learned are between the internal representation in memory of a coerced action and a contiguous verbal utterance in the natural language being learned.

The fundamental importance of this internal representation of a coerced action can be recognized by considering a parallel case of animal learning. When a dog is trained to *Get the paper* or *Get the ball* by being led through the desired action or by some related technique, the residue in memory of what we term the coerced action is surely drastically abstracted from the perceptually rich context of the demonstrated action desired, and it is that abstracted internal representation in memory that must be associated to the verbal stimulus in order for the dog later to perform the desired action upon hearing the verbal command. We are a long way from knowing even the general structure of the dog's internal representation in memory of the action. In this limited sense, life with a robot is much easier, for we ourselves create the form of its internal representation.

Objects, relations and properties. We furthermore assume the robot begins its natural-language learning with all the basic cognitive and perceptual concepts it will have. In other words, our first-language learning experiments are pure language learning. Any learning of new concepts is delayed to another phase. For example, we have assumed that the spatial relations frequently referred to in all, or at least all the languages we consider in detail, are already known to the robot. This is quite contrary to human language learning. For example, probably in no widely used natural language at least, do children at the age of thirty months use or fully understand the relations of left and right. To avoid misunderstanding, we emphasize that we consider it an important future task to have the robot also learn the familiar spatial and temporal relations.

Actions. What was just said about objects and relations applies also to actions, represented in English by such verbs as *pick up, get, place, etc..* The English, of course, must be learned, but not the underlying actions.

Associations and grammatical forms. Before stating any formal principles of learning, we feel it is desirable to describe as informally and intuitively as possible the learning setup we use. Consider the English command *Pick up the screw,* no part of which has as yet been learned by the robot. The learning steps may be roughly schematized as follows:

(i) By coercion, or simulation of coercion, the robot creates in memory an internal representation of the coerced action of picking up the screw;

For statement of learning principles we show this internal representation, not as a LISP expression, but just as a schematic function $I(\ldots)$ of the denoting terms in the LISP expression. Here, by *denoting terms* we mean the names in the internal language of the actions, objects, properties and relations mentioned. The internal representation of *Pick up the screw* is then $I(\$p, \$u, \$s)$, where $\$p = $ the action of picking, $\$u = $ the direction up and $\$s = $ screw.

(ii) By contiguity the robot associates the verbal utterance and the internal representation

$$Pick\ up\ the\ screw\ \sim\ I(\$p, \$u, \$s),$$

where $\sim$ is the symbol we use for association;

(ii) By probabilistic association, the robot associates the internal denotations with the English words, with one possibility the following incorrect result:

$$pick \sim \$s, up \sim \$p, screw \sim \$u.$$

We need to observe the following:

a. We assume from the beginning the robot knows word boundaries, as delineated by the typed input. This is an example of an assumption that is natural for robots, but clearly false for very young children;

b. For our simple example, there are 24 possible ways of associating the three internal symbols to the four denoting words in the English utterance. We initially assign to each of these 24 possibilities equal probability, but as trials continue, modify the probability by dynamic changes in denotational values, as is explained later in detail.

(iv) After the associations are made, by the principle of generalization, which we call the category generalization, each word is assigned the category of its associated internal symbol. In the present case $pick \in O$ – the category of objects, $up \in A$ – the category of actions and $screw \in R$ – the category of relations. A grammatical form is then also generalized from the verbal command:

$$O \; A \; the \; R$$

which, like the assigned categories, is wrong for English, but remember that this is just the starting point of learning. With this grammatical form is associated its internal representation $I(A, R, O)$ which characterizes its meaning.

(v) A new command is presented as the next step, say *Pick up the nut*. By coercion the internal representation $I(\$p, \$u, \$n)$ is created (see (i) above). The robot then first searches its memory to see if any of the words uttered are associated to one of the internal denotations. Here the result is $up \sim \$p$, and also the classification of *the* as a nondenoting word is found. There are then six possibilities of probabilistic association for *pick, the* and *nut*. Note that the earlier incorrect association of *pick* with $\$s$ does not appear here, which means that at this stage of learning it will be changed. So, let us suppose the new associations are

$$pick \sim \$u, nut \sim \$n.$$

We also have as a new grammatical form

$$R \; A \; the \; O$$

which though incorrect, now has only the confusion of the associations of *pick* and *up* as its source. To correct these associations we must separate the constant pairing of *pick* and *up*, which is what we do. In any case, we form at once the association to the internal representation:

$$R \; A \; the \; O \sim I(A, \; R, \; O).$$

(vi) Learning stops whenever the following steps of interpretation can be successfully completed upon giving the robot a verbal command:

a. An association to an internal denotation or a nondenoting classification is found in memory for each word;

b. The category of each word is found in memory;

c. The grammatical form resulting from (b) is found with an associated internal representation in memory;

d. The command is correctly executed on the basis of the internal representation.

1.4　The General Axioms of Association and Denotation

We state the axioms in a general form, but we assume already that each word a of the target natural language has a denotational value $d_n(a)$ on each trial. This value changes from trial to trial according to the two different models presented in the next section.

Probabilistic association. *On any trial n, let a natural language sentence s be associated to σ, its internal representation, let $\{a_i\}$ be the set of words of s not associated to any internal denoting symbol of σ, let $d_n(a_i)$ be the current denotational value of each such a_i and let $\{\alpha_j\}$ be the set of internal denoting symbols not currently associated with any word of s. Then:*

(i) an element α_j is uniformly sampled without replacement from $\{\alpha_j\}$;

(ii) at the same time an element a_i is sampled without replacement from $\{\alpha_j\}$ with the sampling probability

$$p_n(a_i) = \frac{d_n(a_i)}{\sum_{\{a_i\}} d_n(a_i)};$$

(iii) the sampled pairs are associated, i.e., $a_i \sim \alpha_j$;

(iv) sampling continues until either the set $\{a_i\}$ or the set $\{\alpha_j\}$ is empty.

Denotational value computation. *If at the end of a trial a word a in the presented sentence is associated with some internal symbol α, then $d(a)$, the denotational value of a, increases and if a is not so associated $d(a)$ decreases. Moreover, if a word a does not occur on a trial, then $d(a)$ stays the same unless the association of a to an internal symbol α is broken on the trial, in which case $d(a)$ decreases.*

A more detailed denotational axiom is the following:

If, at the end of trial n, a word a_i in the presented verbal stimulus is associated with some denoting internal symbol α_j of the internal representation σ of s at the end of the trial, then

$$d_{n+1}(a_i) = (1 - \theta)d_n(a_i) + \theta, \quad 0 < \theta \le 1$$

and if a_i is not so associated,

$$d_{n+1}(a_i) = (1 - \theta)d_n(a_i).$$

Moreover, if a word a_i does not occur on trial n, then

$$d_{n+1}(a_i) = d_n(a_i).$$

unless the association of a_i to an internal symbol α_j is broken on trial n, in which case

$$d_{n+1}(a_i) = (1 - \theta)d_n(a_i).$$

Table 1: Comprehension Grammars

Chinese, English, German	
1. $O \rightarrow [DA] S$	
2. $O \rightarrow [IA] S$	
3. $S \rightarrow P S$	
4. $S \rightarrow OBJ$	
5. $G \rightarrow R [PO] O$	
6. $D \rightarrow R$	
7. $P \rightarrow P [\&] P'$	
8. $P \rightarrow [\neg] P$	
9. $P \rightarrow P [\vee] P'$	
Chinese, English	**English, German**
10. $A \rightarrow [ADV] A_4 D O$	12. $A \rightarrow [ADV] A_1 O [COP]$
11. $A \rightarrow A_4 O$	13. $A \rightarrow [ADV] A_2 G [COP]$
	14. $A \rightarrow G A_3 O$
	15. $A \rightarrow A_3 O [COP] G$
	16. $A \rightarrow [ADV] A_4 [COP] O D$
	17. $S \rightarrow S [RP] P$
	18. $S \rightarrow S [RP] G$
Chinese	**German**
19. $A \rightarrow [ba] O A_1$	26. $A \rightarrow A_4 [einen] S D [der] P \ ist$
20. $A \rightarrow [xianzai] G [na4] A_2$	27. $A \rightarrow A_4 [nun] S D \ die \ G \ ist$
21. $A \rightarrow zai G A_3 O$	28. $A \rightarrow A_4 \ die \ S R \ von \ dem \ S D [der] G \ ist$
22. $A \rightarrow [ba] O A_3 [dao1] G$	29. $A \rightarrow A_4 \ die \ S R \ der \ S D \ die \ P \ ist$
23. $A \rightarrow [ba] O A_4 D$	
24. $S \rightarrow G [de] S$	
25. $G \rightarrow O [de] R$	

1.5   Comprehension Grammars Generated

Comprehension grammars have been generated so far for English, Chinese, German, French, Dutch and Korean. Mainly because the first languages of the authors are English, German and Chinese respectively, we will concentrate on these three languages with remarks about the others. The corpus used has consisted of 456 commands, the last 60 of more complicated ones. In the first group commands range from the simple *Get the screw.* to *Place the screw on the plate.* In the last 60 a typical more complicated command is the following: *Put a small nut on the screw behind the washer left of the plate.*; *Ba yige xiaode luomu fang zai nage ban zuobian de dianquan houmian de nage luosiding shang.*; *Tu eine kleine Mutter auf die Schraube hinter dem Dichtungsring links von der Platte.*

We now turn to the three grammars generated for English, Chinese and German. The grammatical rules are written in context-free notation, in Table 1, where A is the category of actions, D is that of directions, G of regions, O of objects, P of properties, R of relations and S of sentences. Brackets are used for congruence classes of words which semantically function the same way in comprehension, possibly from different languages. For example, $[DA]$ is the congruence class of definite articles in Chinese, English and German used in our corpus. The full list is given in Table 2. The symbol G denotes the option to omit a word of the congruence class.

Using the axiom on congruence to collapse rules that differ only in the occurrence of semantically equivalent nondenoting words, the number of rules for English is 18, for Chinese 18 and for German 20. What is of perhaps greater interest is the analysis of the structure

Table 2: Congruence Classes

$[DA]_1$	=	*(the; nage, zai, zhege; das, dem, den, der, die)*,
$[IA]_2$	=	*(a; yige; eine, einem, einen, einer)*,
$[PO]_5$	=	*(of, ϵ; ϵ; von, ϵ)*,
$[\&]_7$	=	*(and; he; und)*,
$[\neg]_8$	=	*(not; busi; nicht)*,
$[\vee]_9$	=	*(or; huozhe; oder)*
$[ADV]_{10}$	=	*(now, so, ϵ; ϵ)*
$[ADV]_{12}$	=	*(now, so, ϵ; nun, ϵ)*
$[COP]_{12,13,15,16}$	=	*(ϵ; ist, ϵ)*
$[ADV]_{13}$	=	*(now, so, ϵ; dann, jetzt, nun, ϵ)*
$[ADV]_{16}$	=	*(now, so, ϵ; ϵ)*
$[ADV']_{16}$	=	*(ϵ; jetzt, nun, ϵ)*
$[RP]_{17}$	=	*(that is, which is; der, die, ist die)*
$[RP]_{18}$	=	*(that is, which is, ϵ; der, die, die sich, ϵ)*
$[ba]_{19}$	=	*(ba, xianzai ba, ϵ)*
$[xianzai]_{20}$	=	*(chao, name, xianzai)*
$[na4]_{20}$	=	*(na4, na4li)*
$[ba]_{22}$	=	*(ba, ϵ)*
$[dao1]_{22}$	=	*(dao1, zai, ϵ)*
$[ba]_{23}$	=	*(ba, xiazai ba, ϵ)*
$[de]_{24}$	=	*(de, de nage)*
$[de]_{25}$	=	*(de, ϵ)*
$[der]_{26}$	=	*(der, die)*
$[einen]_{26}$	=	*(einen, jetzt eine)*
$[ist]_{26}$	=	*(ist, ε)*
$[nun]_{27}$	=	*(die, nun die)*

of common rules in comparison with special rules for the particular languages. The basic numerical data are these. There are 9 rules that are common to English, Chinese and German where 'commonality' means that the category such as definite article now includes words from each of the three languages, in this particular case for example, *the, nage, zai, zhege, das, dem, den, der,* and *die*. In addition, there are 2 rules common to English and Chinese that are not in the German grammar, there are 7 rules common to English and German that are not used in Chinese, and no rules common only to Chinese and German.

What is remarkable about rules 1–9 in Table 1 is that none of them are high level rules for the generation of complete commands. The first 4 deal with the generation of object phrases, Rule 5 with the generation of a description of a region, Rule 6 with a direction, and the last 3 with properties. Note that the rules common to English and Chinese are high level rules for generating commands, and that 5 of the 7 rules common to English and German are high-level rules for generating commands. The remaining 2 are for generating object phrases. Five of the 7 rules in Table 1 special for Chinese are high-level rules for generating commands.

The big surprise is that no special grammatical rules are required for English.

In Table 2 the congruence classes for the three languages are shown. The subscript on each class shows the grammatical rule of Table 1 on which it depends. For example, $[ADV]_{10}$ depends on Rule 10, which is relevant only for English and Chinese. The congruence class for Rule 1 is intuitively incorrect. The reason is simple to explain. The Chinese particles *xianzai, dao1, zai,* and *ba* do not generally co-occur with object phrases, as the rules suggest, but with other categories of words. For example, *xianzai* has a meaning that is close to the English adverb *now*, the particle *dao1* occurs with verbs, as does *ba*. What is important is that at the level of the commands we are considering, the nonintuitive Rule 1 is satisfactory for the purposes of comprehension. It would of course lead to very bad Chinese if applied to the production of utterances. We emphasize once again that in considering these Chinese examples, we are generating comprehension by uniform procedures that are the same across all languages. We are not claiming this can be done for production or even that for the ultimate reaches of comprehension it will be satisfactory.

2 Brain Processing of Words and Sentences

In our earlier analyses of brain-wave representations of words and sentences, based on electroencephalographic (EEG) data, we averaged over trials, as well as subjects, and made a discrete fast Fourier transform (FFT) to the frequency domain [6, 7, 8]. We then searched for a filter that optimized correct recognition of the words or sentences being processed. Using filters to eliminate noise from signals is widespread in many kinds of signal processing. When speech or music constitute the signals, such filters not only work well, but are practically necessary, because of the large number of component waves.

The optimal filters we found in our earlier studies usually fell well within the range 2- to 15-Hz. So, using the standard software for discrete fast Fourier transforms with a sampling rate between 600 and 1,000 Hz, we, in fact, were ordinarily using a filter that contained less than 60 discrete frequencies. This relatively small number of frequencies immediately suggests an alternative to filtering for our EEG-recorded brain waves. This is to look at the frequencies with comparatively large amplitudes, select a small set of these, and use their superposition instead of a filter. (The work summarized here comes from Suppes and Han [15].) So, for each observation i the superposition wave S_i is just

$$S_i = \sum_{j=1}^{n} A_j \sin(\omega_j t_i + \varphi_j),$$

where A_j, ω_j and φ_j are the amplitude (in microvolts), frequency (in radians/s) and phase (in radians) of the jth sine wave. Superposition of continuous light waves is familiar from classical optics in the study of diffraction and interference. Here we use discrete wave representations to match our discrete fast Fourier transforms. For simplicity of comparison, we report only relative amplitudes, not microvolt calibrations, for all three experiments, performed with three different EEG systems. We note that the least-squares criterion of fit used in all our analyses is invariant under a change of the units in which amplitude is measured.

For the reason already stated, such superpositions are uncommon in standard signal processing, but in our special low-frequency environment they can work very well. Moreover, to represent a word by a small set of superposed pure sine functions gives a definite sense of the minimum number of parameters needed for the invariant brain waves that seem to characterize rather well the words we have studied. To be explicit, each sine wave j in the superposition is characterized completely by A_j, ω_j and φ_j. For superpositions made up of five frequencies this yields a 15-parameter representation, which is certainly not enough to represent the spectral analysis of any spoken word, and is thereby testimony to the apparently simplifying transformations imposed by the auditory system on a sound-pressure wave as it reaches the cortex.

When we referred to "invariant" brain waves above we had in mind the extensive averaging over trials and subjects used to obtain an invariant result. Such averaging can eliminate more than noise, for the unaveraged signals may well contain much additional information, such as individual associations, not needed to identify the word itself. At present, our success in correct recognition of what word or sentence is represented depends on using such averaging.

Our focus entirely on waves, with no reference to populations of neurons and their intermittent spiking, as the real communication setup physically, may make some readers skeptical of our results. We certainly believe the waves we find in the observed EEG data arise from the spiking activity of many neurons, as their source. Whether or not there is a more fundamental way, at least in principle, to observe more directly, for cognitive purposes, the collective activity of the neurons handling speech, although it is an important question.

2.1 General Methods of Analysis

After applying an FFT to the averaged EEG data, we selected high-energy frequencies for the superposition wave representing a given word. Our criteria for selecting these frequencies were the following:

(i) We excluded any frequencies below 1.5 Hz as being too low, or, in the case of the large direct-current amplitude at 0 Hz, as being irrelevant for classification or prediction;

(ii) Based on our earlier studies, we excluded any frequencies equal to or greater than 20 Hz;

(iii) A frequency selected must be a local maximum in its amplitude;

(iv) When two local maxima were separated by only one frequency in the discrete FFT, we used only the one with higher amplitude, and if the amplitudes were the same, we selected the higher frequency;

(v) For superposition of n frequencies, a frequency selected must be one of the n highest local maxima, subject to the exclusion of (iv).

Subjects were numbered consecutively, and sometimes used in more than one of the earlier studies. The same numbering is used in this article. The sensors referred to follow the nomenclature of the standard EEG 10-20 system.

2.2 Results

In our first experimental study [6], we presented auditorily seven words to subjects in 100 randomized trials for each word. We recorded with the standard 10-20 EEG system subjects' brain waves beginning shortly before the onset of each verbal stimulus. Using the EEG data of the best sensor C4 for recognizing the seven words *first, second, third, yes, no, right* and *left,* we applied the method of superposition described above. First, we averaged together the unfiltered EEG data for subjects S3, S4 and S5, which yielded 300 trials for each stimulus word presented auditorily. These data were the test samples. We next applied an FFT to these averaged data for each word, and we selected, for each word, using the criteria stated earlier, the seven frequencies, i.e., sine waves, with the highest energy from the less than 60 frequencies, computed by the discrete FFT, with 0.662 Hz the difference between successive frequencies. As an example, the frequency-domain graph of the relative amplitude for the auditory word *first* is shown in Fig. 1.

Using now as a prototype for each word the superposition of the seven selected sine waves, we classified the test samples consisting of the unfiltered but averaged data described above. The criterion of fit, as in our previous work, was minimum least squares of all observations over a selected temporal interval. For a number of intervals we correctly classified all seven words.

To test how few superposed sine waves were required, we next systematically reduced the number of frequencies used in the superposition, by deleting first, the highest frequency from the seven for each word. With six superposed sine waves we also correctly classified the test samples for all seven words. Continuing, by deleting always the highest remaining frequency, we continued to classify correctly all seven words, for five, four or three frequencies. Finally, using only the single remaining lowest two frequencies for each word, we correctly classified six of the seven words.

Table 3: Three frequencies selected for each of seven words

Freq in Hz	Relative Amplitude	Phase in degrees	Freq in Hz	Relative Amplitude	Phase in degrees
	First			*Yes*	
2.649	10.86	-10.8	3.974	14.51	-69.79
4.636	13.74	26.8	5.298	10.90	151.69
5.961	14.47	-100.7	7.285	7.50	130.8
	Second			*No*	
3.312	11.01	105.0	2.649	14.02	52.8
5.961	20.16	-110.6	5.961	11.68	-93.9
9.935	4.89	137.1	9.935	4.94	-99.8
	Third			*Right*	
4.636	17.03	12.6	1.987	10.57	-166.1
6.623	17.23	21.1	3.312	13.14	154.8
8.610	12.72	-5.3	4.636	13.00	46.2
				Left	
			3.312	15.70	-170.9
			4.636	9.54	36.9
			7.285	9.57	139.3

In Table 3, we show the superposed three lowest frequencies, their amplitudes and their phases for each of the seven words. The selected frequencies all fall between 1.9 and 10.0 Hz, with more variation in phase than amplitude. The three selected for *first,* as shown in Table 1, are easily identified also in Fig. 1. In Fig. 2, the superposition of the lowest two,

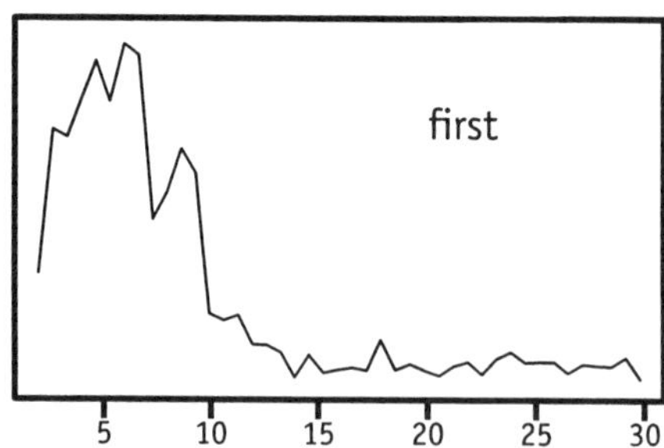

Figure 1: Graph in the frequency domain of the FFT of the averaged data for the word *first,* with discrete frequencies shown on the x axis in hertz and the amplitudes of the frequencies on the y axis in relative amplitude.

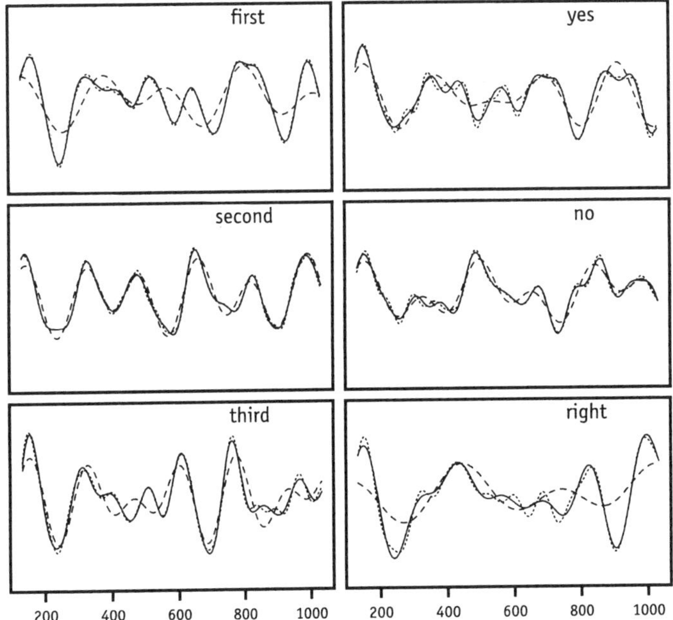

Figure 2: Comparison of superposition of the two lowest frequencies (dashed line), superposition of five frequencies (solid line), and superposition of seven frequencies for six of the seven words (dotted line). The x axis is measured in milliseconds after the onset of the stimulus and the y axis in relative amplitude.

the lowest five and the seven frequencies are shown in panels for six of the seven words. We omit the seventh one, for *left,* only to save space. The graphs for Fig. 2 are for one of the temporal intervals optimal for all superpositions, except the superposition of two frequencies. The temporal interval used was for 897 ms, beginning 132 ms after onset of stimulus. The interval for two frequencies was 750 ms, beginning at 132 ms. x

First, a completely different approach to superposition would be to estimate, purely statistically, for each prototype, a fixed number of best frequencies with optimally fitted amplitudes and phases. But such a purely statistical approach, with no physical constraints, produces frequencies with wildly varying amplitudes that seem obviously implausible, even though, through interference of one frequency with another, very good fits are produced. In contrast, by using frequencies found in the Fourier analysis of the data, the case for their physical reality is clear.

On the other hand, using pure sine waves to represent the fundamental frequencies may seem unrealistic. Because there is a lot of evidence that the images of the stimuli we used last only for a second or so at most, a more obvious mathematical choice would be some kind of wavelet representation with a short temporal domain. There are now many types of wavelets, with accompanying software, available for such purposes [9, 10]. In the end this may turn out to be the right approach, but our earlier efforts [7] to use wavelets did not improve on the Fourier results. For the present, we think the use of pure sine waves is an acceptable approximation for the short temporal length we consider, and, in fact, may be hard to improve on. A different problem that must be dealt with in a more complete model of the brain's processing is the damping out of the activated image. But this extension presents no fundamental difficulty.

### 2.3	*Natural Computations*

The contrast between the style, if not the subject matter, of the first two sections of this paper are stark. The discourse and marshalling of concepts in the first section are very much in the spirit of formal work on languages, as can be found in logic and computer science, even though the target of the machine-learning effort has been on the learning of natural language, but of course a very regimented sublanguage. Such problems do have an empirical aspect, namely, it is impossible to know in advance how well the learning schemes of association, denotation and grammatical generalization will work. Only by running a program can one draw serious conclusions about the learning, at least at the present stage of development of the theory.

The contrast between Section 1 and Section 2 brings out clearly enough how complicated it is to understand thoroughly the processing of natural language by the brain. The work described in Section 2 has been an intense computational effort to get through the complicated data recorded by standard EEG methods to identify well-defined brain-wave representations of words and sentences. The example given here is a simple one, just for reasons of keeping the exposition limited. In many ways the most successful work has been not with words, but with entire sentences, as reported in references here to these 3 articles [7, 8, 11]. In the case of sentences, the contrast with Section 1 is really stark. Subjects were asked to judge the truth or falsity of simple questions about the geography of the world. Typical sentences used in the experiments were *Berlin is not north of Rome, Warsaw in not the capital of Austria,* and *the largest city of Poland is Moscow.* In many ways, the sentences used were as regimented as those in the robotic experiments of Section 1. But here the difference was that we were attempting to identify the processing xof the sentences in EEG-recorded brain waves. We were, for the very best subjects, surprisingly successful. For example, we were able for the best subject, to correctly classify 93 of 100 different sentences [11].

On the other hand, the limitation of this result is also very clear. We are as yet only able to speculate on how we naturally compute the truth value of such sentences. Certainly, the sentences themselves are not sitting in memory. Most of them subjects probably have never heard before, or, if they have, they certainly have forgotten them, and have not stored them; they are long since forgotten. Simple semantic computations must be made by subjects, upon hearing or reading any of these simple sentences about geography. How does a subject compute their truth or falsity, that is, how do subjects make the appropriate semantic computation? It is to be stressed as strongly as possible that formal theories of truth derived from the work of Tarski and others in logic many years ago do not provide any real answer. It was, of course, no part of Tarski's intention to provide a theory of natural semantic computation, and only the most general aspects of that work would seem to have relevance to the actual computation carried out in the brain processing of natural-language sentences.

Something a little more than speculation can be said about these natural computations, and it surely will become a subject of intense focus in this century. The associative processes known to occur naturally and widely in the brain, speculation about which were already an important part of Hobbes' and Hume's theory of the mind, developed especially with considerable thoroughness in Hume's *A Treatise on Human Nature,* [12]. Now a very large psychological literature, a growing linguistic literature and a growing neuroscience literature exist on the processes of association of the brain. The concept of an associative network is the best current framework for these natural semantic computations. It is important to stress that the scientific literature in psychology over the past 50 years on natural processes of association is rich and varied. It does not mainly address the problem of computation that is of focus here, but it provides an excellent basis on which to develop theoretical ideas that can be checked against some observable brain processes. How far those observations will take us with present technology is still an open question. From a direct computational standpoint, in many ways the current work in linguistics is formally more helpful, but it is not possible to try to survey that relevant literature here.

I turn now to my last topic about the features of natural language important for conversation, but which are as yet neither a part of robotic language development nor of empirical brain studies to any serious extent.

3　Semantics of Conversation

It is scarcely news that no robots or computers of any form are capable of conducting, at any time, extensive conversations of a natural sort on any subject. The nihilistic view, and anti-scientific one, is that such conversations will never be possible. This is certainly not the view that I hold, but all the same, what I want to stress here is the depth and difficulty of solving the scientific problems that stand between the present state of affairs and being able to endow robots and other computers with appropriate speaking and listening capabilities.

I mentioned at the beginning that the era of realistic speech recognition has begun. But this does not mean the same thing as having natural conversations with our favorite devices. The kind of work exemplified in the first section on machine learning will take us a long way toward making practical use of speech for information and control.

In this concluding section, I organize my remarks about what I think are the main problems on which we must make serious headway before we can have natural conversations with devices. My remarks fall under two main headings: ambiguity and prosody.

Ambiguity. Perhaps the most striking thing about the semantic quality of natural conversation is the presence of ambiguity in every direction. To begin with, many of the words we use naturally and frequently have a variety of distinct meanings, what is sometimes technically called in linguistics *polysemy*. Famous examples are to be found among some of the most stinging criticisms of artificial intelligence in the past decades. What are the methods by which we determine whether *bank* means riverbank or an institution of finance? And that is only when it is used as a noun, not as a verb. Moreover, once we have fixed in some way the particular meaning of each word in a sentence we find, when we apply standard grammatical parsers, that there are often for sentences of any length, an astronomical number of correct parses, and in a very high percentage of cases, at least two.

The main reason we have great difficulty in thinking about how to solve these problems for robots and other devices so that we can provide them programs, or learning methods for acquiring programs, is our ignorance of how we solve them in human speech.

There is a definite tendency on the part of some logicians, and undoubtedly some computer scientists as well, to think that the right approach to these problems is to live with our ignorance and to simply regard it as a defect of natural language that it is so saturated with ambiguity. But this is a naive strategy for more than one reason. The first is that it is hopeless to have a program of changing the way natural language itself is used by people speaking to each other. A modified program to have people speak in a very regimented sublanguage of their natural language in dealing with their devices will also not be a successful strategy. Competitors will continually be striving to meet the natural demand that the devices talk and speak as well as our normal human companions. I don't mean with this remark to suggest that there won't be good intermediate solutions of the kind I have already referred to. It is just that the ultimate goal here, both scientific and practical, will be to achieve the kind of easy natural discourse so common between us.

The second and deeper reason for claiming that the strategy of changing natural language is mistaken is this. It is not anything like a defect of natural language that it is saturated with ambiguity. The ambiguity provides a kind of instability, as in the case of other unstable systems or chaotic systems, to be taken advantage of in efficient control and communication. In this case it is a matter of communication. What takes place in general terms is a quick reduction of the ambiguity by the context of the talk and the intentions of the speakers and listeners. But the general context, social and physical, as well as the intentions of speakers and listeners, are difficult things to introduce into our formal analysis of language in the tradition of formal semantics. A successful scientific theory must be prepared to deal with a much wider range of subject matter than is traditional in semantics, as conceived by logicians and even most computer scientists. Now it is a familiar move in philosophy to say that the considerations just mentioned belong to pragmatics, not semantics. But that I dismiss as a mere terminological move. However we label it, considerations of context and intentions are central to our own intuitive understanding of speech, and a problem to master at a scientific level.

It is also important to mention that ambiguity is not restricted to the lexicon and grammar. Our production and perception of spoken language, in terms of the sounds meant to represent the phonemes and syllables of speech are often produced or received badly. The problems of ambiguity at this level are at least as plentiful and, at a fundamental level, still difficult to understand in terms of the correction procedures naturally applied in the brain processing of the listener, and often in the correction of mistakes by the speaker. There is much to be said about these phenomena. I will not say more now.

Prosody. Prosody refers to the organization of pitch and loudness in producing variations in the melody and sonority of spoken utterances and, sometimes, longer passages of speech. A description of prosody by phonologists is an important and complicated topic for linguistics in the study of any natural language. My focus here is not on the technical description of how a prosodic contour is produced by a speaker, but what those prosodic contours of pitch and loudness are used to produce in the minds of listeners. The cries of babies and the joyous cheers of older children, as well as the expressions by all of us of sadness, anger or fear, mainly rely upon the prosodic features of our speech. How the prosodic organization of speech, which at one level can be given a near physical description, is processed in a meaningful way in the brains or minds of listeners, is still little understood. As you might expect, there are brain recordings showing immediate response of the brain to unusual prosodic features, for example, sharp changes of pitch or of loudness. This is rather like similar EEG recordings, for many years now, of the effects on the brain's processing of semantic anomalies. The gross evidence of such anomalies, just as in the case of prosody, is still a very great distance from how anything of any subtlety is actually processed in these matters in the cortex, the most likely part of the brain involved.

Irony. The remarks just made on various uses of prosody are superficial, but at least they touch upon familiar subjects. Often, the case that I want to discuss last, is less frequently mentioned, but of great importance in sophisticated conversations of almost every kind. I am thinking of the use of prosody to give a statement, but especially a response to a statement, an ironic turn that often reverses the literal semantic meaning of the words being used. Socratic irony, i.e., the pretense to ignorance, represents a well-known philosophical tradition. Among recent philosophers, one of the best with serious comments on irony is Paul Grice [13], (pp. 53–54). Subtle as the remarks are, they do not go far enough or cover sufficiently the many complexities of the use of irony. Let me give just two examples in conclusion.

The first is the strong contrast between attempts to use irony in written language, as opposed to spoken speech. A novelist famous for the conversations between his characters is Henry Green [14]. But what can he do when he wants to produce some way of noting that a character's remarks are meant ironically? Here is a typical awkward device that is needed.

"Does anyone else know of this?"

"Auntie does."

"Of course," Edge took her up with a heavy irony that was wasted, because the girl did not notice.

Green 1951, p. 157

It is not appropriate here to survey the many ways that novelists mark the use of irony, but to stress the contrast with spoken speech is essential. In this great age of cinema, television and video we all can recognize how easy it is for an actor on the stage or in front of the camera, especially in front of the camera, to portray irony by an easy prosodic variation or by a lifting of eyebrows or a subtle movement of other facial muscles.

My second example is something that is not carried far in Grice's analysis. That is the complexity that arises from the social setting of more than two persons. Two knowing people talking to an innocent third can continually make subtley ironic remarks without the innocent third person being aware of what is going on between the other two. So here, the point that Grice makes about irony, namely the one mentioned above, that in irony, the true intended meaning is usually the opposite of the conventional meaning of the words spoken is often not accurate. The two knowing conversationalists, while in their responses affirming to the third conversationalist, the innocent one, the conventional meaning are denying it between themselves. This play of affirmation and denial, sincerity and irony, ricochets through conversations in all walks of life and about all subject matters.

This subtle use of irony is wonderfully done in modern films, even better than on the stage. A mere knowing lift of an eyebrow while saying something apparently sincere, or even a slight change in prosodic contour, is enough to alert the audience to the sense of irony that has been brought into play.

Whether on the stage, before the camera, or in the chair across the room, a muttered "Indeed" or "You don't say" or "Whatever" can be marked with irony, often to the delight of all and the malice of none. Getting such performances from our digital devices will in the end be a work of art, as well as of serious science.

References

[1] Suppes, P., Böttner, M. and Liang, L. (1995) Comprehension grammars generated from machine learning of natural language. *Machine Learning,* **19**, 133-152.

[2] Suppes, P., Böttner, M. and Liang, L. (1996) Machine learning comprehension grammars for ten languages. *Computational Linguistics,* **22**, 329-350.

[3] Suppes, P. and Liang, L. (1996) Probabilistic association and denotation in machine learning of natural language. In A. Gammerman (ed.), *Computational Learning and Probabilistic Reasoning.* Sussex, England: John Wiley & Sons, Ltd., 87-100.

[4] Skinner, B. F. (1959) *Verbal Behavior.* New York: Appleton.

[5] Chomsky, N. (1959) Review of B. F. Skinner, *Verbal Behavior. Language,* **35**, 26-58.

[6] Suppes, P., Lu, Z.-L. and Han, B. (1997) Brain wave recognition of words. *Proc. Natl. Acad. Sci.* **94**, 14965-14969.

[7]	Suppes, P., Han, B. and Lu, Z.-L. (1998) Brain-wave recognition of sentences. *Proc. Natl. Acad. Sci.* **95**, 15861-15866.

[8]	Suppes, P., Han, B., Epelboim, J. and Lu, Z.-L. (1999) Invariance between subjects of brain wave representations of language. *Proc. Natl. Acad. Sci.* **96**, 12953-12958.

[9]	Daubechies, I. (1992) *Ten Lectures on Wavelets*. Philadelphia: Soc. Indust. Appl. Math.

[10]	Bruce, A. and Gao, H.-Y. (1996) *Applied Wavelet Analysis with S-Plus*. New York: Springer.

[11]	Suppes, P., Wong, D.K., Perreau Guimaraes, M., Uy, E.T. and Yang, W. (to appear) High statistical recognition rates for some persons' brain-wave representations of sentences. *Proc. Natl. Acad. Sci.*

[12]	Hume, D. (1739) *A Treatise on Human Nature*. London: John Noon.

[13]	Grice, P. (1989) *Studies in the Way of Words* Cambridge, MA: Harvard University Press, pp. 53-54.

[14]	Green, H. (1951) *Concluding*. New York: The Viking Press.

[15]	Suppes, P. and Han, B. (2000). Brain-wave representation of words by superposition of a few sine waves. *Proc. Natl. Acad. Sci.* **97**, 8738-8743.

Logic, Artificial Intelligence and Robotics
J.M. Abe & J.I. da Silva Filho (Eds.)
IOS Press, 2001

A New Approach to Type-2 Fuzzy Sets[*]

Helmut Thiele
University of Dortmund, Department of Computer Science I,
D-44221 Dortmund, Germany
Phone: +49 231 755 6152
Fax: +49 231 755 6555
E-Mail: **thiele@ls1.cs.uni-dortmund.de**
WWW: **http://ls1-www.cs.uni-dortmund.de**

Abstract. First, we define type-2 fuzzy sets on the one hand and context dependent fuzzy sets on the other hand and show that there is a bijection between these two kinds of fuzzy sets if we assume a trivial condition.

Secondly, we review some useful applications of the concept of context dependent fuzzy sets, for instance, to make precise the notion of qualitative fuzzy set, to interpret vague concepts, and to develop modal fuzzy approximate reasoning.

Thirdly, starting from a theory of fuzzy modifiers based on concepts of functional analysis, we define so-called external and internal operations with context dependent fuzzy sets.

The last section contains concluding remarks, in particular, on further applications of context dependent fuzzy sets.

Keywords: Type-2 Fuzzy Sets, Context Dependent Fuzzy Sets, External and Internal Operations.

1 Introduction. Fundamental Concepts

Let U be a non-empty (crisp) set called universe. By $\mathbb{R}$ and $\langle 0, 1 \rangle$ we denote the set of all real numbers and the set of real numbers r with $0 \leq r \leq 1$, respectively. An ordinary fuzzy set F on U is a function $F : U \to \langle 0, 1 \rangle$. The crisp power set $\mathbb{P}U$ of U and the fuzzy power set $\mathbb{FP}U$ of U, respectively, are defined by $\mathbb{P}U =_{def} \{X | X \subseteq U\}$ and $\mathbb{FP}U =_{def} \{F | F : U \to \langle 0, 1 \rangle\}$. If $\approx$ is an equivalence relation on a set X then the set of all equivalence classes from X generated by $\approx$ is denoted by $X/_{\approx}$. For $Y \subseteq X$ we define $CL_{\approx}(Y) = \{x | x \in X \wedge \exists y (y \in Y \wedge x \approx y)\}$. Assume that X, Y, Z are arbitrary sets and φ is a function with $\varphi : X \times Y \to Z$. Then for $x \in X$ and fixed $y \in Y$ by $\lambda x \varphi(x, y)$ we denote the function $f_y : X \to Z$ defined by $f_y(x) =_{def} \varphi(x, y)$ for every $x \in X$. *Card* X denotes the cardinal number of the set X. For an integer $n \geq 1$, X^n denotes the set of all n-tuples $\underline{x}$ with components from X.

Now, let V be a further non-empty (crisp) set.

Definition 1 Φ *is said to be a type-2 fuzzy set on* U *with respect to* $V =_{def} \Phi : U \to \mathbb{FP}V$.

---

[*]This research was supported by the Deutsche Forschungsgemeinschaft as part of the Collaborative Research Center "Computational Intelligence" (SFB 531)

This concept was introduced by L.A. ZADEH in [21] and investigated in numerous papers, in particular, under the assumption that $V \subseteq \mathbb{R}$ and V is eventually finite. From these papers, because of lacking space, here we only quote [1–5, 8–11, 20] as examples.

Our new approach to type-2 fuzzy sets applies the fundamental and new concept of (KRIPKE semantics based) context dependent fuzzy set and consists of the following ideas.

In the paper [18] we have introduced the concept (of KRIPKE semantics based) context dependent fuzzy sets as follows. Let W be a further non-empty crisp set called "set of possible worlds". We have defined

Definition 2 Ψ *is said to be a context dependent fuzzy set with respect to W* $=_{def} \Psi : U \times W \to \langle 0,1 \rangle$.

One can easily see that there is a very close interrelation between type-2 fuzzy sets and context dependent fuzzy sets.

If we have a type-2 fuzzy set Φ on U with respect to V, i.e. $\Phi : U \to \mathbb{FP}V$, then by the definition

$$\Psi(u,v) =_{def} \Phi(u)(v) \ (u \in U, v \in V)$$

we obtain a context dependent fuzzy set Ψ on U with respect to the set V of possible worlds.

And vice versa, if we have a context dependent fuzzy set Ψ on U with the set W of possible worlds then for fixed $u \in U$ we define a fuzzy set F_u on W by $F_u(w) =_{def} \Psi(u,w), w \in W$, and put $\Phi(u) =_{def} F_u$. Obviously, Φ is a type-2 fuzzy set on U with respect to W.

We state that the binary relation between type-2 fuzzy sets on U with respect to V and context dependent fuzzy sets on U with respect to W is a bijection if $V = W$.

2 Some Useful Applications of Context Dependent Fuzzy Sets

2.1 Defining "Qualitative" Fuzzy Sets and Interpreting Vague Concepts

Consider the linguistic variable AMOUNT_OF_MONEY. Assume that among its linguistic terms we have HIGH. Let $\mathbb{R}$ be the universe of the linguistic variable introduced above. Then following T.Y. LIN [7] the vague concept "high amount of money" is to interpret by a class C of "equivalent" fuzzy sets on $\mathbb{R}$ generated by an equivalence relation $\approx$ on $\mathbb{FPR}$, i.e. $C \in \mathbb{FPR}/\approx$. Furthermore, the equivalence relation $\underset{L}{\approx}$ used by LIN is defined by applying concepts of general topology, in particular, by using real functions in order to generate topological spaces and by using the concept of continuity. But he took no notice of the following very important fact: In applications one has to *operate* with qualitative fuzzy sets. This means, generally speaking, we have an n-ary $(n \geq 1)$ operation op^n on $\mathbb{FPR}$, i.e. $op^n : \mathbb{FPR}^n \to \mathbb{FPR}$, and an equivalence relation $\approx$ on $\mathbb{FPR}$. Then we are faced with the problem to "lift" the operation op^n with the domain $\mathbb{FPR}$ to a corresponding operation Op^n with the domain $\mathbb{FPR}/\approx$.

The general definition scheme for Op^n has the form

$$Op^n(C_1,\ldots,C_n) =_{def} CL_{\approx}(op^n(F_1,\ldots,F_n))$$

where $F_1 \in C_1,\ldots,F_n \in C_n$.

But from universal algebra we know that this scheme is only correct, i.e. $Op^n(C_1,\ldots,C_n)$ is uniquely defined, if for every $F_1,F_1^{'},\ldots,F_n,F_n^{'} \in \mathbb{FPR}$ and for every $C_1,\ldots,C_n \in \mathbb{FPR}/\approx$

with $F_1 \in C_1, \ldots, F_n \in C_n$ the condition $F_1 \approx F_1', \ldots, F_n \approx F_n'$ implies

$$op^n(F_1, \ldots, F_n) \approx op^n(F_1', \ldots F_n').$$

In a forthcoming paper we shall study this problem, in particular, if op^n describes, for instance the intersection, union, addition, multiplication ect. of fuzzy sets $F_1, \ldots, F_n \in \mathbb{FPR}$.

In the paper [18] we created another way to interpret vague concepts like "high amount of money". Assume we have fixed the set $W = \{w_1, w_2, w_3\}$ of possible worlds, where w_1 is an unemploid person, w_2 is an university professor, and w_3 is an oil sheikh, for instance. Then one could fix that for w_1, w_2, and w_3 the vague concept "high amount of money" is to interpret by a fuzzy set on $\mathbb{R}$ describing about 1 thousand, 1 million, and 1 billion dollars, respectively. So, the linguistic term HIGH of the linguistic variable AMOUNT_OF_MONEY is interpreted by a context dependent fuzzy set on $\mathbb{R}$ with respect to W as defined above.

This observation was the starting point to define, to investigate, and to apply context dependent fuzzy sets (see also [8, 16]).

Before we shall discuss further applications of context dependent sets in the next sections here we present some important concepts used in investigations and applications of context dependent fuzzy sets.

Let F and G be usual fuzzy sets on U, i.e. $F, G : U \to \langle 0, 1 \rangle$. Denote an arbitrary set of bijections of U by Γ.

Definition 3 *F and G are said to be isomorphic with respect to Γ (shortly $F \underset{\Gamma}{\approx} G$) $=_{def}$ There exists a bijection $\beta \in \Gamma$ such that $\forall x(x \in U \to F(x) = G(\beta(x))$ (see [14]).*

Proposition 1 *If Γ is a group with respect to the composition of bijections then the binary relation $\underset{\Gamma}{\approx}$ is an equivalence relation on $\mathbb{FP}U$.*

With respect to our "money example" discussed above we could assume that all fuzzy sets used have the same shape, i.e. they can generate from one of them by translating along the real axis. Obviously, this assumption means that we have to choose Γ as the group of all translations along the real axis.

The example considered gives the occasion to develop the following further concepts.

Assume that W is an arbitrary non-empty set of possible worlds, Φ is a context dependent set on U with respect to W, Γ is a set of bijections of U, and $w, w' \in W$.

Definition 4 *The worlds w and w' are said to be equivalent with respect to Φ and Γ (shortly $w \underset{\Phi,\Gamma}{\approx} w') =_{def} \lambda x \Phi(x, w) \underset{\Gamma}{\approx} \lambda x \Phi(x, w')$.*

Proposition 2 *If Γ is a group then $\underset{\Phi,\Gamma}{\approx}$ is an equivalence relation on W.*

Now, if Γ is a group of bijections of U we can consider the factor set $W/\underset{\Phi,\Gamma}{\approx}$ as an important indicator of the complexity of the context dependent set Φ with respect to Γ.

For instance, our "money example" leads to the

Definition 5 *Φ is said to be simple with respect to $\Gamma =_{def} Card(W/\underset{\Phi,\Gamma}{\approx}) = 1$.*

For applying these concepts we refer to the next two sections on approximate reasoning using context dependent sets and on modal fuzzy approximate reasoning.

Concluding our statements on the concept of context dependent fuzzy sets we finally remark.

The set W of possible worlds can be interpreted in many different ways. So in our "money example" $W = \{w_1, w_2, w_3\}$ is a set of three persons w_1, w_2 and w_3. If W is interpreted as a set of time points or time intervals then $\Phi : U \times W \to \langle 0, 1 \rangle$ describes time variant fuzzy knowledge on U. The same holds if, for instance, W is interpreted as a set of space points or a set of technical (or biological) systems or a set of states of a fixed complex system.

In many applications, for instance, in studying time variant fuzzy knowledge, the following separation property will play an important role.

Let FAC be a binary function with $FAC : \langle 0, 1 \rangle \times \langle 0, 1 \rangle \to \langle 0, 1 \rangle$. This function will be used as "factorization" or "separation" means as follows where $\Phi : U \times W \to \langle 0, 1 \rangle$.

Definition 6

1. *Φ is said to be factorized by FAC $=_{def}$ There are usual fuzzy sets F on U and G on W such that $\forall x \forall w (x \in U \wedge w \in W \to \Phi(x, w) = FAC(F(x), G(w)))$.*

2. *Φ is said to be factorizable $=_{def}$ There exists a function $FAC : \langle 0, 1 \rangle \times \langle 0, 1 \rangle \to \langle 0, 1 \rangle$ such that Φ is factorized by FAC.*

2.2 *Fuzzy Approximate Reasoning with Context Dependent Fuzzy Sets*

Taking into consideration the fact that vague concepts must be interpreted by context dependent fuzzy sets we are faced with the problem to introduce approximate reasoning using context dependent fuzzy sets.

To this end in [19] we generalized the "classical" fuzzy approximate reasoning based on the generalized modus ponens and the compositional rule of inference (see [6,21]) as follows.

Let Φ, Ψ, Φ', Ψ' be context dependent fuzzy sets on U with respect to W, i.e.

$$\Phi, \Psi, \Phi', \Psi' : U \times W \to \langle 0, 1 \rangle.$$

Consider a inference scheme of the form
$$\frac{\Phi \Rightarrow \Psi \qquad \Phi'}{\Psi'}$$

called context dependent generalized modus ponens.

To develop a semantic interpretation of this scheme in [19] we defined

Definition 7 *$\Theta = [\Pi, K, Q]$ is said to be a context dependent semantics for the context dependent generalized modus ponens*
$=_{def}$ *1. $\Pi, K : \langle 0, 1 \rangle \times \langle 0, 1 \rangle \times W \to \langle 0, 1 \rangle$ and*

2. *$Q : \mathbb{P}\langle 0, 1 \rangle \times W \to \langle 0, 1 \rangle$*

On the basis of a given $\Theta = [\Pi, K, Q]$ we define the context dependent generalized compositional rule of inference as follows where $x, y \in U$ and $w \in W$.

Definition 8

1. $FREL(\Pi, \Phi, \Psi)(x, y, w) =_{def} \Pi(\Phi(x, w), \Psi(y, w), w)$

2. $\Psi'(y, w) =_{def} Q(\{K(\Phi'(x, w), FREL(\Pi, \Phi, \Psi)(x, y, w), w)| x \in U\}, w)$

3. $FUNKT_\Theta(\Phi, \Psi, \Phi') =_{def} \Psi'$

Obviously, for fixed $w \in W$ this definition coincides with the definition of the usual generalized compositional rule of inference as defined in [6, 19].

Definition 9

1. *The context dependent generalized modus ponens is said to be correct with respect to the semantics* $\Theta =_{def}$ *For every* $\Phi, \Psi \in \mathbb{FP}(U \times W), FUNKT_\Theta(\Phi, \Psi, \Phi) = \Psi$.

2. *This modus ponens is said to strongly correct with respect to the semantics* Θ *and a topology* TOP *in* $\mathbb{FP}(U \times W) =_{def}$ *This modus ponens is correct with respect to* Θ *and the functional operator* $FUNKT_\Theta$ *is continuous on* $\mathbb{FP}(U \times W)$ *with respect to* TOP.

This definition is fundamental for theoretical investigations and, in particular, for application. In [6, 13, 19] we have presented first results concerning correctness and strong correctness. In a forthcoming paper we shall investigate this problem in detail.

2.3 Modal Fuzzy Approximate Reasoning

In the papers [12, 13, 19] we have defined for $R \subseteq W \times W$ and $S : W \times W \to \langle 0, 1 \rangle$ hard *KRIPKE* frames $\mathfrak{K} = [W, R]$ and soft *KRIPKE* frames $\mathfrak{L} = [W, S]$. Then by adopting ideas of classical modal logic we introduced not only "hard" but also "soft" diamond and box operators and defined their application to context dependent fuzzy sets.

This procedure is generalized to n-ary (n integer, $n \geq 1$) context dependent fuzzy relations χ on U with respect to W, i.e. $\chi : U^n \times W \to \langle 0, 1 \rangle$ as follows where $\mathfrak{x} \in U^n$ and $w \in W$.

Definition 10

1. $\langle \mathfrak{K} \rangle \chi(\mathfrak{x}, w) =_{def} Sup\{\chi(\mathfrak{x}, w')| w' \in W \wedge [w, w'] \in R\}$

2. $[\mathfrak{K}] \chi(\mathfrak{x}, w) =_{def} Inf\{\chi(\mathfrak{x}, w')| w' \in W \wedge [w, w'] \in R\}$

To discuss the "soft" case we fix two functions $K, \Pi : \langle 0, 1 \rangle \times \langle 0, 1 \rangle \times W \times W \to \langle 0, 1 \rangle$ and define

Definition 11

1. $\langle \mathfrak{L}, K \rangle \chi(\mathfrak{x}, , w) =_{def} Sup\{K(\chi(\mathfrak{x}, , w'), S(w, w'), w, w')| w' \in W\}$

2. $[\mathfrak{L}, \Pi] \chi(\mathfrak{x}, , w) =_{def} Inf\{\Pi(\chi(\mathfrak{x}, , w'), S(w, w'), w, w')| w' \in W\}$

To develop a "modal fuzzy approximate reasoning" we carried out the following steps illustrated by the example below [12, 13].

Consider the inference scheme *SCH*:
$$\frac{\begin{array}{c}\Box(A \; \to \; B) \\ \Box A\end{array}}{\Box B}$$

of the classical modal logic where A and B are formulas of the calculus given.

Adopting ZADEH's ideas to formulate the generalized modus ponens we translated the scheme *SCH* into the inference scheme *SCH*$'$:

$$\frac{[\mathfrak{K}](\Phi \;\Rightarrow\; \Psi) \qquad [\mathfrak{K}]\Phi'}{[\mathfrak{K}]\Psi'}$$

where $\mathfrak{K}$ is a hard *KRIPKE* frame and Φ, Ψ, Φ', Ψ' are context dependent fuzzy sets on U with respect to W. The we defined Ψ' (see here definitions 7 and 8) as follows

Definition 12 $\Psi'(y,w) =_{def} Q(\{K([\mathfrak{K}]\Phi'(x,w), [\mathfrak{K}]\Pi(\Phi(x,w), \Psi(y,w), w)|x \in U\}, w)$

and, finally, we constructed $[\mathfrak{K}]\Psi'$.

3   External and Internal Operations with Context Dependent Fuzzy Sets

In [17] we have studied mappings $Mod : \mathbb{FP}U \to \mathbb{FP}U$ called modifiers on U. Besides other concepts we have defined

Definition 13

1. *Mod is said to be an external local modifier on U*
   $=_{def}$ *There exist a mapping ext* $: \mathbb{R} \to \mathbb{R}$ *such that* $\forall x \forall F (x \in U \wedge F \in \mathbb{FP}U \to Mod(F)(x) = ext(F(x)))$.

2. *Mod is said to be an internal local modifier on U*
   $=_{def}$ *There exist a mapping int* $: U \to U$ *such that* $\forall x \forall F (x \in U \wedge F \in \mathbb{FP}U \to Mod(F)(x) = F(int(x)))$.

For operating with context dependent fuzzy sets we generalize this approach as follows.

For an integer $n \geq 1$ let ext_n be an n-ary "external" operation, i.e. $ext_n : \langle 0,1 \rangle^n \to \langle 0,1 \rangle$. By the following definition we "lift" the n-ary external operation ext_n on $\langle 0,1 \rangle$ to a corresponding n-ary "external defined" operation Ext_n on $\mathbb{FP}(U \times W)$. Assume $\Phi_1, \ldots, \Phi_n : U \times W \to \langle 0,1 \rangle, x \in U$, and $w \in W$.

Definition 14 $Ext_n(\Phi_1, \ldots, \Phi_n)(x,w) =_{def} ext_n(\Phi_1(x,w), \ldots, \Phi_n(x,w))$

Obviously, by this definition an arbitrary property of ext_n is lifted to a corresponding property of Ext_n, for instance, if $n = 2$ the commutativity and the associativity of ext_2.

Now, we are going to define so-called n-ary internal operations with context dependent fuzzy sets. Therefore, let int_n be a so-called n-ary internal operation, i.e. $int_n : W^n \to W$. Then using ZADEH's extension principle we lift the internal operation int_n on W to a corresponding n-ary "internal defined" operation Int_n on $\mathbb{FP}(U \times W)$ where again $\Phi_1, \ldots, \Phi_n : U \times W \to \langle 0,1 \rangle, x \in U$ and $w \in W$.

Furthermore, for defining this internal lifting procedure we need an $(n+n)$-ary operation K on $\langle 0,1 \rangle^n \times W^n$, i.e. $K : \langle 0,1 \rangle^n \times W^n \to \langle 0,1 \rangle$. Then we define

Definition 15 $Int_n(\Phi_1, \ldots, \Phi_n)(x,w)$
$=_{def} Q(\{K(\Phi_1(x,w_1), \ldots, \Phi_n(x,w_n), w_1, \ldots, w_n)|w_1, \ldots, w_n \in W \wedge w = int_n(w_1, \ldots, w_n)\}, w)$

We underline that lifting properties of int_n to corresponding properties of Int_n is impossible, in many cases. The possibility of lifting strongly depends on the structure of W and the functional properties of int_n, K, and, last but not least, on the structure of $\Phi_1, \ldots, \Phi n$.

In case of $W = \langle 0,1 \rangle$, in the papers [9,10] one can find old but interesting results if $n = 2$, int_2 is a usual operation in $\mathbb{R}$ (min, max, addition, multiplication ect.), and K is a suitable t- or t-conorm.

4 Conclusion. Further applications

Because of lacking space we could only present the concept of context dependent fuzzy set and some applications of this concept. The idea of external and internal operations with such fuzzy sets is new, as far as we know.

Due to close relations to type-2 fuzzy sets, many results on context dependent fuzzy sets can be immediately applied to type-2 fuzzy sets. So, a lot of work is to be done on this field.

For example, we describe only the following three sets of problems which should be investigated.

Firstly, assume that W is equipped with a set of n-ary operations, i.e. W is an universal algebra. By considering these operations as internal operations and lifting them to $\mathbb{FP}(U \times W)$ we obtain a (new) universal algebra with the carrier $\mathbb{FP}(U \times W)$. We plan to investigate the structure of this universal algebra and to compare it with the structure of W. Important special cases are studied in [9, 10], for instance.

Secondly, let U be the set of formulas of a logical calculus. Assume $\varphi : U^n \to U$, i.e. φ is an n-ary rule to generate a new formula $\varphi(u_1, \ldots, u_n)$ from $u_1, \ldots, u_n \in U$. A context dependent fuzzy set $\Phi : U \times W \to \langle 0, 1 \rangle$ is said to be an external and internal, respectively, evaluation with respect to Op_n if for every $u_1, \ldots, u_n \in U$ and $w \in W$ the equation

$$\Phi(\varphi(u_1, \ldots, u_n), w) = Op_n(\Phi(u_1, w), \ldots, \Phi(u_n, w))$$

holds and Op_n is an n-ary external (internal) Operation on $\mathbb{FP}(U \times W)$. Calculi with only external evaluations are studied in many papers, with respect to internal evaluation one can only find some ideas, for instance in [2].

Thirdly, studying fuzzy systems with context dependent components is very important for theory as well as for practice, in particular, if W is a set of time points (or time intervals) for describing and handling time variant knowledge or if W is a set of spatial coordinates (or spatial regions) for describing and handling spatial dependent knowledge or if W is a set of real numbers and the context dependent fuzzy sets considered can be interpreted as usual type-2 fuzzy sets. Among the components of such fuzzy systems context dependent inference machines play an important role. If such inference machine is given by a (context dependent!) fuzzy IF-THEN rule base we are faced with the problem to interpret such bases by using the principles FATI or FITA (see [6]) and to combine context dependent fuzzy sets occurring in such bases by internal or external operations.

Acknowledgement

The author would like to thank VOLKHER KASCHLUN for his help in preparing the manuscript.

References

[1] R. E. Bellmann and L. A. Zadeh. Local and fuzzy logics. In J. M. Dunn and G. Epstein, editors, *Modern Uses of Multiple-Valued Logic — Invited Papers of 5th ISMVL Symposium 1975*, pages 103–165. Reidel, Dordrecht, 1977.

[2] Didier Dubois and Henri Prade. Operations in a fuzzy-valued logic. *Information and Control*, 43(2):224–240, November 1979.

[3] R. I. John. Type 2 fuzzy sets: an appraisal of theory and applications. *International Journal of Uncertainty, Fuzziness and Knowledge-Based Systems*, 6(6):563–576, December 1998.

[4] N. N. Karnik, J. M. Mendel, and Q. Liang. Type-2 fuzzy logic systems. *IEEE-FS*, 7(6):643, December 1999.

[5] N. N. Karnik and J. M. Mendel. Operations on type-2 fuzzy sets. *Fuzzy Sets and Systems*. to appear.

[6] Stephan Lehmke, Bernd Reusch, Karl-Heinz Temme, and Helmut Thiele. On interpreting fuzzy IF-THEN rule bases by concepts of functional analysis. Technical Report CI-19/98, University of Dortmund, Collaborative Research Center 531, February 1998.

[7] T.Y. Lin. Neighborhood systems - a qualitative theory for fuzzy and rough sets. In *Second Annual Joint Conference on Information Science*, Wrightsville Beach, North Carolina, USA, September 28-October 1, 1995. Conference Proceedings, pages 255–258.

[8] J.M. Mendel. Computing with words when words can mean different things to different people. In *3rd Annual Symposium on Fuzzy Logic and Applications*, Rochester, New York, USA, June 22-25, 1999.

[9] M. Mizumoto and K. Tanaka. Some properties of fuzzy sets of type 2. *Information and Control*, 31(4):312–340, August 1976.

[10] M. Mizumoto and K. Tanaka. Fuzzy sets under various operations. In *4th International Congress of Cybernetics and Systems*, Amsterdam, The Netherlands, August 21-25, 1978.

[11] J. Nieminen. On the algebraic structure of fuzzy sets of type-2. *Kybernetica*, 13(4): 261-273, 1977.

[12] B. Reusch and H. Thiele. Modales approximatives Schließen (On modal approximate reasoning. In *20th Chair Workshop "Interdisziplinäre Methoden der Informatik"*, Haus Nordhelle, Meinerzhagen-Valbert, Germany, September 25-28, 2000. Forschungsbericht No 749, Feb. 2001, pages 9-19, Department of Computer Science, University of Dortmund.

[13] B. Reusch and H. Thiele. On modal fuzzy approximate reasoning. In *Joint 9th IFSA World Congress and 20th NAFIPS International Conference*, Vancouver, Canada, July 25-28, 2001. To appear in Conference Proceedings.

[14] Helmut Thiele. On fuzzy quantifiers. In Z. Bien and K. C. Min, editors, *Fuzzy Logic and its Applications to Engineering, Information Sciences, and Intelligent Systems. Selected Papers of the 5th IFSA World Congress, Soul, Korea, July 1993*, pages 343–352. Kluwer Academic Publishers, 1995.

[15] Helmut Thiele. On logical systems based on fuzzy logical values. In *EUFIT '95 — Third European Congress on Intelligent Techniques and Soft Computing*, volume 1, pages 28–33, Aachen, Germany, August 28–31, 1995.

[16] Helmut Thiele. On semantic models for investigating 'computing with words', Keynote Address. In *Second International Conference on Knowledge-Based Intelligent Electronic Systems*, Adelaide, Australia, April 21–23, 1998. Extended Version: On Semantic Models for Investigating 'Computing with Words'. Technical Report CI-32/98, University of Dortmund, Collaborative Research Center 531 (Computational Intelligence), April 1998, 13 pages.

[17] Helmut Thiele. Interpreting linguistic hedges by concepts of functional analysis and mathematical logic. In *EUFIT '98 — Sixth European Congress on Intelligent Techniques and Soft Computing*, Aachen, Germany, September 7–11, 1998. Conference Proceedings, vol 1, pages 114–119.

[18] Helmut Thiele. On the concept of qualitative fuzzy set. *The Twenty-Ninth International Symposium on Multiple-Valued Logic* (ISMVL '99), Freiburg im Breisgau, Germany, May 20–22, 1999. Proceedings, pages 282–287.

[19] Helmut Thiele. On approximate reasoning with context-dependent fuzzy sets. In *WAC 2000 — Fourth Biannual World Automation Congress*, Wailea, Maui, Hawaii, USA, June 11–16, 2000.

[20] M. Wagenknecht and K. Hartmann. Application of fuzzy sets of type 2 to the solution of fuzzy equation systems. *Fuzzy Sets and Systems*, 25:183–190, 1988.

[21] Lotfi A. Zadeh. The concept of a linguistic variable and its application to approximate reasoning — I. *Information Sciences*, 8:199–249, 1975.

Logic, Artificial Intelligence and Robotics
J.M. Abe & J.I. da Silva Filho (Eds.)
IOS Press, 2001

A Concurrent Algorithm for Logical Subsumption

Isabel Tonin and Guilherme Bittencourt
Departamento de Automação e Sistemas
Universidade Federal de Santa Catarina
88040-900 - Florianópolis - SC - Brazil
E-mail: {isabel | gb}@lcmi.ufsc.br

Abstract.
This paper presents an algorithm to eliminate subsumed clauses from a set of first-order logic clauses. The algorithm proposes a method, based on the use of a hypercube data structure, to reduce the number of subsumption tests needed to determine the set of subsumed clauses. Besides reducing the number of subsumption tests, the adopted hypercube data structure is suitable for a parallel implementation. The algorithm has been implemented and compared with an implementation of a naïve algorithm.
Keywords: Subsumption, Automated Reasoning, First-order logic.

1 Introduction

The exponential growth of the search space during the inference process is a common problem faced by resolution-based automated theorem provers. Since their introduction, great effort has been made to minimize this problem. As the growth is due to the number of clauses generated (and mainly retained) during the inference process, the deletion of those clauses, that are irrelevant to the inference process, plays an important role among the techniques developed to reduce the search space. One of the most important and effective techniques is based on subsumption[1, 13], that is, the deletion of clauses subsumed by other.

The elimination of subsumed clauses is cited as one of the "basic research problems to solve" in the Handbook of Logic in Artificial Intelligence and Logic Programming [16]. Because it must be repeated many times, the use of subsumption becomes very expensive. In fact, some authors state that due to the high processing costs involved, the use of subsumption based simplification does not improve the efficiency of automated theorem provers [15]. Nevertheless, many successful theorem provers apply subsumption tests. The problem of reducing the subsumption processing cost can be approached in two different ways: by improving the subsumption test itself or by reducing the number of necessary subsumption tests[10].

In this paper we adopt the first approach and introduce an algorithm that reduces the number of necessary tests for subsumed clauses elimination. Because of its underlying data structure, this algorithm is naturally concurrent and suitable for a parallel implementation. The proposed algorithm does not introduce any improvement to the subsumption test itself, rather it adopts the most efficient tests available in the literature, e.g. [10], [9] and [13].

This paper is organized as follows. In Section 2, the subsumption problem is introduced and the basic definitions are given. In Section 3, the proposed algorithm is described in the following sequence: its general idea, basic data structure adopted, its formal definition, an example, and, some comments about its behavior. In Section 4 some experimental results are presented. Finally, in Section 5, this work is concluded.

2 Definitions

Consider the first-order language $L(P, F, C)$, where P, F and C are finite or countable sets of predicate, function and constant symbols, respectively. Following the usual definition of *terms, atomic formulas*, and *formulas* (e.g., [8]), and given the formulas $X_1, X_2, ..., X_n$, a *generalized disjunction* is defined as $[X_1, X_2, ..., X_n] \equiv X_1 \vee X_2 \vee ... \vee X_n$, and a *generalized conjunction* as $\langle X_1, X_2, ..., X_n \rangle \equiv X_1 \wedge X_2 \wedge ... \wedge X_n$. A *literal* is an atomic formula, or the negation of an atomic formula, or one of the constants *True* or *False*. A *clause* is a generalized disjunction $[X_1, X_2, ..., X_n]$ in which each member is a literal. A first-order formula W_c is either in *conjunctive normal form*, or in *clause form* or is a *set of clauses* when it is a generalized conjunction $\langle C_1, C_2, ..., C_n \rangle$ in which each member is a clause.

A clause C_1 *subsumes*[1] a clause C_2, noted $C_1 \succ C_2$, if there is a substitution θ such that: $C_1\theta \subset C_2$. A clause C is *condensed* [9], if it cannot be split into two subclauses, C_1 and C_2, such that C_1 subsumes C_2 with a substitution θ that does not have any effect on C_2.

Given a set W of clauses, $\lfloor W \rfloor$ is a set that contains (i) only condensed versions of the clauses in W, and (ii) only clauses which are not subsumed by any other clause of the set.

3 The Concurrent Algorithm

The proposed algorithm is based on the following fact: if a clause subsumes another, the set of predicate symbols which occur in it must be a subset of the set of predicate symbols which occur in the subsumed clause. The idea is to represent explicitly the partial order defined by the *pertinence* relation on sets of predicate symbols associated with clauses. This is done using a hypercube, as underlying data structure, where, for any given clause, all clauses subsumed by it lie either in the same hypercube vertex or in some upper vertex.

3.1 The Hypercube

In general, a hypercube of dimension n is a graph where the vertices are the set of all n-tuples whose elements belong to the set $\{0, 1\}$ and the edges connect any two vertices whose representations differ only at one position [12]. For the proposed algorithm this definition is specialized as follows.

Given a set of clauses W, let Π_W be the set of all predicates that occur in W, a hypercube is defined as $H_W = \{(k_1, ..., k_n) \mid k_i \in \{0, 1\}\}$ where $n = \mid \Pi_W \mid$. For all $v \in H_W$, we also define the functions *pred* and *succ* as:

$$succ(v) = \{u \in H_W \mid \#_1(u) = \#_1(v) + 1\}$$

---

[1]There is no terminological agreement about the definition of *subsumption* in the standard literature. For instance in [11], a clause C subsumes a clause D if C implies D. We adopt the definition in [7], which, with the additional assumption that the number of literals in C is less than or equal to the number in D, is called θ-subsumption in [11].

$$pred(v) = \{u \in H_W \mid \#_1(u) = \#_1(v) - 1\}$$

where, $\#_1(\omega)$ stands for the number of coordinates of a vertex ω of the hypercube H_W that are different from zero.

Each hypercube vertex v is associated with two clause sets: the *local clause set* C_v and the *transmitted clause set* Γ_v. For each clause $C \in W$, the *target* vertex of C, notated $T(C)$, is the n-tuple $(k_1, \ldots, k_n)$ such that, $k_i = 1$, if P_i occurs in C and $k_i = 0$ otherwise, where P_i the i-th element of Π_W.

3.2   The Algorithm

The proposed algorithm consists basically of three steps: (i) the initial distribution of the clauses in the hypercube, according to their targets, (ii) the execution, at each hypercube vertex, of the subsumed clauses elimination process, and (iii) the collection of the remaining non subsumed clauses. This can be formalized by the following procedure.

Concurrent(W)
1. Create hypercube H_W with dimension $n = \mid \Pi_W \mid$.
2. Add each clause $C \in W$, such that $v = T(C)$, to the local clause set C_v.
3. Activate a copy of the procedure *Subsumption* at each $v \in H_W - \{\vec{0}, \vec{1}\}$.
4. Send an empty clause set from vertex $\vec{0}$ to vertices in $succ(\vec{0})$.
5. Wait for the arrival, at vertex $\vec{1}$, of all transmitted clause sets associated with
   vertices in $pred(\vec{1})$.
6. Return $\lfloor W \rfloor = \cup_{v \in H_W} C_v$

To complete the description of the algorithm, it is necessary to describe the procedure *Subsumption*, referred to at step 3 of procedure *Concurrent*. This procedure is defined as follows.

Subsumption(v)
1. **Wait until** $\{\Gamma_u\}$ **arrive from** $u \in pred(v)$
2. $C_v \leftarrow \lfloor C_v \rfloor$
3. $\Gamma_v \leftarrow \cup_{u \in pred(v)} \{\Gamma_u\}$
4. **for all** $\gamma \in \Gamma_v$ **do**
   for all $\gamma' \in C_v$ **do**
   if $\gamma \succ \gamma'$ **then** $C_v \leftarrow C_v - \{\gamma'\}$
5. **Send** $\Gamma_v \cup C_v$ **to all** $u \in succ(v)$

where the parameter v is the hypercube vertex where the procedure is executed, C_v is the local clause set, Γ_v is the transmitted clause set, that consists of the union of all transmitted clause set Γ_u received from lower dimension vertices u, and γ and γ' are clauses.

The behavior of this procedure can be summarized in five steps. In step 1, the procedure waits until all transmitted clause sets Γ_u arrive from lower dimension vertices. In step 2 and 3, the condensed form of the vertex local clauses $\lfloor C_v \rfloor$ and the transmitted clause set $\Gamma_v = \cup \{\Gamma_u\}$, respectively, are calculated. Step 4 removes from C_v all clauses that are subsumed by some other clause belonging to Γ_v. Finally, step 5 emits the new transmitted clause set to its successors vertices. This transmitted set consists of the union of the remaining local clause set and the original transmitted clause set.

There is no global control. The computation is started by the empty clause set sent out of vertex $\vec{0}$, at step 4 of procedure *Concurrent*, and proceed asynchronously through the hypercube vertices until all the transmitted clause sets arrive at the vertex $\vec{1}$. At this point, the procedure *Concurrent* resumes in step 5, and the simplified clause set can be obtaining by collecting the remaining clauses at all the vertex local clause sets:

$$\lfloor W \rfloor = \cup_{v \in H_W} C_v.$$

3.3 An Example

Consider the following set of clauses.[2]

$$W = \langle\, [P_0(f_1 a), P_1(f_1 a)], [P_0(f_1 a), P_1(f_1 a), P_2(x_0)],$$
$$[P_1(f_2 a)], [P_1(f_2 a), P_2(x_1)], [P_0(f_3 a)], [P_0(f_3 a), P_2(x_1)]\,\rangle$$

One can see that the set W contains (i)only condensed clauses and (ii) some subsumed clauses that can be removed. The goal of the algorithm is to calculate $\lfloor W \rfloor$. The set of predicates that occur in W is $\Pi_W = \{P_0, P_1, P_2\}$ and therefore $n = |\Pi_W| = 3$. Initially, it is necessary to generate the three dimensional hypercube H_W (procedure *Concurrent*, step 1) and assign to its vertices the associated clauses (step 2). The result of this operation is shown in the figure 1.

Next, the propagation process begins with an empty clause set sent from vertex $\vec{0}$ to the copies of the procedure *Subsumption* associated with the vertices of dimension one (step 4). As the clause sets arrive, the procedure *Subsumption*, local to each vertex, is executed, updating the local clause set and sending a new transmitted clause set to the vertices of higher dimension. The propagation continues until the vertex $\vec{1}$ is reached. This behavior is shown in the table 1. The table columns refer to: vertex labels, clause packages arrived in the vertex, vertex local clauses, remaining vertex local clauses after subsumption elimination, and clause packages sent to upper dimension vertices.

Finally (procedure *Concurrent*, step 5), collecting the remaining clauses in all hypercube vertices, results in $\lfloor W \rfloor = \langle\, [P_0(f_1 a), P_1(f_1 a)], [P_1(f_2 a)], [P_0(f_3 a)]\,\rangle$ as one can verify in the table.

3.4 Comments

The behavior of the processes in the hypercube, although complex, is deterministic. At each moment, there are a set of global computations occurring in parallel. Each of these computations is assigned with one of the hypercube vertices. All the processes execute the same procedure. This procedure is a data driven process, depending, at each vertex, on information propagated from lower vertices. The processes communicate with each other through messages that contain *clause sets*. The communication paths are restricted to hypercube edges.

It is interesting to observe that nothing was said about how to perform the subsumption tests itself required in steps 2 and 4 of procedure *Subsumption*. In fact, the subsumption test performed either between clauses or to obtain the condensed clauses, should follow the techniques given in the literature, e.g. [10], [9] and [13]. The goal of the proposed algorithm is

---

[2]This set was generated as described in the section 4 with $n = 3$, $m = 2$, and $k = 1$.

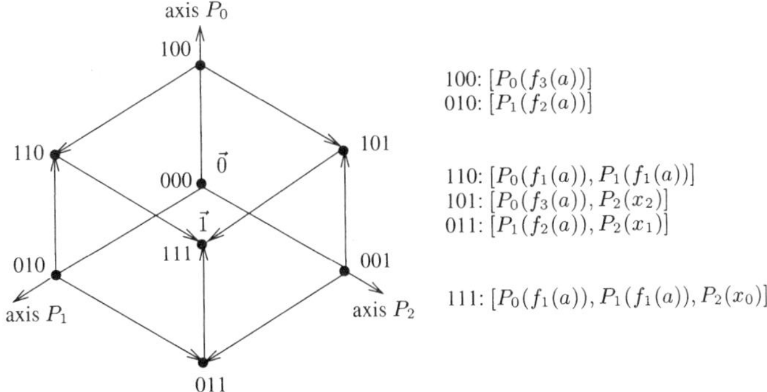

Figure 1: Hypercube and associated clauses with targets

v	*arrived*	*local*	*remained*	*sent*
100	$\{\}$	$\{[P_0(f_3(a))]\}$	$\{[P_0(f_3(a))]\}$	$\{[P_0(f_3(a))]\}$
010	$\{\}$	$\{[P_1(f_2(a))]\}$	$\{[P_1(f_2(a))]\}$	$\{[P_1(f_2(a))]\}$
001	$\{\}$	$\{\}$	$\{\}$	$\{\}$
110	$\{\{[P_0(f_3(a))]\},$ $\{[P_1(f_2(a))]\}\}$	$\{[P_0(f_1(a)), P_1(f_1(a))]\}$	$\{[P_0(f_1(a)),$ $P_1(f_1(a))]\}$	$\{[P_0(f_3(a))],$ $[P_1(f_2(a))],$ $[P_0(f_1(a)),$ $P_1(f_1(a))]\}$
011	$\{\{[P_1(f_2(a))]\},\{\}\}$	$\{[P_1(f_2(a)), P_2(x_1)]\}$	$\{\}$	$\{[P_1(f_2(a))]\}$
101	$\{\{[P_0(f_3(a))]\},\{\}\}$	$\{[P_0(f_3(a)), P_2(x_1)]\}$	$\{\}$	$\{[P_0(f_3(a))]\}$
111	$\{\{[P_0(f_3(a))],$ $[P_1(f_2(a))],$ $[P_0(f_1(a)), P_1(f_1(a))]\}$ $\{[P_1(f_2(a))]\}$ $\{[P_0(f_3(a))]\}\}$	$\{[P_0(f_1(a)), P_1(f_1(a)),$ $P_2(x_0)]\}$	$\{\}$	-

Table 1: Subsumption local vertex process behavior

to decrease the number of subsumption tests to be performed, restricting them to the clauses present in each vertex.

The complexity of the algorithm has not yet been calculated, but it is possible to highlight some of its costs, based on its characteristics. In general, the larger the number of clauses to be tested for subsumption is, the larger the cost will be. As each dimension of the hypercube represents one predicate present in a set of clauses W, for sets containing only one predicate, the hypercube collapses into one vertex that contains all the clauses. In this worst case, the algorithm cost is equal to the cost of a normal subsumption test, which would depend on the chosen technique. The hypercube generation adds a cost of 2^n vertices creation, where $n = | \Pi_W |$ is the number of predicates present in the clause set W. But, as all clauses inferred from W have the same predicates already present in W, the hypercube can be generated only once at the beginning of the inference process and be reused along its cycles.

The number of subsumption tests that can be avoided also depends on how the clause can be initially distributed in the hypercube vertices. The best a set of clauses can be initialy distributed, the best is the cost of the algorithm. For sets that can not be well distributed, the associated hypercube vertices either contain a large number of clauses, or are empty (or almost). Boths configurations are bad: the first because it increases the number of subsumption tests, and the second because, even empty, a vertex has to send its transmitted clause sets. To minimize this problem, the same algorithm can be adapted to work on a lattice extracted from the hypercube. This lattice contains the hypercube $\vec{0}$ and only those vertices that initially have clauses assigned to them. With the lattice, one can avoid excessive propagation of empty packages through the hypercube vertices, but one should also not forget that the lattice has to be redefined at each inference cycle.

4 Experimental Results

To test the subsumption algorithms, a generic form of a theory was defined. This generic theory with n clauses, each with maximum m literals, which is given by the following expression:

$$\langle [P_{(1+1)\bmod w}(f_1(a)), P_{(1+2)\bmod w}(f_1(a)), \dots, P_{(1+m)\bmod w}(f_1(a)),],$$
$$[P_{(2+1)\bmod w}(f_2(a)), P_{(2+2)\bmod w}(f_2(a)), \dots, P_{(2+m)\bmod w}(f_2(a)),],$$
$$\vdots \qquad\qquad\qquad \vdots$$
$$[P_{(n+1)\bmod w}(f_n(a)), P_{(n+2)\bmod w}(f_n(a)), \dots, P_{(n+m)\bmod w}(f_n(a)),]\rangle$$

where w is $(2m - 1)$. In this expression, the presence of a literal in a clause, is conditioned to its predicate index being smaller than m. This condition causes the absence of some literals in the generated clauses which allows a better distribution of the generated theory on the associated hypercube. For each clause C given by the expression above, k additional clauses are introduced. Each one of these additional clauses is subsumed by C, and is constructed by adjoining, to an instance of C, an additional open literal that do not occur in C.

Being $C = [P_j(f_i(a)), P_k(f_i(a)), \dots, P_l(f_i(a))]$, the resulting additional clauses are:

$$[P_j(f_i(a)), P_k(f_i(a)), \dots, P_l(f_i(a))],$$
$$[P_j(f_i(a)), P_k(f_i(a)), \dots, P_l(f_i(a)), P_{m+1}(z_i)],$$
$$\vdots \qquad\qquad\qquad \vdots$$
$$[P_j(f_i(a)), P_k(f_i(a)), \dots, P_l(f_i(a)), P_{m+1}(z_i), P_{m+k}(z_i)]$$

n	m	k	sub	pred	naïve (secs)	concurrent (secs)	speed up
40	3	0	0	3	.42	.08	5.25
50	3	0	0	3	.98	.23	4.26
100	3	0	0	3	22.57	2.24	10,07
40	3	1	40	4	9.62	.38	25.31
50	3	1	50	4	24.16	.78	30.97
100	3	1	100	4	315.29	13.15	23.97
40	3	2	80	5	57.35	2.2	26.07
50	3	2	100	5	136.00	5.24	25.95
100	3	2	200	5	1879.53	99.4	18.90

Table 2: Performance Results

Each value of n, m and k defines a different theory.

To compare the efficiency of the proposed algorithm, besides the procedure *concurrent*, a naïve algorithm was also implemented. This algorithm performs the subsumption test on all pair of clauses of the initial set W. The parallelism of the *concurrent* procedure was not implemented. The hypercube vertices were processed sequentially. The experimental results obtained with the naïve and the *concurrent* implementations are shown in table 2. There the columns *sub* and *pred* refer to the number of subsumed clauses and predicate symbols, respectively, present in the input theory. The results were obtained with a compiled version of the programs in the *CMU Common Lisp 18b* running on a *Linux Red Hat 7.1* system with *800MHz AMD K7 (Athelon)* machine with 256MB main memory.

5   Conclusion

A concurrent algorithm to improve the process of elimination of subsumed clauses present in first-order logic theories is presented. The proposed algorithm uses an hypercube as underlying data structure. The dimensions of this hypercube are defined by the predicate symbols that occur in the logical theory. The algorithm performance depends on how well the theory clauses can be distributed among the hypercube vertices.

The proposed algorithm has been implemented and the results of the tests performed with it are presented. A sequential implementation of the algorithm performs well in comparison with a *naïve* algorithm. Providing a method for distributing the subsumption deletion process gives the proposed algorithm two interesting features. Firstly, the algorithm is suitable for a parallel implementation. Secondly, it can be combined with different techniques for improving the subsumption test itself.

This paper is part of a knowledge representation project undertaken by the authors in recent years (e.g. [2], [3]) where the hypercube data structure plays an important role [5, 4, 6, 14].

Acknowledgments

The authors express their thanks to the Brazilian research support agency "Fundação Coordenação de Aperfeiçoamento de Pessoal de Nível Superior (Capes)" for the partial support of this work.

Logic, Artificial Intelligence and Robotics
J.M. Abe & J.I. da Silva Filho (Eds.)
IOS Press, 2001

On Interpolation and Modularity for Ultrafilter Logic

Paulo A. S. VELOSO
Praça Eugênio Jardim, 6/apt. 501; 22061-040 Rio de Janeiro RJ; Brazil

Abstract. We show that ultrafilter logic shares with classical first-order logic interpolation and related properties, such as joint consistency and modularity (preservation of conservativeness). By adapting a strategy used for classical first-order logic, we establish a joint model existence lemma, from which these properties, as well as Löwenheim-Skolem and completeness, follow.

1. Introduction

In this paper we show that ultrafilter logic shares with classical first-order logic interpolation and related properties, such as joint consistency and modularity (preservation of conservativeness). We adapt a strategy used for the classical case, establishing a joint model existence lemma, from which we can obtain these properties, as well as Löwenheim-Skolem and completeness.

Ultrafilter logic intends to capture directly the intuition of a property holding for a large [1] (or important [2]) set of elements and to serve as a precise basis for generic reasoning [3, 4, 5, 6]. For this purpose, one extends (conservatively) classical first-order logic by a new generalized quantifier ∇.

An interesting strategy for establishing interpolation relies on formulating this metalogical property in terms of semantic consequence and derivability [7]. When we manage to establish a joint model existence lemma, we can then obtain from it Löwenheim-Skolem, (strong) completeness, (Craig-Robinson) interpolation, modularity (preservation of conservativeness) [8] and (Robinson) joint consistency.

To establish such a joint model existence result, we adapt a method used for the classical case [9], based on the introduction of witnesses [10]. Here, we apply it to the case of ultrafilter logic, and it will be readily seen how to apply it to other extensions by generalized quantifiers [11], capturing versions of 'generally', such as 'many' and 'most' [12].

To make the paper self contained we briefly review ultrafilter logic [3, 4, 5].

2. Ultrafilter Logic

Ultrafilter logic extends classical first-order logic by a generalized quantifier ∇, whose intended interpretation is (very) 'important' or 'large'. In this section we briefly review ultrafilter logic: its syntax, semantics and axiomatics.

We consider a fixed signature (logical type) σ with a repertoire of symbols for predicates, functions and constants. We also consider a denumerably infinite set V of new symbols for variables. We let $L(\sigma)$ be the usual first-order language (with equality 3) of signature σ, closed under the propositional connectives, as well as under the quantifiers $\forall$ and $\exists$.

2.1 Syntax of ∇

The syntax of our logic is obtained by extending the usual first-order syntax by the new quantifier. We use $L^\nabla(\sigma)$ for the extension of the usual first-order language $L(\sigma)$ obtained by adding the new operator ∇.

The formulas of $L^\nabla(\sigma)$ are built by the usual formation rules and the new (variable-binding) formation rule: for each variable $v \in V$, if φ is a formula in $L^\nabla(\rho)$ then so is $\nabla v \varphi$.

We shall also use the notations, for a formula φ in $L^\nabla(\sigma)$:

- $occ(\varphi)$ ($fr(\varphi)$) for the set of variables with (free) occurrences in φ,
- $\varphi(\underline{v} := \underline{t})$ for the result of simultaneously substituting each term t_i for all the free occurrences of variable v_i in formula φ (for given sets $\underline{v} := \{ v_1, \ldots, v_n \}$ of variables and $\underline{t} := \{ t_1, \ldots, t_n \}$ of terms), which we sometimes simplify to $\varphi(\underline{t})$, when safe.

Other syntactic notions, such as sentence, substitutable, etc., [13] can be appropriately adapted.

2.2 Ultrafilter semantics of ∇

The semantics of ultrafilter logic is provided by enriching first-order structures with ultrafilters and extending the usual definition of satisfaction to the generalized quantifier ∇.

An *ultrafilter structure* $A^{\mathcal{U}} = (A, \mathcal{U})$ for signature σ consists of a usual (first-order) structure A for signature σ together with a proper ultrafilter $\mathcal{U}$ over the universe A of A.

We extend the usual definition of *satisfaction* of a formula φ in a structure under an assignment $s:V \to A$ to variables as follows (where, as usual, $s(v \propto a)$ is the assignment agreeing with s on every variable but v, and $s(v \propto a)(v)=a$):

(p^∇) for a formula $\nabla v \varphi$, we define $A^{\mathcal{U}} p \, \nabla v \varphi [s]$ iff the set

$$\{ a \in A : A^{\mathcal{U}} p \, \varphi [s(v \propto a)] \}$$ belongs to the ultrafilter $\mathcal{U}$.

As usual, satisfaction of a formula depends only on the realizations assigned to its symbols. In particular, satisfaction for purely first-order formulas (without ∇) does not depend on the ultrafilter, i. e. for a formula φ of $L(\sigma)$, we have $A^{\mathcal{U}} p \, \varphi [s]$ iff $A p \, \varphi [s]$. Also, satisfaction of a formula hinges only on the values assigned to its free variables[1].

Other semantic notions, such as reduct, model ($A^{\mathcal{U}} p \, \Sigma$), etc., are as usual [9]. The corresponding notion of *ultrafilter consequence* is as expected: $\Sigma p^{\mathcal{U}} \tau$ iff $A^{\mathcal{U}} p \, \tau$ whenever $A^{\mathcal{U}} p \, \Sigma$, for every ultrafilter structure $A^{\mathcal{U}}$.

2.3 Axiomatics of ultrafilter logic

We will now formulate a deductive system for our logic by adding schemata coding properties of ultrafilters to a calculus for classical first-order logic.

Consider the following sets of formulas of $L^\nabla(\sigma)$:

$[\nabla\exists] := \{\nabla v\varphi \to \exists v\varphi : \varphi \in L^{\nabla}(\sigma)\}$;

$[\neg\nabla] := \{\neg\nabla v\varphi \to \nabla v\neg\varphi : \varphi \in L^{\nabla}(\sigma)\}$;

$[\nabla\wedge] := \{(\nabla v\psi \wedge \nabla v\theta) \to \nabla v(\psi \wedge \theta) : \psi,\theta \in L^{\nabla}(\sigma)\}$;

$[\nabla v] := \{\nabla v\varphi \to \nabla u\varphi[v \coloneqq u] : \varphi \in L^{\nabla}(\sigma), u \notin occ(\varphi)\}$.

Now, consider the set $A^u(\sigma)$ consisting of the generalizations of the formulas in the union $B^u(\sigma) := [\nabla\exists] \cup [\neg\nabla] \cup [\nabla\wedge] \cup [\nabla v]$.

We can now set up a deductive system for our logic by adding the schemata in $A^u(\sigma)$ to a sound and complete deductive calculus $A(\sigma)$ for classical first-order logic, with Modus Ponens as the sole inference rule [14].

Thus, for a set Σ of sentences and a formula φ in $L^{\nabla}(\sigma)$, we have

$$\Sigma \, o^u \, \varphi \text{ iff } \Sigma \cup A^u(\sigma) \, o \, \varphi \tag{o^u}.$$

We also have substitutivity of equivalents: $\Sigma \, o^u \, \nabla v\psi \leftrightarrow \nabla v\theta$ whenever $\Sigma \, o^u \, \psi \leftrightarrow \theta$. Also, within equivalence, the generalized quantifier ∇ provably commutes with negation ($o^u \neg \nabla v\varphi \leftrightarrow \nabla v\neg\varphi$) and distributes over the binary propositional connectives $\wedge$, $\vee$, $\to$ and $\leftrightarrow$ (e. g., $o^u \nabla v(\psi \wedge \theta) \leftrightarrow (\nabla v\psi \wedge \nabla v\theta)$ and $o^u \nabla v(\psi \vee \theta) \leftrightarrow (\nabla v\psi \vee \nabla v\theta)$).

Other usual deductive notions, such as (maximal) consistent sets, conservative extension, etc., can be appropriately adapted [14].

3. Metamathematical Properties of Ultrafilter Logic

We now establish some metamathematical properties of our logic.

3.1 Soundness

We first examine the soundness of our deductive system with respect to ultrafilter structures. As usual, soundness is easily established.

Indeed, the axioms in $A^u(\sigma)$ code properties of ultrafilters, so they hold in all ultrafilter structures[2]. We thus have soundness of our deductive system with respect to ultrafilter consequence ($o^u \subseteq p^{\mathcal{U}}$), since Modus Ponens preserves validity.

3.2 Deductive Properties

We will use some familiar constructions and properties, which can be established as for classical first-order logic [13, 14], by relying on the connection (o^u) in 2.3.

We first recall the addition of witnesses for existential sentences.

A *witness axiom* of h for an existential sentence $\exists v\varphi$ is a sentence $\omega[\exists v\varphi \backslash h]$ of the form $\exists v\varphi \to \varphi[v \coloneqq h]$, where h is a <u>new</u> constant: not occurring in $\exists v\varphi$.

Given a signature σ, consider a set K of new constants <u>not</u> in σ with cardinality $|L^{\nabla}(\sigma)|$, and form the expansion $\sigma[K] := \sigma \cup K$ of signature σ obtained by adding the new constants from K. Considering a subset $H \subseteq K$ with the same cardinality $|H| = |K|$, we can construct a theory $\Omega\{\sigma[K]\backslash H\}$ of witnesses from H for signature $\sigma[K]$, where every existential sentence of $L^\nabla(\sigma[K])$ has a witness from set H, as in the classical Henkin proof[3] [10].

This witness theory $\Omega\{ \sigma[K] \setminus H \}$ can be used to construct a conservative extension $\Sigma\{ K \setminus H \} := \Sigma \cup \Omega\{ \sigma[K] \setminus H \}$ of a theory $\Sigma \subseteq L^\nabla(\sigma)$. Also, for each formula φ of $L^\nabla(\sigma)$ and constants $\underline{k}$ from K, such that $\Sigma\{ K \setminus H \} \, o^\mu \, \varphi[\underline{v} := \underline{k}]$, we have both $\Sigma \, o^\mu \, \underline{\forall} \underline{v} \varphi$ and $\Sigma \, o^\mu \, \underline{\exists} \underline{v} \varphi$.

3.3 Other Metamathematical Properties

We will now consider other metamathematical properties of our logic.

Consider sets Ψ and Θ of sentences. We say that a sentence ρ *separates* Ψ from Θ iff $\Psi \, o^\mu \, \rho$ and $\Theta \, o^\mu \, \neg\rho$. We call Ψ and Θ *separable* over a given signature σ iff some sentence of $L^\nabla(\sigma)$ separates Ψ from Θ; and *inseparable* over signature σ otherwise[4].

Given signatures μ and ν, consider the following signatures
- $\lambda := \mu \cap \nu$, consisting of the common extra-logical symbols; and
- $\kappa := \mu \cup \nu$, consisting of the extra-logical symbols of either.

We shall establish the following joint model existence lemma.

Joint Model Existence: Given signatures μ and ν, consider the signatures $\lambda := \mu \cap \nu$ and $\kappa := \mu \cup \nu$. If sets of sentences Γ, of $L^\nabla(\mu)$, and Δ, of $L^\nabla(\nu)$, are inseparable over the common signature λ, then their union $\Gamma \cup \Delta$ has an ultrafilter model $M^\mathcal{U} = (M , \mathcal{U})$ with cardinality at most $|L^\nabla(\kappa)|$ ($|M| \leq |L^\nabla(\kappa)|$).

From this joint model existence property, we can, as usual, obtain several metamathematical properties.

First, we have the Löwenheim-Skolem property.

Löwenheim-Skolem: Each consistent set Σ of sentences of $L^\nabla(\sigma)$ has an ultrafilter model $M^\mathcal{U} = (M , \mathcal{U})$ with cardinality at most $|L^\nabla(\sigma)|$[5].

Next, we have the completeness of our deductive system with respect to ultrafilter consequence: $p^\mathcal{U} \subseteq o^\mu$.

Completeness: Each ultrafilter consequence τ of a set Σ of sentences of language $L^\nabla(\sigma)$ ($\Sigma \, p^\mathcal{U} \, \tau$) is ultrafilter-derivable from Σ ($\Sigma \, o^\mu \, \tau$)[6].

We have a sound and complete deductive system for ultrafilter logic. As usual, such a result transfers the finitary character of derivability o^μ to the compactness of the semantical consequence $p^\mathcal{U}$.

We also have the property of (distributed) interpolation.

Interpolation: Given signatures μ and ν, consider the signatures $\lambda := \mu \cap \nu$ and $\kappa := \mu \cup \nu$. Consider sets of sentences Γ, of $L^\nabla(\mu)$, and Δ, of $L^\nabla(\nu)$. Then, each sentence τ of $L^\nabla(\nu)$ such that $\Gamma \cup \Delta \, o^\mu \, \tau$ has an interpolant: a sentence ρ of the common signature λ, such that $\Gamma \, o^\mu \, \rho$ and $\{ \rho \} \cup \Delta \, o^\mu \, \tau$[7].

As usual, from the distributed interpolation property we can easily obtain modularity: preservation of conservative extension [15, 16].

Modularity: Given signatures μ and ν, consider the signatures $\lambda := \mu \cap \nu$ and $\kappa := \mu \cup \nu$. Consider sets of sentences Φ, of $L^\nabla(\lambda)$, Γ, of $L^\nabla(\mu)$, and Δ, of $L^\nabla(\nu)$. For any set of sentences $\Phi \subseteq L^\nabla(\lambda)$, if Γ extends Φ conservatively then so does $\Gamma \cup \Delta$ extends conservatively $\Phi \cup \Delta$[8].

Finally, we have the joint consistency property, as it follows from modularity also as usual [8].

Joint Consistency: Given signatures μ and ν, consider the signatures $\lambda := \mu \cap \nu$ and $\kappa := \mu \cup \nu$. If sets of sentences Γ, of $L^\nabla(\mu)$, and Δ, of $L^\nabla(\nu)$, are consistent extensions of a maximal consistent set Φ, of $L^\nabla(\lambda)$, then their union $\Gamma \cup \Delta$ is consistent[9].

4. The Joint Model Existence Property for Ultrafilter Logic

We now outline the proof of the joint model existence result. The proof and the constructions used are an adaptation of the analogous ones for classical first-order logic [7, 9]. We emphasize sets of formulas and construct the model at the very end providing an adequate ultrafilter by means of witnesses [5].

Given sets of sentences Γ, of $L^\nabla(\mu)$, and Δ, of $L^\nabla(\nu)$, that are inseparable over the common signature λ, we shall extend them to maximally inseparable sets Γ_* and Δ_*, over a common signature with witnesses, and use these sets to construct an ultrafilter model for their union $\Sigma := \Gamma_* \cup \Delta_*$. We proceed to outline how this can be done.

4.1 Addition of Witnesses

We first extend the sets of sentences Γ, of $L^\nabla(\mu)$, and Δ, of $L^\nabla(\nu)$, to sets of sentences Γ' and Δ' with witnesses, preserving inseparability.

Given signatures μ and ν, we construct the signatures $\lambda := \mu \cap \nu$ and $\kappa := \mu \cup \nu$, and consider the language $L^\nabla(\kappa)$ of the union signature. We now select two disjoint sets C and D of new constants (<u>not</u> in κ), each one with cardinality $|L^\nabla(\kappa)|$, and form their union $E := C \cup D$.

Now, consider the expansions of the original signatures by the set E of new constants, namely $\lambda' := \lambda \cup E$, $\mu' := \mu \cup E$, $\nu' := \nu \cup E$, and $\kappa' := \kappa \cup E$. Notice that the four expanded languages have the same (infinite) cardinality, namely $|L^\nabla(\kappa)|$. Thus, we have theories of witnesses for signatures μ' and ν', i. e. theory $\Omega\{\mu[E]\backslash C\}$ of witnesses from C for signature $\mu[E] = \mu'$, and theory $\Omega\{\nu[E]\backslash D\}$ of witnesses from D for signature $\nu[E] = \nu'$, which we use to extend conservatively $\Gamma \subseteq L^\nabla(\mu)$ and $\Delta \subseteq L^\nabla(\nu)$ to

$$\Gamma' := \Gamma \cup \Omega\{\mu[E]\backslash C\} \subseteq L^\nabla(\mu') \text{ and } \Delta' := \Delta \cup \Omega\{\nu[E]\backslash D\} \subseteq L^\nabla(\nu').$$

We can notice that these sets of sentences Γ', of $L^\nabla(\mu')$, and Δ', of $L^\nabla(\nu')$, remain inseparable over their common signature $\lambda' = \lambda \cup E$ (in view of the properties of the extensions $\Gamma' \supseteq \Gamma$ and $\Delta' \supseteq \Delta$)[10].

4.2 Maximally Inseparable Pair of Extensions

We next extend the inseparable sets of sentences Γ', of $L^\nabla(\mu')$, and Δ', of $L^\nabla(\nu')$, with witnesses, to maximally inseparable sets of sentences Γ_* and Δ_*, over the common signature $\lambda' = \lambda \cup E$ (by applying Zorn's lemma to the (nonempty) family of inseparable pairs of extensions, partially ordered by inclusion of pairs)[11].

Now, we can see that such a maximally inseparable pair $<\Gamma_*,\Delta_*>$ consists of pair of maximally consistent theories over their languages[12].

Next, we can also see that a maximally inseparable pair $<\Gamma_*,\Delta_*>$ has the following crucial properties (due to maximal consistency and inseparability)[13]:

$$\text{for each sentence } \psi \text{ of } L^\nabla(\mu'): \text{ if } \psi \in \Delta_* \text{ then } \psi \in \Gamma_*;$$
$$\text{for each sentence } \theta \text{ of } L^\nabla(\nu'): \text{ if } \theta \in \Gamma_* \text{ then } \theta \in \Delta_*.$$

We can also note that the intersection $\Gamma_* \cap \Delta_*$ of a maximally inseparable pair $<\Gamma_*,\Delta_*>$ is maximally consistent theory over its language $L^\nabla(\lambda')$[14].

4.3 Ultrafilter Model

We finally construct an ultrafilter structure for signature κ' and then argue that it will be an ultrafilter model for the union $\Sigma := \Gamma_* \cup \Delta_*$.

Considering the set T of variable-free terms of $L(\kappa')$, the canonical structure H has universe $H := T/\approx$, where $t'\approx t''$ iff $\Sigma \vdash^u t'3t''$. So, its cardinality $|H|$ is at most that of the languages: $|L^\nabla(\kappa')| = |L^\nabla(\kappa)|$. We introduce the realizations of predicates and functions as usual, e. g., for an m-ary predicate of signature $\kappa' = \kappa \cup E$, we set

$$p^H := \{ \,<t_1/\approx,\ldots,t_m/\approx> \in H^m : p(t_1,\ldots,t_m) \in \Sigma \,\}.$$

An ultrafilter over universe H can be obtained as follows. We consider the set represented within Σ by formula φ of $L^\nabla(\kappa')$ with single free variable v, namely $\varphi^\Sigma := \{ \, t/\approx \in H : \varphi[v \coloneqq t] \in \Sigma \,\}$, and form the family of provably important represented subsets, i. e. $_\nabla\Sigma := \{ \, \varphi^\Sigma \subseteq H : \nabla v\varphi \in \Sigma, fr(\nabla v\varphi) = \varnothing \,\}$. Now, in view of our axioms, this family $\Sigma_\nabla \subseteq \wp(H)$ has the finite intersection property, and so, it can be extended to a proper ultrafilter $\mathcal{U} \subseteq \wp(H)$. We use such ultrafilter $\mathcal{U}$ to expand the canonical structure H to an ultrafilter structure $H^\mathcal{U} := (H, \mathcal{U})$ for $L^\nabla(\kappa')$.

We can now show, by induction, that

$$H^\mathcal{U} \vdash_p \psi \text{ iff } \psi \in \Gamma_*, \text{ for each sentence } \psi \text{ of } L^\nabla(\mu');$$
$$H^\mathcal{U} \vdash_p \theta \text{ iff } \theta \in \Delta_*, \text{ for each sentence } \theta \text{ of } L^\nabla(\nu').$$

The basis steps for the atomic formulas follow from the definition of H[15]. The inductive steps for propositional connectives as well as the quantifiers $\forall$ and $\exists$ are as in the classical proof. Now, the inductive step for the new quantifier ∇, namely

$$H^\mathcal{U} \vdash_p \nabla u\psi \text{ iff } \nabla u\psi \in \Gamma_*, \text{ for } \psi \text{ in } L^\nabla(\mu'),$$
$$H^\mathcal{U} \vdash_p \nabla z\theta \text{ iff } \nabla z\theta \in \Delta_*, \text{ for } \theta \text{ in } L^\nabla(\nu');$$

can be seen to follow from the following property of family Σ_∇

$$\varphi^\Sigma \in \Sigma_\nabla \text{ iff } \varphi^\Sigma \in \mathcal{U}, \text{ for each } \varphi \text{ in } L^\nabla(\mu') \cup L^\nabla(\nu')[16].$$

Hence, we can conclude that, $H^\mathcal{U} \vdash_p \tau$, for each sentence $\tau \in \Gamma \cup \Delta \subseteq \Gamma_* \cup \Delta_*$. Thus, we have an ultrafilter model $H^\mathcal{U} = (H, \mathcal{U})$ of the union $\Gamma \cup \Delta$ with cardinality $|H| \leq |L^\nabla(\kappa)|$.

5. Conclusion

Ultrafilter logic was introduced to capture directly the intuition of a property holding for a large or important set of elements [1, 4] and to serve as a precise basis for generic reasoning [3, 5]. For this purpose, one extends (conservatively) classical first-order logic by adding a new generalized quantifier ∇ giving rise to generalized formulas, which are interpreted as having extension in a given ultrafilter.

We have established that ultrafilter logic shares with classical first-order logic interpolation and related properties, such as joint consistency and modularity (preservation of conservativeness). By adapting a strategy used for the classical case [7], we have shown a joint model existence lemma, from which these properties, as well as Löwenheim-Skolem and completeness, can be seen to follow.

As a logic with generalized quantifiers, ultrafilter logic is connected to such extensions of first-order logic [11, 17]. Also, ultrafilter logic is a proper extension of classical first-order logic [18, 19] with compactness and Löwenheim-Skolem properties[17]. This feature may confer to our logic some independent logical interest.

Ultrafilter logic belongs to a family of logics for qualitative reasoning [18], including the so-called modulated logics [12]. These logics appear to have some connections with fuzzy logics, as well as with empirical reasoning [3, 6], suggesting the possibility of interesting applications.

Notes

[1] Thus, we can employ the familiar notation $A^{\mathcal{U}} \, p \, \varphi \, [\, \underline{a} \,]$ (for satisfaction of a formula φ - with at most m free variables - by $\underline{a} \in A^m$); such a formula defines an m-ary relation: $A^{\mathcal{U}}[\varphi] := \{\, \underline{a} \in A^m : A^{\mathcal{U}} \, p \, \varphi \, [\, \underline{a} \,] \,\}$. Similarly, we can introduce the extension $A^{\mathcal{U}} \, [\, \varphi(\underline{a}, v)\,] := \{b \in A : A^{\mathcal{U}} \, p \, \varphi(\underline{u}, v) \, [\, \underline{a}, b\,]\}$; with this notation, satisfaction of a generalized formula $\nabla v \varphi(\underline{u}, v)$ becomes: $A^{\mathcal{U}} \, p \, \nabla v \, \varphi(\underline{u}, v) \, [\, \underline{a} \,]$ iff the extension $A^{\mathcal{U}} \, [\, \varphi(\underline{a}, v)\,]$ belongs to the ultrafilter $\mathcal{U}$.

[2] Clearly, the axioms in $[\nabla \exists] \cup [\neg \nabla] \cup [\nabla \wedge]$ code properties of ultrafilters, so they hold in every ultrafilter structure. As for $[\nabla v]$, if variable u does not occur in φ ($u \notin \mathrm{occ}(\varphi)$), we have $A^{\mathcal{U}} \, p \, \varphi \, [\, s\,]$ iff $M^{\mathcal{U}} \, p \, \varphi \, [v := u] \, [\, s(u \oes(v))\,]$.

[3] We can construct such a theory $\Omega\{\, \sigma[K] \setminus H \,\}$ of witnesses from H for signature $\sigma[K]$, where every existential sentence of $L^{\nabla}(\sigma[K])$ has a witness from set H, by enumerating these sentences and H, as in the classical Henkin-style proof [9]. (Note that, for each subset $S \subseteq K$ with $|S| < |K|$ and existential sentence $\exists v \varphi$ of $L^{\nabla}(\sigma[K])$, we have a constant $h_\varphi \in H$ not occurring in $S \cup \{\, \varphi \,\}$, which can be used as a witness in the axiom $\omega[\, \exists v \varphi \setminus h_\varphi\,]$.)

[4] Clearly, if Ψ and Θ are inseparable they must be consistent. Also, Ψ and Θ are inseparable iff their sets of theorems are inseparable.

[5] Indeed, with $\mu := \nu := \sigma$, consistent set Σ is inseparable from Σ over σ, so Σ has an ultrafilter model $M^{\mathcal{U}} = (M, \mathcal{U})$ with cardinality $|M| \leq |L^{\nabla}(\sigma)|$.

[6] Indeed, otherwise $\Sigma \cup \{\, \neg \tau \,\}$ would be a consistent set in $L^{\nabla}(\sigma)$, and we would have an ultrafilter model $M^{\mathcal{U}}$ of Σ such that $M^{\mathcal{U}} \, P \, \tau$.

[7] Indeed, otherwise Γ and $\Delta \cup \{\, \neg \tau \,\}$ would be inseparable over signature λ, and we would have an ultrafilter model $M^{\mathcal{U}}$ of $\Gamma \cup \Delta$ such that $M^{\mathcal{U}} \, P \, \tau$.

[8] Indeed, each sentence τ of $L^{\nabla}(\nu)$ such that $\Gamma \cup \Delta \, \sigma^\mu \, \tau$ has an interpolant sentence ρ of signature λ; and $\Gamma \, \sigma^\mu \, \rho$, by conservativeness, gives $\Phi \, \sigma^\mu \, \rho$, whence $\{\, \rho \,\} \cup \Delta \, \sigma^\mu \, \tau$ yields $\Phi \cup \Delta \, \sigma^\mu \, \tau$.

[9] Recall that a consistent extension Σ of a maximal consistent set Φ must be conservative. (If $\Phi \, O^\mu \, \tau$, then $\Phi \, \sigma^\mu \, \neg \tau$, whence $\Sigma \, \sigma^\mu \, \neg \tau$ and $\Sigma \, O^\mu \, \tau$.) Thus, from the conservativeness of Γ over Φ we have the conservativeness from Δ to $\Gamma \cup \Delta$, whence the consistency of $\Gamma \cup \Delta$ follows from that of Δ.

[10] Indeed, a sentence of $L^{\nabla}(\lambda')$ is of the form $\rho(\underline{e})$, with $\rho(\underline{w})$ in $L^{\nabla}(\lambda)$ and $\underline{e}$ from E. We thus see that

Γ $\vdash^\mu \rho(\underline{c})$ would yield Γ $\vdash^\mu \underline{\forall w}\rho(\underline{w})$, and Δ' $\vdash^\mu \neg\rho(\underline{c})$ would yield Δ $\vdash^\mu \underline{\exists w}\neg\rho(\underline{w})$, thus $\underline{\forall w}\rho(\underline{w}) \in L^\nabla(\lambda)$ would separate Γ from Δ.

[11] In the family $\mathcal{E} := \{ \ <\Psi, \Theta> $ inseparable over $\lambda' : \Gamma \subseteq \Psi \subseteq L^\nabla(\mu')$ and $\Delta' \subseteq \Theta \subseteq L^\nabla(\nu) \ \}$, partially ordered by inclusion of pairs, any chain has the pairwise union as an upper bound. (Indeed, if $\cup_{i \in I}\Psi_i \vdash^\mu \rho$ and $\cup_{i \in I}\Theta_i \vdash^\mu \neg\rho$, then, for some $j \in I$, $\Psi_j \vdash^\mu \rho$, and, for some $k \in I$, and $\Theta_k \vdash^\mu \neg\rho$, whence for $m \geq \max\{j,k\}$, $\Psi_m \vdash^\mu \rho$ and $\Theta_m \vdash^\mu \neg\rho$.)

[12] Clearly, Γ_* and Δ_* are theories. For a sentence ψ of $L^\nabla(\mu')$ undecided by Γ_*, both $\Gamma_* \cup \{\psi\}$ and $\Gamma_* \cup \{\neg\psi\}$ would be separable from Δ_*, so we would have separating sentences ρ and τ over the common signature λ', which would yield sentence $\rho \vee \tau$ in $L^\nabla(\lambda')$ separating Γ_* from Δ_*. Similarly for Δ_*.

[13] Indeed, given $\psi \in L^\nabla(\mu')$ with $\psi \in \Delta_* \subseteq L^\nabla(\nu)$, if $\psi \notin \Gamma_*$, then $\neg\psi \in \Gamma_* \subseteq L^\nabla(\mu')$, by maximal consistency. So, $\neg\psi \in L^\nabla(\lambda')$ would separate Γ_* from Δ_*. Similarly, for $\theta \in L^\nabla(\nu)$, from $\theta \in \Gamma_* \subseteq L^\nabla(\mu')$ we have $\theta \in \Delta_*$. We will need these properties for atomic sentences.

[14] Indeed, given $\rho \in L^\nabla(\lambda')$, if $\rho \notin \Gamma_* \cap \Delta_*$, then $\rho \notin \Gamma_*$ or $\rho \notin \Delta_*$. In the former case, as $\rho \in L^\nabla(\lambda') \subseteq L^\nabla(\mu')$, the preceding property yields $\rho \notin \Delta_*$. In the latter case, we have similarly, $\rho \notin \Gamma_*$. Thus, in either case, $\rho \notin \Gamma_*$ and $\rho \notin \Delta_*$, whence maximal consistency, yields $\neg\rho \in \Gamma_*$ and $\neg\rho \in \Delta_*$, thus $\neg\rho \in \Gamma_* \cap \Delta_*$.

[15] For instance, consider an atomic sentence $p(\underline{t})$ of $L^\nabla(\mu')$. On the one hand, if $p(\underline{t}) \in \Gamma_* \subseteq \Sigma$, then $H^\mathcal{U} p\,p(\underline{t})$, by construction. On the other hand, if $p(\underline{t}) \notin \Gamma_*$, then the previous crucial property yields $p(\underline{t}) \notin \Delta_*$, whence $p(\underline{t}) \notin \Sigma$, and thus $H^\mathcal{U}P\,p(\underline{t})$, by construction.

[16] This property follows from maximal consistency. Clearly, $\varphi^\Sigma \in \mathcal{U}$, whenever $\varphi^\Sigma \in \Sigma_\nabla \subseteq \mathcal{U}$, for each φ in $L^\nabla(\mu') \cup L^\nabla(\nu)$. Conversely, for $\psi(u)$ in $L^\nabla(\mu')$, if $\psi(u)^\Sigma \notin \Sigma_\nabla$ then $\nabla u\,\psi(u) \notin \Gamma_*$, so $\neg\nabla u\,\psi(u) \in \Gamma_*$ and $\nabla u\neg\psi(u) \in \Gamma_*$, whence $\neg\psi(u)^\Sigma \in \Sigma_\nabla \subseteq \mathcal{U}$, thus $\psi(u)^\Sigma \notin \mathcal{U}$. Similarly for $\theta(z)$ in $L^\nabla(\nu)$.

[17] The apparent conflict with Lindström's theorems [20] is explained because of our notion of models (with ultrafilters).

References

[1] A. B. Slomson, Some Problems in Mathematical Logic. Doctoral Dissertation, Oxford University, Oxford, 1967.

[2] W. A. Carnielli and A. M. Sette, Default operators. In Abstracts of the Workshop on Logic, Language, Information and Computation, Recife, 1994.

[3] W. A. Carnielli and P. A. S. Veloso, Ultrafilter logic and generic reasoning. In G. Gottlob, A. Leitsch, and D. Mundici (eds.) Computational Logic and Proof Theory (5th Kurt Gödel Colloquium) {LNCS 1289}, Springer-Verlag, Berlin, 1997, pp. 34-53.

[4] A. M. Sette, W. A. Carnielli and P. A. S. Veloso, An alternative view of default reasoning and its logic. In E. H. Haeusler and L. C. Pereira (eds.) Pratica: Proofs, Types and Categories, PUC-Rio, Rio de Janeiro, 1999, pp. 127-158.

[5] P. A. S. Veloso and W. A. Carnielli, An ultrafilter logic for generic reasoning and some applications. Res. Rept. ES-437/97, COPPE-UFRJ, Rio de Janeiro, 1997.

[6] P. A. S. Veloso, On ultrafilter logic as a logic for 'almost all' and 'generic' reasoning. Res. Rept. ES-488/98, COPPE-UFRJ, Rio de Janeiro, 1998.

[7] L. Henkin, Metamathematics: class notes for course Math 225. University of California, Berkeley, 1972.

[8] P. A. S. Veloso, On pushout consistency, modularity and interpolation for logical specifications, *Information Processing Letters* **60**(1996) 59-66.

[9] C. C. Chang and H. J. Keisler, Model Theory. North-Holland, Amsterdam, 1973.

[10] L. Henkin, The completeness of the first-order functional calculus, *Journal of Symbolic Logic* **14**(1949) 159-166.

[11] J. Barwise and S. Feferman., Model-Theoretic Logics. Springer-Verlag, New York, 1985.

[12] M. C. G. Grácio, Lógicas Moduladas e Raciocínio sob Incerteza. Doctoral Dissertation, UNICAMP, Campinas, 1999.

[13] J. R. Shoenfield, Mathematical Logic. Addison-Wesley, Reading, 1967.

[14] H. B. Enderton, A Mathematical Introduction to Logic. Academic Press; New York, 1972.

[15] P. A. S. Veloso, A new, simpler proof of the modularisation theorem for logical specifications, *Bulletin of the IGPL* 1(1993) 3-12.

[16] P. A. S. Veloso and T. S. E. Maibaum, On the modularization theorem for logical specifications, *Information Processing Letters* 53(1995) 287-293.

[17] H. J. Keisler, Logic with the quantifier 'there exist uncountably many', *Annals of Mathematical Logic* 1(1970) 1-93.

[18] P. A. S. Veloso, On 'almost all' and some presuppositions. In L. C. P. D. Pereira and M. B. Wrigley (eds.) Logic, Language and Knowledge: essays in honour of Oswaldo Chateaubriand Filho, *Manuscrito* **XXII**(1999) 469-505.

[19] P. A. S. Veloso, On the power of ultrafilter logic, *Bulletin of the Section of Logic* 29(2000) 89-97.

[20] P. Lindström, On extensions of elementary logic, *Theoria* 35(1966) 1-11.

Logic, Artificial Intelligence and Robotics
J.M. Abe & J.I. da Silva Filho (Eds.)
IOS Press, 2001

On a Logical Framework for 'Generally'

Sheila R. M. VELOSO[+] and Paulo A. S. VELOSO[+*]
[+]*Systems and Computer Engin. Program, COPPE-UFRJ, Brazil;*
[+]*Computer Sci. Dept., Inst. Mathematics, UFRJ, Brazil*
[*]*Dept. of Informatics, PUC-Rio, Brazil (on leave)*

Abstract. We examine logical systems with generalized quantifiers for expressing and reasoning about 'generally'. The primary motivation is a qualitative approach to assertions and arguments involving a version of such vague notions occurring often. The idea of 'generally' "as all but for a 'negligible' set of exceptions" is rendered precise by means of filters. This gives a monotonic and conservative extension of classical first-order logic, with which it shares several properties. Its sorted version captures relative notions with appropriate behavior. Our filter logic, though related to default logics, is quite different, belonging to a family of connected systems with generalized quantifiers for vague notions.

1. Introduction

We examine logical systems with generalized quantifiers, for expressing and reasoning about 'generally'. Assertions and arguments involving notions, such as 'generally', 'most', 'many', etc., occur often both in ordinary language and in some branches of science. The primary motivation is a qualitative, rather than quantitative, approach to such vague notions.

We wish to express assertions, such as "birds 'generally' fly", and reason about them in a formal manner. One usually understands "birds 'generally' fly" as "all birds, but for 'negligible' set of exceptions, fly", which suggests the paraphrases "most birds fly" or "the set of flying birds is an 'important' set". To <u>express</u> such 'generally' assertions formally, we introduce the new operator ∇ and express "birds 'generally' fly" by $\nabla v F(v)$. To give precise <u>meaning</u> to such assertions, we extend the usual notions, by providing a family $\mathcal{F}$ of 'important' sets, and stipulate that $\nabla v F(v)$ means the set $\{b \in B : F(b)\}$ is in the family $\mathcal{F}$ as a rigorous counterpart for "the set of flying birds is an 'important' set". To <u>reason</u> about such 'generally' assertions in a formal manner, we will set up a deductive system by extending (conservatively) the classical first-order predicate calculus. This logic is related to default logic [1] and its variants [2, 3], as indicated by benchmark examples, which was one of motivations for the introduction of some related systems [4, 5]. But, they are quite different logical systems, both technically[1] and in terms of intended interpretation [6, 7].

The structure of this paper is as follows. We begin by motivating the usage of filters for capturing an intuitive idea of 'generally'. Next, we consider a basic (unsorted) framework: we introduce our logic for 'generally', in section 3, and examine its properties, such as completeness, in section 4. Some interesting situations may require assertions relative to several universes, involving "most birds" and "most penguins" for instance, which we take up in section 5, where we motivate ideas concerning 'relative generally' and introduce our sorted framework for reasoning about versions of 'generally' relative to various universes.

2. On 'Generally' and 'Negligibly Few'

We will now indicate how one can arrive at filters [9] as capturing an idea of 'generally' . The approach is based on the familiar intuition of 'generally' as "all but for a 'negligible' set of exceptions", but we also employ some related, and more basic notions [10].

We will first motivate and outline our approach to making precise a notion of 'generally'. Since we shall be dealing with local qualitative notions[2], we will prefer to use names like 'important' and 'negligible'. We shall try to explain them by relying on a relation, which we will call 'almost as important as' and denote by $\approx$[3]. In the sequel, we will indicate that filters [9] are appropriate for giving precise counterparts for such notions.

The intuition of 'generally' as "all but for a 'negligible' set of exceptions" suggests understanding "objects 'generally' have a given property" as "the exceptional objects (failing to have this property) form a 'negligible' set. If we understand 'negligible' as "fit to be neglected or discarded", it appears reasonable to say that two sets are 'almost as important' when their difference (i. e. the part where they differ) is negligible. The difference is the so called symmetric difference: $X \Delta Y := (X-Y) \cup (Y-X)$. We can now put forward some postulates about these notions (based on common sense and ordinary understanding) [10].

(Δ) $X \approx Y$ iff $X \Delta Y \in \mathcal{N}${"Sets with negligible symmetric difference are almost as important"}

($\subseteq$) $X \in \mathcal{N}$ whenever $X \subseteq N \in \mathcal{N}${"Subsets of negligible sets are negligible"}.

($\approx$) $X \in \mathcal{N}$ if $X \approx N \in \mathcal{N}${"Sets almost as important as negligible ones are negligible"}.

($\in$) $\varnothing \in \mathcal{N}${"The empty set $\varnothing$ is negligible"}.

($\notin$) $V \notin \mathcal{N}${"The universe V is not negligible"}.

In virtue of these postulates, the family $\mathcal{N}$ of negligible sets forms a proper ideal [9]. Conversely, each proper ideal is a family of subsets satisfying our five postulates. Thus, the interpretation of " objects generally have a given property φ" as "the set of objects failing to have property φ is negligible" can be seen to amount to "the set of objects having φ belongs to a given filter" (the family $\mathcal{F}$ of sets with negligible complements).

3. A Logic for Generally

Our logic for 'generally' adds to classical first-order logic a generalized quantifier ∇ for generally. We now examine this logic $\mathcal{L}_{\omega\omega}(\mathcal{F})$ - its syntax, semantics and axiomatics. Given a signature ρ, we let $L(\rho)$ be the usual first-order language (with equality $\equiv$) of signature ρ.

We use $L^{\nabla}(\rho)$ for the extension of the first-order language $L(\rho)$ by the new operator ∇. The formulas of $L^{\nabla}(\rho)$ are built by the usual formation rules and the new variable-binding formation rule for generalized formulas: for each variable v if φ is a formula in $L^{\nabla}(\rho)$ then so is $\nabla v\varphi$. Other syntactic notions, such as substitution ($\varphi(t)$) can be easily adapted [11, 12]. The next example illustrates the expressive power of such languages with ∇.

Example Consider a signature λ with a binary predicate L (standing for 'loves').
Some assertions expressed by means of ∇ are: "people generally love somebody" by $\nabla x \exists y L(x,y)$, "somebody loves people in general" by $\exists x \nabla y L(x,y)$, "everybody loves people in general" by $\forall x \nabla y L(x,y)$, "people generally love everybody" by $\nabla x \forall y L(x,y)$, and "people generally love each other" by $\nabla x \nabla y L(x,y)$.   □

The semantic interpretation for our logic $\mathcal{L}_{\omega\omega}(\mathcal{F})$ is provided by enriching first-order structures with filters and extending the definition of satisfaction to the new quantifier ∇. A *filter structure* $A^{\mathcal{F}} = (A, \mathcal{F})$ for signature ρ consists of a usual structure A for ρ together with a proper filter $\mathcal{F}$ over the universe A of A. We extend the Tarskian definition of *satisfaction* of a formula in a structure under assignment $\underline{a}$ to its (free) variables, using the *extension* $A^{\mathcal{F}}[\varphi(\underline{a},v)] := \{\, b \in A : A^{\mathcal{F}} \models \varphi(\underline{u},v)\,[\,\underline{a},b\,]\,\}$, as follows:

($\models^{\nabla}$) for a formula $\nabla v\varphi(\underline{u},v)$, we define $A^{\mathcal{F}} \models \nabla v\varphi(\underline{u},v)\,[\,\underline{a}\,]$ iff $A^{\mathcal{F}}[\,\varphi(\underline{a},v)\,]$ is in $\mathcal{F}$.

Satisfaction of a formula hinges only on the realizations assigned to its symbols[4]. Other semantic notions, such as reduct, model ($A^{\mathcal{F}} \models \Gamma$) and validity are as usual [11,12]; also, the notion of *filter consequence* is as expected: $\Gamma \models^{\mathcal{F}} \tau$ iff $A^{\mathcal{F}} \models \tau$ whenever $A^{\mathcal{F}} \models \Gamma$[5].

We will now formulate a deductive system for our logic by adding schemata, coding properties of filters, to a calculus for classical first-order logic.

We set up a deductive system for filter logic by taking a sound and complete deductive calculus for classical first-order logic, with Modus Ponens as the sole inference rule (as in [11]), and extending its set $\Lambda(\rho)$ of axiom schemata by adding a set $\Phi^{f}(\rho)$ of new axiom schemata, to form $\Lambda^{f}(\rho):=\Lambda(\rho)\cup\Phi^{f}(\rho)$. This set $\Phi^{f}(\rho)$ consists of all the generalizations of the following five schemata (where φ, ψ and θ are formulas of $L^{\nabla}(\rho)$):

$[\forall\nabla]:\forall v\varphi \to \nabla v\varphi;$ $\qquad\qquad\qquad\qquad$ $[\nabla\exists]:\nabla v\varphi \to \exists v\varphi;$

$[\nabla\wedge]:(\nabla v\psi \wedge \nabla v\theta) \to \nabla v(\psi\wedge\theta);$ $\qquad$ $[\to\nabla]:\forall v(\psi\to\theta)\to(\nabla v\psi\to\nabla v\theta);$

$[\nabla v]:\nabla v\varphi(v) \to \nabla u\varphi(u)$, for a new variable u $\qquad\qquad\qquad\qquad$ {variant}.

These schemata express properties of filters, the last one covering alphabetic variants. Other usual deductive notions, such as (maximal) consistent sets, witnesses and conservative extension [11, 12], can be easily adapted. Filter derivability amounts to first-order derivability from the filter schemata: $\Gamma \vdash^{f} \varphi$ iff $\Gamma \cup \Lambda^{f}(\rho) \vdash \varphi$[6].

Example Consider the following facts about a universe of people.
"People generally oppose those in conflict with any one with whom they sympathize" expressed by $\nabla x\forall y\nabla z[S(x,y)\wedge C(z,y)\to O(x,z)]$ and "People generally sympathize with Bill" expressed by $\nabla yS(y,b)$. Then, one can conclude the sentence $\forall x\nabla z[C(z,b)\to O(x,z)]$ {i. e. "People generally oppose those in conflict with Bill"}.   □

4. Filter Logic

We shall now establish some properties of filter logic, including soundness and completeness of the deductive system with respect to filter consequence.

Our deductive system provides a sound and complete deductive calculus for reasoning about assertions involving 'generally': $\Gamma \vdash^{f} \tau$ iff $\Gamma \models^{\mathcal{F}} \tau$.

Soundness ($\vdash^{f} \subseteq \models^{\mathcal{F}}$) is easily established as usual[7].

For completeness ($\models^{\mathcal{F}} \subseteq \vdash^{f}$) we can adapt Henkin's well-known proof for classical first-order logic [11, 12, 13], by providing an adequate filter by means of witnesses. We proceed to outline how this can be done (cf. [6, 14]). Given a consistent set Γ in $L^{\nabla}(\rho)$, extend it to a maximal consistent set Σ in $L^{\nabla}(\rho*)$, with witnesses for the existential sentences of $L^{\nabla}(\rho*)$ in set C of new constants, where $\rho*:=\rho\cup C$. We form the canonical structure H for signature $\rho*$ with universe H as usual[8], and provide a filter by means of formulas of $L^{\nabla}(\rho*)$ with a single free variable, as follows. Considering the set $\varphi(v)^{\Sigma} := \{\, t/\equiv \,\in H : \varphi(t)\in\Sigma\,\}$ represented within Σ by formula $\varphi(v)$ of $L^{\nabla}(\rho*)$, we form the family of provably important subsets of H: $_{\nabla}\Sigma := \{\, \varphi(v)^{\Sigma} \subseteq H : \nabla v\varphi(v)\in\Sigma\,\}$. By our axioms, this family $_{\nabla}\Sigma$ has the finite intersection property, so its closure $\mathcal{F}_{\Sigma} \subseteq \wp(H)$ under supersets is a proper filter. We use this filter $\mathcal{F}_{\Sigma}$ on

H to expand the canonical structure H to a filter structure $H^{\mathcal{F}_\Sigma} := (H, \mathcal{F}_\Sigma)$ for $\rho*$. We now show, by induction, that $H^{\mathcal{F}_\Sigma} \vDash \tau$ iff $\Sigma \vdash^f \tau$, for each sentence τ of $L^\nabla(\rho*)$. The inductive step for ∇, namely $H^{\mathcal{F}_\Sigma} \vDash \nabla v\varphi(v)$ iff $\Sigma \vdash^f \nabla v\varphi(v)$, follows from the property $\varphi(v)^\Sigma \in \mathcal{V}^\Sigma$ iff $\varphi(v)^\Sigma \in \mathcal{F}^\Sigma$. (by schema $[\to\nabla]$)[9].

We thus have a Löwenheim-Skolem Theorem for our logical system.

Löwenheim-Skolem Theorem for filter logic: Each $\vdash^f$-consistent set Γ of sentences of $L^\nabla(\rho)$ has a filter model $M^{\mathcal{F}}$ with cardinality at most that of its language: $|M| \leq |L^\nabla(\rho)|$.   □

Hence, we have the desired result for filter logic.

Theorem Completeness of filter derivability with respect to filter consequence

The deductive system $\vdash^f$ is complete with respect to $\vDash^{\mathcal{F}}$: $\Gamma \vdash^f \tau$ whenever $\Gamma \vDash^{\mathcal{F}} \tau$.   □

We now examine other metamathematical properties of the filter logic $\mathcal{L}_{\omega\omega}(\mathcal{F})$ (cf. [14]). We have a sound and complete deductive system for $\mathcal{L}_{\omega\omega}(\mathcal{F})$. As usual, such a result transfers the finitary character of $\vdash^f$ to the compactness of $\vDash^{\mathcal{F}}$. Thus, our logic is a proper extension of classical first-order logic $\mathcal{L}_{\omega\omega}$ with compactness and Löwenheim-Skolem properties[10]. Also, $\mathcal{L}_{\omega\omega}(\mathcal{F})$ has some other connections with classical first-order logic: its conservativeness over $\mathcal{L}_{\omega\omega}$ and the universality of ∇-consequences of first-order theory.

Proposition *Filter logic and classical logic*

Consider a set Δ of sentences and formula θ of $L(\rho)$.

   a) Conservativeness of $\mathcal{L}_{\omega\omega}(\mathcal{F})$ over $\mathcal{L}_{\omega\omega}$: $\Delta \vdash \theta$ iff $\Delta \vdash^f \theta$.

   b) Generalized consequences: $\Delta \vdash^f \nabla v\theta$ iff $\Delta \vdash \forall v\theta$ and $\Delta \vdash^f \neg\nabla v\theta$ iff $\Delta \vdash \neg\exists v\theta$.

Proof outline. Any nonempty set can be extended to some proper filter.   □

Item (b) corroborates that 'generally' requires explicit information, otherwise it reduces to classical quantification (only the universe can be guaranteed to be in every filter)[11].

Example Consider consistent theories with information about a universe of birds.

a. Consider a consistent purely first-order theory Δ. Assume that one knows that: "some birds fly", "all birds have beaks", "every bird is a biped", "flying birds have wings"[12]. Then, one does not know that "birds generally do not fly": $\Delta \nvdash^f \nabla v\neg F(v)$. Also, not knowing that "all birds fly" ($\Delta \nvdash \forall vF(v)$), one does not know that " birds generally fly": $\Delta \nvdash^f \nabla vF(v)$.

b. Consider a consistent theory Γ with generalized information. Assume that one knows that "all feathered winged birds fly", "birds generally have wings", "birds generally have feathers"[13]. Then, one concludes that "birds generally fly": $\Gamma \vdash \nabla vF(v)$.   □

5. Relative Generally

We shall now examine the idea of having a notion of 'generally' relative to a universe: how it arises and is formulated, as well as some related issues (cf. [14]). We will first indicate how the proper expression of "relative generally" assertions brings about the idea of a notion of important with respect to each universe, leading to its natural formulation in a sorted version of filter logic. Then, the need for establishing some connections while blocking others leads to comparing such relative concepts. Finally, these ideas are incorporated into a sorted framework for reasoning about relative generally.

Our generalized quantifier ∇ may exhibit somewhat unexpected behavior in some cases. We shall now examine these undesirable side-effects and propose a way to overcome this difficulty. The generalized quantifier ∇ is meant to capture the idea of holding generally, i. e. for most objects of the universe. Sometimes we wish to express the idea of holding generally over a given subset of the universe, i. e. for most objects of a given sub-universe. We now examine the expression of such relative generally assertions.

On a universe of birds, we express "birds generally fly" by $\nabla v F(v)$. How are we to express relative generalized assertions, like "eagles generally have wings" or "penguins generally have beaks"? By analogy with the classical quantifiers, relativization is an apparently natural suggestion, i. e. expressing "M's generally are N's" by $\nabla v[M(v) \rightarrow N(v)]$. Unfortunately, relativization fails to be adequate for expressing "relative generally" assertions, due to the behavior of the quantifier ∇. The next example illustrates this issue.

Example Consider expressing the following facts on birds by relativization. "All penguins are winged birds" by $\forall v[P(v) \rightarrow W(v)]$ and "penguins generally do not fly" by $\nabla v[P(v) \rightarrow \neg F(v)]$. From these two sentences, one concludes $\nabla v[W(v) \rightarrow \neg F(v)]$ {i. e. "winged birds generally do not fly"}. Now, the two given premises appear to express reasonable facts. On the other hand, the conclusion, <u>as read</u>, is not so reasonable. This unexpected conclusion indicates that relativization fails to express the intended idea. The reason comes from neglecting the relative aspect[14].　　　　　　　　　　　　　　　　　　　□

For a formula $\nabla v[M(v) \rightarrow N(v)]$ the reading "M's generally are N's" is not appropriate. For, one must bear in mind that what this does assert is "for most <u>objects</u> a, if M(a) then N(a)"[15]. A natural approach to overcome this problem, thus expressing 'relative generally' assertions, rests on relative notions of important subsets: each given universe has its own notion of important subsets. This idea may be formulated by providing a filter $\mathcal{F}_S$ over each universe S. With relative notions of importance, we can paraphrase "M's generally are N's" as "most M's are N's" meaning that the set $\{a \in M : N(a)\}$ is 'almost as important as' the universe M, i. e. $N \cap M \approx M$ or $N \cap M \in \mathcal{F}_M$[16].

A many-sorted approach will provide an appropriate framework for formulating the idea of distinct notions of important relative to the universes, by assigning filters corresponding to these relative notions of important. We shall now examine sorted versions of filter logic. The basic idea is that the previous (unsorted) concepts now become relativized to sorts.

We consider many-sorted signatures, where the extra-logical symbols, as well as variables, come classified according to sorts [11]. Quantifiers are relativized to sorts, as expressed in the formation rules: for each variable v over sort s, if φ is a formula in $L^\nabla(\rho)$, then so are $(\forall v{:}s)\varphi$, $(\exists v{:}s)\varphi$ and $(\nabla v{:}s)\varphi$. A *filter structure* $A^\mathcal{F}$ for S-sorted signature ρ is an expansion of an S-sorted first-order structure A for ρ, obtained by assigning to each sort s of signature ρ a filter $\mathcal{F}_s$ over the universe A[s] of sort s (giving the important subsets of A[s]). The extension of *satisfaction* becomes relativized to sorts accordingly, namely $A^\mathcal{F} \vDash (\nabla y{:}s)\varphi(\underline{u},v)[\underline{a}]$ iff the extension $\{\, b \in A[s] : A^\mathcal{F} \vDash \varphi(\underline{u},v)[\underline{a},b] \,\}$ is in the filter $\mathcal{F}_s$. The *filter axiom schemata* in the set $\Phi^f(\rho)$ become sorted as well[17].

As in classical first-order logic, the sorted and unsorted versions are similar. So, the results in section 4 carry over, by relativizing to sorts the previous arguments[18].

We now examine the proposal of employing distinct notions of important subsets in the sense of how the need for establishing some connections while blocking others leads to comparing relative notions of important sets. The next example illustrates how some (undesired) conclusion are blocked.

Example (Birds and penguins with unconnected important sets)
Given that "All penguins are birds" ($P \subseteq B$), consider the assertions σ: "birds generally fly" (the flying birds form a large set of birds, i. e. $F \cap B \in \mathcal{F}_B$) and τ: "penguins generally fly" (the flying penguins form a large set of penguins, i. e. $F \cap P \in \mathcal{F}_P$). Now, neither σ entails τ (since we may even have $P \cap F = \varnothing$), nor does τ entail σ (since $P \subseteq B$ may very well be a negligible set of <u>birds</u>), if the notions of important subsets are not connected.　　　□

This example illustrates the idea of independent notions of important subsets. If the set of penguins is not an important set of <u>birds</u> (P⊆B not almost as important as B), then a set X⊆P may be an important set of <u>penguins</u> without being an important set of <u>birds</u>. It is this independence that blocks the undesired conclusion[19]. The next example illustrates how some (desired) conclusions can be achieved.

Example (Birds and winged birds with connected important sets)
Given that "All birds have wings", i. e. W⊆B, consider the assertions σ: "birds generally fly" (as before $F \cap B \in \mathcal{F}_B$ or $F \cap B \approx B$) and ρ: "winged birds generally fly" ($F \cap W \in \mathcal{F}_W$ or $F \cap W \approx W$). Given also that "birds generally have wings" ($W \approx B$), the set $B - W$ of exceptional wingless birds is a negligible set (of <u>birds</u>). So, it appears intuitively plausible that the important subsets of W are the relativizations $Y \cap W$ of the important subsets Y of B[20]. So, we shall also assume the coherence principle: for any set Y⊆B, $Y \cap W \approx W$ iff $Y \approx B$. Now, in the presence of this principle, assertions σ and ρ become equivalent.   □

The two preceding examples illustrate the following idea. Given S⊆T and a proper filter $\mathcal{F}_S$ over S, consider the *relativizable complex* $^T\mathcal{F}_S := \{Y \subseteq T : Y \cap S \in \mathcal{F}_S\}$. In case $(T-S) \in \mathcal{F}_T$ or $S \notin \mathcal{F}_T$, then we need an independent notion $\mathcal{F}_T$ of important subsets of T. If $S \in \mathcal{F}_T$, then the relativizable complex $^T\mathcal{F}_S$ is a filter over T, which we may take as $\mathcal{F}_T$ if we wish to enforce coherent inheritance[21], i. e. for every subset Y⊆T: $Y \cap S \in \mathcal{F}_S$ iff $Y \in \mathcal{F}_T$.

We shall now consider comparison of universes, with distinct notions of important subsets, in a sorted framework. We shall examine how to formulate some ideas related to sub-universes as well as coherent inheritance in a many-sorted approach.

In our sorted framework, sorts are unrelated: we have equality only over a sort, rather than between distinct sorts. Nevertheless, we can express some relationships among sorts by means of appropriate injections. The idea is that an injection i:s→t establishes a bijection from s onto its image i(s), the latter being a real subset of t. To express that sort s is a subsort of sort t, we can resort to a unary function i from s to t together with an axiom asserting its injectivity[22] [15]. This yields transitivity of subsorts.

We now formulate our previous coherent transfer principle for an injection i:s→t, where the image i(s) is an important subset of t. Then, the non-image t−i(s) is a negligible subset of t, where the distinction between a set Z⊆t and its pre-image $i^{-1}(Z)$ is confined. So, we consider Z⊆t as an important subset of t iff $i^{-1}(Z)$ is an important subset of s[23].

Now, given i:s→t and formula φ(z) with variable z over t, we can express: "most objects of t are in the image", "most objects of t have property φ(z)", and "most objects of s give objects in t with property φ(z)". This leads to the *coherent connection schema* [∇i:s⊆t], with the instances (∇i:s⊆t/φ): $(\nabla z{:}t)(\exists x{:}s)z \equiv i(x) \rightarrow [(\nabla z{:}t)\varphi \leftrightarrow (\nabla x{:}s)\varphi(i(x)))]$.

Let us now examine our preceding examples in this sorted formulation.

Example (Sorted birds, winged birds and penguins)
Consider three sorts: b (for birds), w (for winged birds) and p (for penguins) and a unary predicate F (for flies) over sort b, with j:w→b and k:p→b.
a. Considering all winged birds as birds, assume $(\nabla z{:}b)(\exists x{:}w)z \equiv j(x)$ {"birds generally have wings"}. Then, from the instance (∇j:w⊆b/F(z)) of the connection schema, we conclude the equivalence between $(\nabla z{:}b)F(z)$ {"birds generally fly"} and $(\nabla x{:}w)F(j(x))$ {"winged birds generally fly"}. We thus see that, since the winged birds form an important set of birds, "generally flying" is inherited both downwards and upwards.
b. Considering all penguins as birds, assume $(\nabla z{:}b)F(z)$ {"birds generally fly"} and the coherent connection schema. If we have $(\nabla z{:}b)(\exists y{:}p)z \equiv k(y)$ {"birds generally are penguins"}, instance (∇k:p⊆b/F(z)) yields $(\nabla y{:}p)F(k(y))$ {"penguins generally fly"}; but otherwise this conclusion is not forced upon us. In fact, if we have $(\nabla x{:}p)\neg F(k(y))$ {"penguins generally do

not fly"}, instance (∇k:p$\subseteq$b/$\neg$F(z)) yields $\neg$(∇z:b)($\exists$y:p)z$\equiv$k(y) {"it is not the case that birds generally are penguins"}. $\qquad\square$

6. Conclusion

We have examined monotonic logical systems with generalized quantifiers over filters, which provide rigorous bases for qualitative reasoning with vague notions such as 'generally' or 'most' [8]. The unsorted logical system is a conservative extension of classical first-order logic, with which it shares several properties. Some situations, however, require assertions relative to several universes, leading to the ideas of 'relative most' and to our sorted framework for them.

We can similarly introduce generalized quantifiers for the dual notion of 'negligible'. Modal versions of these logics can also be contemplated.

This logical system, though related to default logics, is quite different, both technically and in terms of intended interpretation [7]. Our filter logic belongs to a family of closely related systems with generalized quantifiers for reasoning about qualitative notions [8], including an ultrafilter logic for 'almost all' [14, 10]. These systems appear to have some interesting connections with fuzzy logic [16] (e. g. expressing 'very tall' by 'taller than most'), as well as with empirical reasoning [17], which suggest the possibility of other applications [6, 14]. They are deservedly undergoing further investigation [18].

Notes

[1] Our logic is a monotonic and conservative extension of classical logic (in sharp contrast to non-monotonic approaches) and caters to a positive view (representing 'generally' explicitly, rather than interpreting it as "in the absence of information to the contrary").

[2] The metaphor of a filter congregating the 'large' sets might suggest a quantitative notion, such as viewing as 'small' the unlikely sets (with low probability), but this is a non-local notion: a set with the same size as a 'large' one would be 'large'. The notion we employ is, in contrast, local. Also, we wish a qualitative account, dealing with properties more of a topological rather than metrical nature.

[3] As an example, consider a set consisting of a horse and an ox and another one consisting of a horse and a dog. These sets may be just as important to a conservationist. Yet, the former may be more important to a farmer, whereas the latter might be preferred by an English gentleman, keen on fox hunting. So, sets with the same size may fail to be equally important. Also, sets with different sizes may be just as important.

[4] In particular, satisfaction of purely first-order formulas (without ∇) does not depend on the filter.

[5] The behavior of ∇ is intermediate between those of the classical quantifiers (cf. schemata [$\forall\nabla$] and [$\nabla\exists$]), but the behavior of iterated ∇'s contrasts sharply with the commutativities of each classical $\forall$ and $\exists$: the formula ∇x∇yL(x,y)$\rightarrow$$\nabla$y$\nabla$xL(x,y) fails to be valid.

[6] Hence we have monotonicity of $\vdash^f$ and substitutivity of equivalents: $\Gamma \vdash^f \nabla$v$\psi\leftrightarrow\nabla$v$\theta$, if $\Gamma \vdash^f \psi\leftrightarrow\theta$.

[7] The axioms in $\Lambda^f(\rho)$ hold in all filter structures and Modus Ponens preserves satisfaction.

[8] It has universe H:=T/$\sim^\Sigma$ where T is the set of variable-free terms of L(ρ*) and t'$\sim^\Sigma$t" iff $\Sigma\vdash^f$t'$\equiv$t".

[9] The other steps are as in Henkin's proof.

[10] Conflict with Lindström's theorems [19] is only apparent (since we have filters in the models),

[11] Similar results hold for a set of simply generalized sentences (of the form ∇vφ, for some purely first-order formula φ). For a set Γ of simply generalized sentences, $\Delta\cup\Gamma \vdash^f \theta$ iff $\Delta \cup \{ \exists$uψ(u) $\} \vdash \theta$, where ψ(u) is some finite conjunction ψ_1(u)$\wedge...\wedge\psi_n$(u), with ∇v$_1\psi_1$(v$_1$),...,∇v$_n\psi_n$(v$_n$)$\in \Gamma$ and a new variable u. Also,

$\Delta \cup \Gamma \vdash^f \nabla$v$\theta$ iff $\Delta \vdash \forall$u(ψ(u)$\rightarrow\theta$(u)), for such a finite conjunction ψ(u) and new variable u.

[12] More precisely, $\Delta\vdash \exists$vF(v), $\Delta\vdash \forall$vK(v), $\Delta\vdash \forall$vD(v) and $\Delta\vdash \forall$v[F(v)$\rightarrow$W(v)].

[13] More precisely, $\Gamma\vdash \forall$v[W(v)$\wedge$T(v)$\rightarrow$F(v)], $\Gamma\vdash \nabla$vW(v) and $\Gamma\vdash \nabla$vT(v).

[14] One can consistently hold that "winged birds generally fly", "all penguins are winged birds" and " penguins generally do not fly" (or even "no penguin flies"). One feels that the penguins, forming a negligible set of winged birds, fail to amount to important exceptions to the belief that winged birds generally fly.

[15] Indeed, given the (classical) meaning of the conditional, formula $\nabla v[P(v) \rightarrow \neg F(v)]$ means that the set $P \cap F$ of flying penguins is a negligible set of <u>birds</u> (rather than of <u>penguins</u>).

[16] Thus, we can distinguish "eagles generally fly" ($F \cap E \in \mathcal{F}_E$) from "penguins generally fly" ($F \cap P \in \mathcal{F}_P$).

[17] For instance, $[\nabla v]_S : (\nabla v{:}s)\psi(v) \rightarrow (\nabla u{:}s)\psi(u)$, for a new variable $u{:}s$.

[18] For completeness, the witnesses introduced for the existential quantifiers inherit the corresponding sorts.

[19] Given filters $\mathcal{F}_T$, over T, and $\mathcal{F}_S$, over $S \subseteq T$, if $S \notin \mathcal{F}_T$ then, for every $X \subseteq S$: $X \notin \mathcal{F}_T$ (even for $X \in \mathcal{F}_S$).

[20] For a set $Y \subseteq B$, its sets of exceptions $W - Y$ and $B - Y$ are connected by $W - Y \subseteq B - Y \subseteq (W-Y) \cup (B-W)$. If 98% of the birds have wings, the wingless birds have negligible impact on the likelihood of birds flying.

[21] Notice that such (non-logical) questions fall outside the realm of our logic for 'generally'.

[22] More precisely $(\forall x', x''{:}s)[i(x') \equiv i(x'') \rightarrow x' \equiv x'']$.

[23] This connection can be explained by resorting to the family $^i\mathcal{F}_S := \{Z \subseteq t : i^{-1}[Z] \in \mathcal{F}_S\}$.

References

[1] R. Reiter, A logic for default reasoning. *J. Artificial Intelligence* **13**(1):81-132, 1980

[2] G. Antoniou, *Nonmonotonic Reasoning*. MIT Press, Cambridge, MA, 1997.

[3] G. Brewka, J. Dix and K. Konolige, *Nonmonotonic Reasoning: an overview*. CSLI, Stanford, 1997.

[4] W. A. Carnielli and A. M. Sette, Default operators (Abstr.). *Workshop on Logic, Language, Information and Computation*, Recife, Brazil, 1994.

[5] K. Schlechta, Defaults as generalized quantifiers. *J. Logic and Computation* **5**(4):473-494, 1995.

[6] W. A. Carnielli and P. A. S. Veloso, Ultrafilter logic and generic reasoning. In: G. Gottlob, A. Leitsch, and D. Mundici (eds.) Computational Logic and Proof Theory (5th Kurt Gödel Colloquium), Springer-Verlag, Berlin, 1997, pp. 34-53.

[7] A. M. Sette, W. A. Carnielli and P. A. S. Veloso, An alternative view of default reasoning and its logic. In: E. H. Haeusler and L. C. Pereira (eds.) Pratica: Proofs, Types and Categories, PUC-Rio, Rio de Janeiro, 1999, pp. 127-158.

[8] M. C. G. Grácio, *Lógicas Moduladas e Raciocínio sob Incerteza*. D. Sc. diss., UNICAMP, Campinas, Oct. 1999.

[9] J. L. Bell and A. B. Slomson, *Models and Ultraproducts: an introduction*. North-Holland, Amsterdam, 1971 (2nd rev. pr.).

[10] P. A. S. Veloso, On 'almost all' and some presuppositions. In: L. C. P. D. Pereira and M. B. Wrigley (eds.) Logic, Language and Knowledge: essays in honour of Oswaldo Chateaubriand Filho, *Manuscrito* **XXII** (2): 469-505, 1999.

[11] H. B. Enderton, *A Mathematical Introduction to Logic*. Academic Press, New York, 1972.

[12] J. R. Shoenfield, Mathematical Logic. Addison-Wesley, Reading, 1967.

[13] L. Henkin, The completeness of the first-order functional calculus. *J. of Symbolic Logic* **14**: 159-166, 1949.

[14] P. A. S. Veloso, On ultrafilter logic as a logic for 'almost all' and 'generic' reasoning. Res. Rept. ES-488/98, COPPE-UFRJ, Rio de Janeiro, Dec. 1998.

[15] M. C. Meré and P. A. S. Veloso, Definition-like extensions by sorts. *Bull. IGPL* **3**(4): 579-595, 1995.

[16] W. Turner, *Logics for Artificial Intelligence*. Ellis Horwood, Chichester, 1984.

[17] Hempel, C. *Aspects of Scientific Explanation and Other Essays in the Philosophy of Science*. Free Press, New York, 1965.

[18] P. A. S. Veloso and W. A. Carnielli, Logics for qualitative reasoning. In preparation.

[19] P. Lindström, On extensions of elementary logic. *Theoria* **35**: 1-11, 1966.

Author Index

Abe, J.M.	1,13,53,174,208,215	López-Escobar, E.G.K.	147
Akama, S.	13	Maia, E.C.	101
Ávila, B.C.	23	Martini, A.	148
Bianconi, R.	31	Massad, E.	156
Bittencourt, G.	263	Murai, T.	166,200
Cao, T.H.	43	Nakamatsu, K.	174
da Costa, N.C.A.	52	Noanaka, H.	62
da Rocha Costa, A.C.	68	Ohsuga, S.	186
da Silva Filho, J.I.	53,215	Ortega, N.R.S.	156
da Silva, P.R.O.	224	Resconi, G.	166,200
Da-te, T.	62	Roisenberg, M.	35
de Almeida Costa, C.I.	135	Sato, Y.	166
Dimuro, G.P.	68	Scalzitti, A.	208,215
Dubois, D.M.	76	Shioya, H.	62
Ebecken, N.F.F.	84	Simão, J.M.	224
Gonzaga, H.	135	Smrž, P.	232
González, C.G.	101	Stadzisz, P.C.	224
Haeusler, E.H.	148	Suppes, P.	238
Hembecker, F.	23	Suzuki, A.	174
Hruschka, E.R.	84	Thiele, H.	255
Iséki, K.	109,123	Tonin, I.	263
Kondo, M.	129	Veloso, P.A.S.	270,279
Künzle, L.A.	224	Veloso, S.R.M.	279
Lambert-Torres, G.	135	Wolter, U.	148
Lima De Campos, G.A.	35	Yoshikawa, T.	62
Lima De Campos, L.M.	35